ELEMENTS
OF
HUMAN
CANCER

——————The Jones and Bartlett Series in Biology——————

ELEMENTS
OF
HUMAN
CANCER

Geoffrey M. Cooper

Dana-Farber Cancer Institute
Harvard Medical School

JONES AND BARTLETT PUBLISHERS
BOSTON   LONDON

Editorial, Sales, and Customer Service Offices

Jones and Bartlett Publishers   Jones and Bartlett Publishers International
One Exeter Plaza   PO Box 1498
Boston, MA 02116   London W6 7RS
                                  England

Library of Congress Cataloging-in-Publication Data

Cooper, Geoffrey M.
    Elements of human cancer / Geoffrey M. Cooper.
      p. cm.
    Includes bibliographical references and index.
    ISBN 0-86720-191-6
    1. Cancer. 2. Oncology. I. Title.
    [DNLM: 1. Neoplasms—etiology. 2. Neoplasms—genetics.
  3. Neoplasms—therapy. QZ 200 C776e]
    RC261.C66 1992
    616.99′4—dc20
    DNLM/DLC
    for Library of Congress   91-35336
                                             CIP

Sponsoring Editor: Joseph E. Burns
Production: Hoyt Publishing Services
Cover design: Susan RZ Slovinsky
Artist: Kramer Design Associates
Typesetter: Huron Valley Graphics
Printer: R.R. Donnelley & Sons
Cover printer: Henry N. Sawyer

The cover photograph shows two cells that have just divided and are still joined at the midbody. DNA in the nucleus is stained blue. Tubulin is stained green, displaying microtubules in the cytoplasm. The photograph was generously supplied by Dr. David Albertini of Tufts University.

Printed in the United States of America
95 94 93   10 9 8 7 6 5 4 3 2

This book is dedicated to my cousin Sue and her family.

Contents

Preface

CANCER IS ONE of our most common health problems. Since it strikes about one out of every three Americans, almost all of us have family members or friends who have suffered from this disease. Thus, understanding cancer is not only important to scientists and physicians; it is a subject that touches the lives of everyone.

Many of us have a number of basic questions about cancer. What is cancer? How many different kinds of cancer are there? What causes cancer? Can it be prevented? How can cancer be treated, and how well do these treatments work? What is going on in cancer research? Will there ever be a cure for cancer? The answers to these questions are both of general interest and, in some cases, of immediate practical consequence.

Cancer is an old problem, which has been with us throughout the history of the human race. However, our understanding of cancer has undergone major changes in recent years. Cancer is fundamentally a disease in which some of the cells in the body begin to grow in an uncontrolled manner. Within the last decade, scientists have begun to unravel the basic mechanisms that control the growth of normal cells, and we are beginning to understand how defects in these mechanisms can lead to cancer. This fundamental understanding of cancer forms a central part of this book. I have attempted to integrate it into consideration of the causes of cancer and the practical aspects of cancer prevention and treatment. Since many of the most important recent advances have come directly from studies of human cancer, it has been possible to focus this book explicitly on cancer in human beings, drawing from experimental animal studies only as needed to illustrate particular points.

Elements of Human Cancer is intended to meet the needs of readers wishing to obtain a broad understanding of the cancer problem. Since cancer is a topic of general interest, the book is written so as to be accessible to a wide range of readers. The relevant scientific concepts are reviewed and explained as they are discussed. A number of additional features have been incorporated to help make the book easy to follow. An extensive glossary is provided; each chapter is previewed by a detailed outline of its contents; and key words are highlighted in the text and listed at the end of each chapter. For those readers who may wish to delve more deeply into the scientific or medical literature, each chapter is also followed by a list of references and further readings, which is keyed to individual chapter sections.

Elements of Human Cancer thus serves both as a college-level text, suitable for students with a wide range of backgrounds, and as a resource for other readers, including medical students, graduate students, and members of the public who are interested in the current status of cancer research, prevention, and treatment. For such readers, the book provides an overview and perspective of both the scientific and the clinical aspects of cancer, as well as the background needed to understand continuing advances in our attempts to deal with this disease.

Acknowledgments

I AM GRATEFUL to a number of individuals who contributed to this project.

My wife, Ann Kiessling (Faulkner Centre for Reproductive Medicine, Harvard Medical School), played a major role in stimulating me to undertake the preparation of a book providing a general overview of cancer, as well as critiquing the manuscript throughout its preparation.

The manuscript was reviewed by Robert Bast (Duke Comprehensive Cancer Center), Dorothea Becker (University of Pittsburgh Cancer Center), Joseph Bertino (Memorial Sloan-Kettering Cancer Center), Charles Boone (National Cancer Institute), Hung Fan (University of California, Irvine), Florence Haseltine (National Institute of Child Health and Human Development), Carol McClure (University of California, Riverside), Paul Neiman (Fred Hutchinson Cancer Research Center), Albey Reiner (University of Massachusetts, Amherst), Machelle Seibel (Faulkner Centre for Reproductive Medicine, Harvard Medical School), and Bert Vogelstein (The Johns Hopkins Oncology Center). Their comments were of great help and contributed significantly to both the content and presentation of the final manuscript.

It is a pleasure to thank Joe Burns, Heather Stratton, Paula Carroll, and Judy Songdahl of Jones and Bartlett Publishers for their interest, support, and encouragement throughout the various stages of both preparation of the manuscript and production of the book. The editorial and production services of David Hoyt are also greatly appreciated, as are the care and effort expended by Kris Kramer on the artwork.

PART I

THE NATURE AND CAUSES OF CANCER

THE FIRST CHAPTER of this book provides an introduction and initial overview of the cancer problem. The remaining chapters of Part I consider the different kinds of cancer, how cancers develop, and the variety of agents that cause human cancer.

Chapter 1
Basic Facts About Cancer

CANCER MAY BE the most feared disease of our time. It is second only to heart disease as a leading cause of death in the United States, and it is estimated that about one out of every three Americans will develop cancer at some point in life. In spite of major progress in cancer treatment, about half of patients with cancer ultimately die of their disease. Moreover, there is something intrinsically frightening about the very nature of cancer itself. Cancer results from abnormal growth of otherwise healthy cells. The cancer cells continue to grow and divide without restraint, eventually spreading throughout the body, interfering with the function of normal tissues and organs, and progressively leading to death. Some of the horror of cancer may be the feeling that a part of one's own body has revolted against the whole, leading to destruction from within.

WHAT IS CANCER?

There Are Many Kinds of Cancer

Cancer is a disease entity characterized by uncontrolled cell proliferation. There are many kinds of cancer; it is really a family of more than a hundred different diseases. The distinctions among the various kinds of cancer are of great practical importance, since they are treated differently and can have quite distinct outcomes for the patient. Not only are there numerous kinds of cancer, but individual cancers of the same type sometimes behave very differently from each other. For the patient and family members facing a diagnosis of cancer, it is critical to realize that such a diagnosis is not a death sentence. Some patients with cancer can be readily cured; in other cases, life can be prolonged for many years by effective therapies.

Cell Growth and Division

In spite of this considerable diversity among individual cancers, the fundamental defect in all forms of cancer is the uncontrolled growth and division of cancer cells. The human body is composed of approximately 50 trillion (5×10^{13}) individual cells, which make up tissues and organs such as the liver, heart, and brain. The growth and division of normal cells are carefully controlled to meet the needs of the whole organism. Since the entire body originates from a single cell—the fertilized egg—there is obviously a great deal of cell growth and division during normal development. The behavior of individual cells is programmed as part of the overall developmental scheme, so that each cell grows and divides as required to form the tissues and organs of the developing embryo. In the adult, a few kinds of cells (such as nerve cells) are no longer capable of division, but most types of cells continue to divide as required to replace cells that have been lost due to cell injury or death. Skin cells, for example, are continually shed and must be replaced by cell division. Some types of cells, such as the blood-forming cells, the cells that line the intestine, and the cells that form hair, divide particularly rapidly throughout adult life. In these cases, frequent cell division is needed to replace mature cells that have short lifespans. For example, approximately 5×10^{11} blood cells die each day in a normal adult, and these must be replaced by division of the blood-forming cells in the bone marrow. The system is rigorously controlled, so that the rate of division of the blood-forming cells precisely matches the rate of death of the mature blood cells. The division rates of other cell types are likewise carefully controlled, so that the different tissues and organs are maintained at a functional steady state level in the adult organism.

Unregulated Proliferation of Cancer Cells

This careful regulation of normal cell growth and division is lacking in cancer cells. Cancer cells continue to grow when they should not, apparently oblivious to the factors that control the growth of their normal counterparts. Cancer begins when a single cell begins to proliferate abnormally. This altered cell divides to form two abnormally proliferating cells, which in turn divide to form four abnormal cells, and so on (Fig. 1.1). Since each cancer cell divides to form two new cancer cells, the total number of such cells continues to increase exponentially. Thus, in this simple model, 20 cell divisions result in the formation of about a million cancer cells from one original, abnormal cell. After another 20 cell divisions, the number of cancer cells in such a hypothetical tumor would be approximately 1 trillion (10^{12})—which would correspond to about 1 pound of tissue. Rapidly proliferating cells in optimal conditions can divide as often as once a day, so a single cancer cell dividing at this rate could develop in a little over a month (40 cell divisions = 40 days) to a tumor 1 pound in size. The size of most tumors in the body does not increase this rapidly, however; it usually takes several months or years for a cancer this large to develop. In addition, as will be discussed in later chapters, cancers actually develop in a much more gradual way, so lengthy periods of time are generally required before an initially altered cell becomes a full-fledged cancer.

Overly simplistic though it may be, the above example illustrates the salient feature that is central to our understanding of cancer. Cancer is fundamentally a disease at the cellular level, in which cell proliferation has gone out of control. Consequently, cancer cells continue to grow and divide, yielding an ever-increasing mass of cancer cells. Unless checked, the cancer cells invade surrounding normal tissues, enter the circulation, and spread throughout the body, eventually interfering with the function of normal cells and leading to the death of the patient.

HOW FREQUENT IS CANCER?

The Prevalence of Cancer in Current Society

It is estimated that approximately one out of every three Americans will develop cancer at some point during life. Currently, in spite of intensive research and

Figure 1.1 Proliferation of cancer cells. Each cancer cell divides to form two new cancer cells, so the number of cancer cells doubles with each cell division.

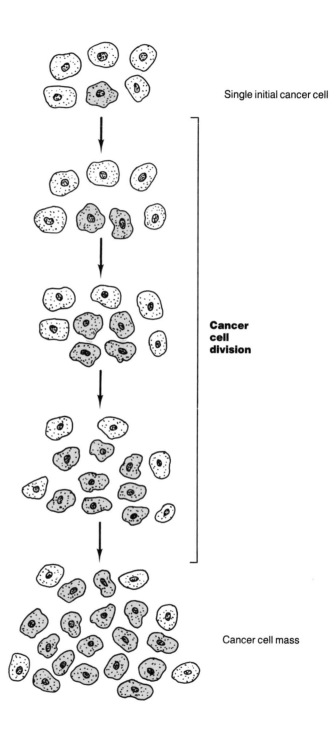

Single initial cancer cell

Cancer cell division

Cancer cell mass

some major advances in treatment, cancer claims the life of nearly one out of every four Americans (22%). Cancer is thus second only to heart disease, which is responsible for about 35% of deaths, as a cause of mortality in this country. Other causes of death, such as accidents, murders, and AIDS, each account for less than 5% of all deaths in the United States. About 1 million cases of cancer are diagnosed each year in the United States, and about 500,000 Americans die annually of the disease. Moreover, the number of cancer deaths continues to increase steadily. For example, approximately 510,000 Americans died of cancer in 1990. The corresponding number was approximately 500,000 in 1989, 490,000 in 1988, 480,000 in 1987, and 470,000 in 1986.

Cancer Was Less Common in Previous Centuries

Although cancer is clearly a major affliction of present society, this has not always been the case. Cancer has been with us throughout the history of mankind, but it has become a leading cause of death only in the last century. Prior to 1900, cancer was a comparatively rare disease that accounted for a relatively small percentage of deaths. At that time, most deaths were due to infectious diseases, such as influenza, pneumonia, and tuberculosis, and life expectancy was less than 50 years. Now, due both to general improvements in public health (such as sanitation, nutrition, and personal hygiene) and to the development of vaccines and antibiotics, infectious diseases have been virtually eliminated as major causes of death. Consequently, life expectancy has increased to over 70 years, and the major causes of death in our society have shifted to heart disease and cancer. The prevalence of cancer in current society is thus largely a consequence of the elimination of other diseases that constituted major killers in the past. The triumph of medical science against infectious disease has brought new health problems—cancer and heart disease—to the forefront of our present concerns.

THE COMMON KINDS OF CANCER

Major Cancer Sites

Although there are over one hundred different kinds of cancer, only a few occur frequently. In fact, cancers of only 11 different sites account for about 80% of all cancers in the United States (Table 1.1). The most frequent is skin cancer, which accounts for over 600,000 cases a year. However, the vast majority of skin cancers are highly curable and are therefore not included in Table 1.1. The next four most common are cancers of the lung, the colon and rectum, the breast, and the prostate. Together, these four account for over half of the total cancer incidence. Lung cancer, with approximately 157,000 cases per year, is the most frequent lethal cancer; it accounts for over one-fourth of all cancer deaths. About one-half

Table 1.1   *Most Frequent Cancers in the United States*

Cancer Site	Cases per Year		Deaths per Year	
Lung	157,000	(15%)	142,000	(28%)
Colon/rectum	155,000	(15%)	61,000	(12%)
Breast	151,000	(14%)	44,000	(9%)
Prostate	106,000	(10%)	30,000	(6%)
Bladder	49,000	(5%)	10,000	(2%)
Uterus	47,000	(5%)	10,000	(2%)
Lymphomas	43,000	(4%)	20,000	(4%)
Oral cavity	31,000	(3%)	8,000	(2%)
Pancreas	28,000	(3%)	25,000	(5%)
Leukemias	28,000	(3%)	18,000	(4%)
Skin	28,000	(3%)	9,000	(2%)
	823,000	(79%)	377,000	(74%)
All Sites	1,040,000	(100%)	510,000	(100%)

Data are for the year 1990. Nonmelanoma skin cancers (approximately 600,000 cases per year) and carcinomas of the uterine cervix diagnosed *in situ* (approximately 50,000 cases per year) are not included in incidence figures. (From American Cancer Society, *Cancer Facts and Figures*, 1990.)

of all cancer deaths are caused by three kinds of cancer—those of the lung, the breast, and the colon and rectum.

Changes in Cancer Rates over Time

The frequency of many kinds of cancer has remained relatively constant over the last 50 years, but some have shown significant changes in incidence or mortality (Fig. 1.2). The most striking change is in the frequency of lung cancer, which has increased more than tenfold since 1930. This continuing increase in the frequency of lung cancer accounts for the steady rise in the overall incidence of cancer in the United States. As will be discussed in chapter 3, this rise in lung cancer incidence is directly attributable to increased use of tobacco, particularly cigarette smoking. It follows that lung cancer could be effectively prevented by cessation of tobacco use, which would eliminate a major fraction of cancer deaths.

In contrast to the increasing mortality from lung cancer, there have been significant decreases in deaths from cancers of the stomach and uterine cervix (Fig. 1.2). In 1930, stomach cancer was the most common cause of cancer death, and cervical cancer was second. However, the incidence of stomach cancer has now declined more than fivefold. The reason for this substantial decrease in stomach cancer is probably related to changes in dietary practices—possibly the use of refrigeration correlated with decreased consumption of cured meats and increased consumption of fresh fruits and vegetables—as discussed further in

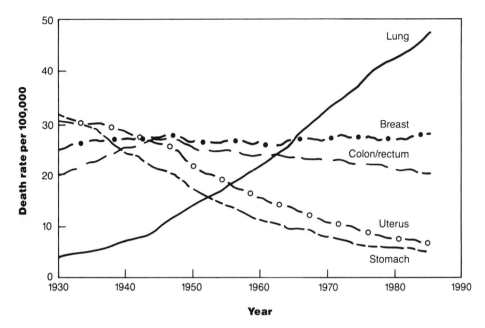

Figure 1.2   Death rates for representative cancers since 1930. Annual age-standardized death rates are shown for the United States population. Rates for lung, colon/rectum, and stomach cancers are for both sexes. Rates for cancers of the breast and uterus are for females only. Cancers of the uterus include both endometrial and cervical carcinomas. (From American Cancer Society, *Cancer Facts and Figures*, 1990.)

chapters 3 and 12. Interestingly, the incidence of stomach cancer remains very high in other countries. For example, stomach cancer is the most common cancer in Japan, with an incidence about eight times higher than in the United States. As discussed in detail in chapter 3, such differences between countries suggest the importance of environmental factors (such as differences in diet between Japan and the United States) as causes of cancer.

The Impact of Early Detection on Cervical Cancer

Whereas the decline in mortality from stomach cancer is due to a decreased incidence of the disease, the decline in mortality from cervical cancer is due, at least in part, to improved diagnosis and treatment. In particular, cancer of the uterine cervix can be diagnosed at an early stage by microscopic examination of a sample of cells from the uterus, readily obtained as part of a routine physical examination. This is the **Pap test,** named after its originator, George Papanicolau. Abnormal cervical cells can be reliably identified in such samples, allowing detection of this cancer at an early stage of the disease, when it can be effectively and easily treated. Over 50,000 cases of cervical cancer are diagnosed and cured in this way each year in the United States. The Pap test is thus the classic success story of an early **screening** test.

CANCER AND AGE

The Incidence of Cancer Increases with Age

Cancer can occur at all ages, but it becomes much more common as we grow older. This is illustrated in Figure 1.3 for the three most common cancers—those of the lung, the breast, and the colon and rectum. The incidence of colon and rectum cancer, for example, increases more than tenfold between the ages of 30 and 50, and another tenfold between 50 and 70. Such dramatic increases in cancer incidence with age are, of course, directly related to the prevalence of cancer in modern society. As previously discussed, elimination of infectious diseases resulted in a substantial increase in the average lifespan, leading to a larger fraction of older individuals and a correspondingly increased incidence of cancer in our population.

Multiple Abnormalities Are Needed to Generate Most Cancer Cells

The increasing incidence of cancer with age reflects a fundamental feature of the biology of cancer cells. As will be discussed in detail in later chapters, the conversion of a normal cell to a cancer cell does not occur as a single one-step event. Rather, the loss of growth control that characterizes cancer cells is the end result of accumulated damage to several different mechanisms that regulate normal cell growth. Development of cancer thus involves a series of progressive changes that gradually convert a normal cell into one that has lost control of its proliferation. Many years are required to accumulate the multiple abnormalities needed to generate most cancer cells, so the majority of cancers develop late in life.

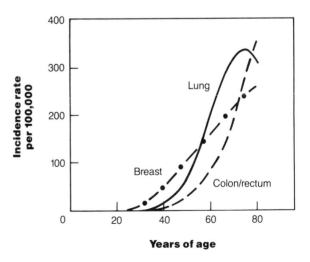

Figure 1.3 Relation of cancer incidence to age. Annual incidence rates are shown for both sexes in the United States. (Data are plotted from National Cancer Institute, *Cancer Statistics Review,* 1989.)

Childhood Cancers Are Relatively Rare

Not all cancers, however, are confined to advancing age. Indeed, the most tragic cancers are those of childhood. Fortunately, cancer is comparatively rare in children. Of the approximately 1 million cases of cancer diagnosed yearly in the United States, less than 8,000 affect children. Nonetheless, cancer is responsible for about 10% of deaths in children under age 15, making it second only to accidents (which cause about 40% of childhood deaths) as a cause of childhood mortality. The common adult cancers are rare in children. Instead, cancers of the blood and lymph systems, **leukemias** and **lymphomas,** account for about half of all childhood cancers. The other kinds of cancers that are common in children, including cancers of the brain, nervous system, bone, and kidney, are rare in adults.

TREATMENT OF CANCER

As will be discussed in detail in chapter 14, cancer is treated by surgery, radiation, and **chemotherapy.** The success of these treatments varies considerably according to the kind of cancer and how early it is detected. As previously noted, the common skin cancers (basal and squamous cell carcinomas) and cancer of the uterine cervix can be detected at very early stages, at which time they can be readily cured. As will be discussed in later chapters, early detection and treatment of breast cancer and of colon and rectum cancers are also of major importance to the outcome of these diseases.

Survival Rates for Common Cancers

The success of treatment of most cancers is usually measured as the fraction of patients who survive for five years without evidence of disease. Most patients surviving this long can be considered to have been cured of their cancer, although in some cases the cancer may recur even after this time.

The overall five-year survival rate for all cancers is now about 50%. Survival rates for some of the common adult cancers are illustrated in Figure 1.4. The most common cancer of adults, lung cancer, is difficult to detect before the disease has reached an advanced stage, and only about 10% of patients with lung cancer survive five years after diagnosis. However, the five-year survival rates for the other major adult cancers are more encouraging: approximately 75% for breast cancer, 70% for prostate cancer, and 50% for colon and rectum cancer. These survival rates are substantially influenced by the time at which the cancer is detected and treatment is initiated. For example, the five-year survival rate for breast cancer is over 90% if the cancer is detected early, but it declines to only about 20% if the cancer has progressed to an advanced stage and spread (**metastasized**) to distant body sites by the time of diagnosis. At the other extreme is pancreatic cancer, which is usually not detected until an advanced stage and is associated with a five-year survival rate of only 3%.

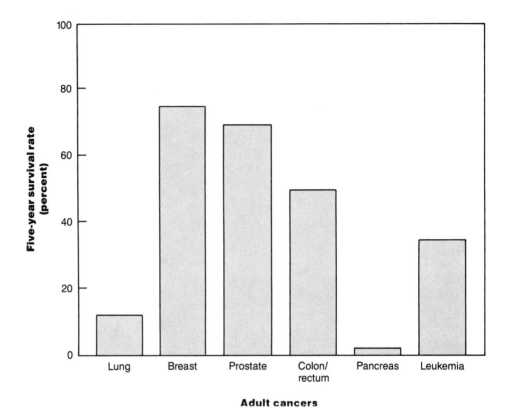

Figure 1.4   Survival rates for representative adult cancers. Five-year survival rates for all stages of the indicated cancers are shown for the United States. (From American Cancer Society, *Cancer Facts and Figures,* 1990.)

Treatment of Childhood Leukemias and Lymphomas

The development of effective therapies for some of the childhood leukemias and lymphomas has been a major, and particularly gratifying, advance in cancer treatment. Chemotherapy now leads to cures for up to 75% of children with acute lymphocytic leukemia, the most common childhood cancer. In contrast, this disease was fatal for more than 95% of children diagnosed in 1960. Chemotherapy of lymphomas has also been highly effective, with survival rates of up to 90% for Hodgkin's disease and 60% for non-Hodgkin's lymphomas. Survival rates for other childhood cancers are about 50% for bone, brain, and nervous system cancers and over 80% for kidney cancer (Wilms' tumor).

Substantial advances in the treatment of cancer have clearly been made, and the diagnosis of cancer is no longer a hopeless one. In addition, in many cases, appropriate therapy can prolong life for many years, even if a cure is not achieved. Nonetheless, the fact remains that many patients with cancer eventually die of their disease, so current treatments are ultimately unsuccessful. In addition, the survival rates for the most common adult cancers (lung, breast, and colon/rectum) have improved only slightly over the last 30 years.

The Difficulty in Finding a Cure for Cancer

The prospect of a cure for cancer—in the sense that penicillin is a cure for many bacterial infections—remains a distant hope. The difficulty lies in the nature of cancer as compared to the infectious diseases. Penicillin is an effective antibiotic because it kills the bacteria that cause disease without damaging the normal cells of the body. This works because there are major differences between bacterial cells and our own. In particular, penicillin prevents synthesis of the bacterial cell wall. Since animal cells do not have a cell wall, the normal cells of the human body are unaffected. The success of antibiotics is thus based on fundamental biochemical differences between bacteria and human cells.

Cancer, on the other hand, is due to uncontrolled growth of otherwise normal cells. There are, therefore, no readily apparent targets (like the bacterial cell wall) for a "magic bullet" against cancer. As will be discussed in subsequent chapters, most of the drugs currently used in cancer therapy are directed against all rapidly proliferating cells. They consequently affect not only cancer cells but also some normal cells, particularly the rapidly dividing cells that line the intestine, form hair, and form blood cells. Since such drugs kill normal cells in addition to cancer cells, they are quite toxic to the patient, and this toxicity severely limits their effectiveness. Consequently, much of the present research on cancer is devoted to understanding the mechanisms that control normal cell growth and to elucidating the abnormalities in cancer cells that result in loss of normal growth control. The long-range hope is that understanding the basis of normal and abnormal growth at the cellular and molecular levels will eventually lead to new strategies for selectively blocking the growth and division of cancer cells.

SUMMARY

Cancer is second only to heart disease as a cause of death in the United States, and it is expected to affect approximately one out of every three Americans. Although there are many different kinds of cancer, they all share a common fundamental basis: abnormal growth and division of cancer cells, which eventually spread through the body, invading and interfering with the function of normal tissues and organs. Cancer is thus fundamentally a disease at the cellular level, in which the cancer cell fails to respond to the controls that regulate normal cell growth and division. Such loss of growth control usually requires the accumulation of damage to several different cellular regulatory mechanisms, so most cancers develop late in life. Substantial progress has been made in the treatment of cancer, but in most cases current therapies ultimately fail, and about 50% of patients with cancer eventually die of their disease. Since cancer cells closely resemble normal cells, the fundamental problem in cancer treatment is selectively interfering with the growth of cancer cells without adverse side effects to the patient.

KEY TERMS

cancer cell
cell proliferation
growth control
Pap test
early detection
screening
chemotherapy
five-year survival
metastasis

REFERENCES AND FURTHER READING

American Cancer Society. 1990. *Cancer facts and figures—1990.* American Cancer Society, Atlanta.

Cairns, J. 1978. *Cancer: science and society.* W.H. Freeman, New York.

DeVita, V.T., Jr., Hellman, S., and Rosenberg, S.A., eds. 1989. *Cancer: principles and practice of oncology.* 3rd ed. J.B. Lippincott, Philadelphia.

Jandl, J.H. 1991. *Blood: pathophysiology.* Blackwell Scientific Publications, Boston.

National Cancer Institute. 1989. *Cancer statistics review.* National Institutes of Health, Bethesda.

Page, H.S., and Asire, A.J. 1985. *Cancer rates and risks.* 3rd ed. National Institutes of Health, Bethesda.

Pitot, H.C. 1986. *Fundamentals of oncology.* 3rd ed. Marcel Dekker, New York.

Ruddon, R.W. 1987. *Cancer biology.* 2nd ed. Oxford University Press, New York.

Silverberg, E., Boring, C.C., and Squires, T.S. 1990. Cancer statistics, 1990. *CA—A Cancer Journal for Clinicians* 40:9–26.

Chapter 2

Classification and Development of Neoplasms

SUMMARY
KEY TERMS
REFERENCES AND FURTHER READING

AS NOTED IN THE PRECEDING CHAPTER, there are more than a hundred distinct kinds of cancer. Although they all arise from fundamentally similar abnormalities in the control of cell growth and division, different kinds of cancer vary in many important characteristics, including response to treatment and the course of disease. Cancers originate from different types of normal cells and vary in their rates of growth and ability to spread to other parts of the body. In addition, individual cancers develop in a gradual, progressive fashion, so they become more rapidly growing and increasingly malignant over time. This chapter considers the classification of cancers according to both behavior and origin, as well as the ways in which cancers progressively develop and spread through the body.

BENIGN AND MALIGNANT NEOPLASMS

Neoplasms and Tumors

The central issue in cancer pathology is the distinction between benign and malignant neoplasms. A **neoplasm** (literally, "new growth") is any abnormal growth of cells. The term **tumor** is usually used synonymously with neoplasm. Neoplasms or tumors may be either benign or malignant—a difference that is critical to all aspects of dealing with the disease.

Benign Tumors

A **benign neoplasm** is a growth that remains confined to its original location, usually enclosed in a fibrous capsule, and neither invades surrounding normal tissue nor spreads to other body sites. A common skin wart is an example of a benign neoplasm. Since benign tumors remain localized to their site of origin, they can almost always be completely removed by surgery. Therefore, benign tumors are generally not life-threatening, except for those that occur in inoperable locations, such as some brain tumors.

Malignancy and Metastasis

In contrast, a **malignant neoplasm** is capable both of invading adjacent tissue and of spreading to other tissues and organs. Only malignant neoplasms are properly considered **cancers,** and it is their ability to invade normal tissues and spread throughout the body (**metastasize**) that makes cancer so dangerous. Once metastasis has occurred, the cancer can no longer be successfully dealt with by localized treatment such as surgery.

Benign and malignant neoplasms also differ in other characteristics. For example, benign neoplasms generally grow more slowly and more closely resemble their normal tissue of origin. However, the critical difference, which constitutes cancer's principal health hazard, is the ability of malignant neoplasms to invade and metastasize.

CLASSIFICATION OF NEOPLASMS ACCORDING TO ORIGIN

The Main Types of Cancer

Both malignant and benign tumors are classified according to the type of cell and tissue from which they arise. Malignant tumors are divided into three main types: carcinomas, sarcomas, and the leukemias and lymphomas. **Carcinomas,** which constitute approximately 90% of all human cancers, are malignancies of the **epithelial cells** that cover the surface of the body and line the internal organs. **Sarcomas,** which are rare in humans, are solid tumors of connective tissues, such as muscle and bone. **Leukemias** and **lymphomas,** which constitute about 8% of all human cancers, are malignancies of the blood and lymph systems.

This classification of tumors is generally correlated with the embryonic origin of the tissue from which they develop. As will be discussed in more detail in chapter 6, three layers of cells (called the germ layers) become apparent at an early stage of embryonic development: the **ectoderm,** the **mesoderm,** and the **endoderm** (Fig. 2.1). The ectoderm gives rise to the skin and nervous system; the mesoderm gives rise to supporting tissues such as bone, muscle, and blood; and

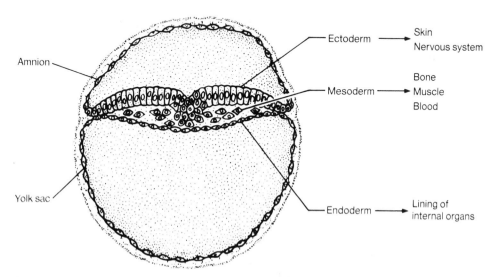

Figure 2.1   Formation of the germ layers. A cross-section through a 15–16 day embryo is shown. Cells of the three germ layers (ectoderm, mesoderm, and endoderm) subsequently give rise to the indicated tissues of the adult.

the endoderm gives rise to internal organs such as the liver, pancreas, and the epithelial lining of the stomach, intestine, and lungs. Cancers arising in tissues of ectodermal or endodermal origin are carcinomas, and cancers arising in tissues of mesodermal origin are generally termed sarcomas. Leukemias and lymphomas are formally considered subtypes of sarcomas, since these cancers of the blood and lymph systems are of mesodermal origin. It should be noted, however, that a few kinds of epithelial cells, in particular those lining the organs of the urogenital tract (such as the kidney and ovary), are actually of mesodermal origin, although cancers of these epithelia are still classified to as carcinomas.

Further Classification of Cancers

Tumors are further classified according to (1) site of origin, such as lung, breast, or colon carcinomas, and (2) cell type, such as **adenocarcinoma** (a carcinoma arising from glandular epithelium), **squamous cell carcinoma** (a carcinoma arising from flat epithelial cells), **rhabdomyosarcoma** (a sarcoma arising from muscle cells), or **acute lymphocytic leukemia** (a leukemia arising from immature lymphocytes). Table 2.1 illustrates the classification of a variety of representative cancers. For example, lung carcinomas can be divided into four major types, depending on the morphology of the cancer cells: adenocarcinomas, squamous cell carcinomas, large cell carcinomas, and **small cell carcinomas.** These different types of lung cancer grow and metastasize at different rates, so they respond to treatment somewhat differently. Squamous cell carcinomas, for example, grow more slowly and remain localized longer than the other types, so they are more likely to be successfully treated by surgery.

   Skin cancer provides a dramatic illustration of the potential differences in behavior exhibited by different histological types of cancer. Skin cancers are divided into three types: **basal cell carcinomas, squamous cell carcinomas,** and **melanomas.** Basal and squamous cell carcinomas are the most frequent, accounting for over 90% of skin cancers. These cancers very rarely metastasize; consequently, they are readily treated, with cure rates of over 99%. On the other hand, melanomas, which are cancers of the pigment-forming cells, can metastasize rapidly; they therefore represent a much more serious type of skin cancer, which is fatal to about 20% of patients.

Principles and Exceptions to Cancer Nomenclature

Several exceptions to this nomenclature further confuse an already complicated classification. As already noted, cancers of the blood (leukemias) and lymph system (lymphomas) are types of sarcomas, since they arise from tissues of mesodermal origin. However, these cancers are almost always referred to simply as leukemias and lymphomas. They are further classified according to the type of cell involved (Table 2.1). Cancers that resemble embryonic tissues are frequently designated by the suffix "-blastoma." For example, **neuroblastoma** is a childhood cancer of neuronal cells, and **retinoblastoma** is a childhood eye cancer. Finally, some cancers are named after their discoverers, such as **Wilms' tumor** (a childhood kidney cancer) and **Hodgkin's disease** (a type of lymphoma).

Table 2.1 Representative Human Cancers

Origin	Common Types of Cancer
Carcinomas	
Bladder	Transitional cell carcinoma
Breast	Adenocarcinoma: ductal, lobular, medullary, Paget's disease, inflammatory
Colon/rectum	Adenocarcinoma
Esophagus	Squamous cell carcinoma
Gall bladder	Adenocarcinoma
Kidney	Renal cell adenocarcinoma, Wilms' tumor
Larynx	Squamous cell carcinoma
Liver	Hepatocellular carcinoma
Lung	Adenocarcinoma, squamous cell carcinoma, small cell carcinoma, large cell carcinoma
Oral cavity	Squamous cell carcinoma
Ovary	Adenocarcinoma: serous, mucinous, endometrioid, clear cell
Pancreas	Adenocarcinoma
Pharynx	Squamous cell carcinoma
Prostate	Adenocarcinoma
Skin	Basal cell carcinoma, squamous cell carcinoma, melanoma
Stomach	Adenocarcinoma
Thyroid	Papillary carcinoma, follicular carcinoma, medullary carcinoma, anaplastic carcinoma
Uterus	
Cervix	Squamous cell carcinoma
Endometrium	Adenocarcinoma
Sarcomas	
Bone	Osteosarcoma, giant cell tumor, Ewing's sarcoma
Cartilage	Chondrosarcoma
Muscle	Rhabdomyosarcoma
Fibrous tissue	Fibrosarcoma
Fat cells	Liposarcoma
Blood vessels	Hemangiosarcoma, Kaposi's sarcoma
Leukemias and Lymphomas	
All blood cells	Chronic myelogenous leukemia
Lymphocytes	Acute lymphocytic leukemia, chronic lymphocytic leukemia, Hodgkin's disease, non-Hodgkin's lymphomas, myeloma
Erythrocytes	Acute erythroid leukemia
Granulocytes	Acute myelocytic leukemia, acute promyelocytic leukemia
Monocytes	Acute monocytic leukemia
Megakaryocytes	Acute megakaryocytic leukemia
Nervous System Tumors	
Brain	Astrocytoma, glioblastoma, ependymoma, medulloblastoma, meningioma, Schwannoma
Eye	Retinoblastoma
Peripheral nervous system	Neuroblastoma, neurofibrosarcoma
Germ Cell Tumors	
Testis	Seminoma, choriocarcinoma, embryonal carcinoma, yolk sac carcinoma
Ovary	Dysgerminoma, choriocarcinoma, yolk sac carcinoma, teratoma

Benign tumors are generally designated by the suffix "-oma," following a prefix that indicates the tissue of origin. Thus, a **lipoma** is a benign tumor of fat cells, whereas a **liposarcoma** is its malignant counterpart. Likewise, a colon **adenoma** is benign (a **polyp**), whereas a colon adenocarcinoma is malignant. However, this simple nomenclature is not universally followed. For example, melanomas are highly malignant skin cancers, in spite of the fact that the name "melanoma" suggests a benign tumor.

Thus, while there are general principles that govern the classification and nomenclature of both benign and malignant neoplasms, the system is complicated by many exceptions. For this reason, Table 2.1 presents a fairly comprehensive classification of representative neoplasms according to their origin. The different histologic types of many of the common cancers will be further discussed in chapters 15–17.

TUMOR INITIATION AND PROGRESSION

Tumor Clonality

A fundamental feature of the development of neoplasms, already noted in chapter 1, is that an individual tumor develops from a single initially altered cell that begins to proliferate abnormally. This is referred to as tumor **clonality,** meaning that all of the cells in a tumor arise by the growth and division of a single progenitor cell. Thus, tumors are clonal growths in which all of the neoplastic cells are the descendants of a single cell of origin. Since tumor cells inevitably divide to form more tumor cells, the continued proliferation of a single initially altered cell eventually gives rise to a clonally derived tumor cell population.

The fact that a tumor originates from a single cell does not, however, imply that the initially altered cell that ultimately gives rise to a malignant neoplasm has acquired all of the characteristics of a cancer cell. On the contrary, the development of cancer (**carcinogenesis**) is a gradual process. Normal cells become converted to cancer cells not in a single step, but through a series of changes that lead to increasingly abnormal cell growth, culminating in malignancy. As noted in chapter 1, most cancers are much more frequent in older people. This increasing incidence of cancer with age is a reflection of the fact that the generation of a cancer cell is a stepwise process that requires the accumulation of multiple alterations.

Stages of Carcinogenesis

At the cellular level, the development of a malignant neoplasm is commonly viewed as a multistep process involving mutation and selection for more rapidly growing cells with increasing invasiveness and metastatic potential (Fig. 2.2). **Tumor initiation,** the first step in development of a neoplasm, is the result of a genetic alteration (**mutation**) leading to the abnormal proliferation of a single cell. During the second stage of tumor development, **tumor promotion,** cell division leads to the formation of an actively proliferating tumor cell population.

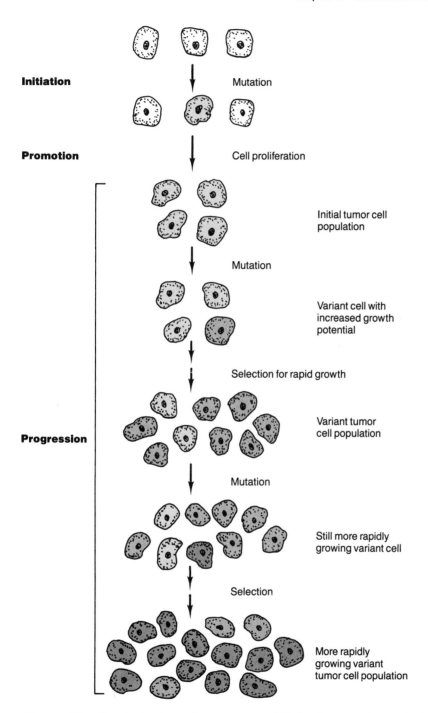

Initiation   Mutation

Promotion   Cell proliferation

                                         Initial tumor cell
                                         population

                                         Mutation

                                         Variant cell with
                                         increased growth
                                         potential

                                         Selection for rapid growth

Progression   Variant tumor
                                         cell population

                                         Mutation

                                         Still more rapidly
                                         growing variant cell

                                         Selection

                                         More rapidly
                                         growing variant
                                         tumor cell population

Figure 2.2   Clonal selection and tumor progression. The development of a malignant neoplasm occurs by a series of steps, each of which involves mutation and selection for more rapidly growing cells within the tumor cell population.

However, this is not the end of the story. **Tumor progression** continues as additional mutations occur within cells of this proliferating population. Some of these mutations may have deleterious effects on the cells, but eventually one will result in more rapid cell growth. Because of their increased growth potential, the descendants of a cell bearing such a mutation enjoy a selective advantage and outgrow the other cells in the tumor. The process is called **clonal selection,** since a new clone of tumor cells with increased proliferative potential has taken over the tumor cell population. The clonal selection process is repeated multiple times during tumor progression, resulting in the formation of more and more rapidly growing tumor cells. As will be discussed further in chapter 7, the genetic material of cancer cells may undergo more frequent alterations than that of normal cells, and this genetic instability could accelerate mutation and selection during tumor development. Tumor progression is thus viewed as a series of steps, each of which involves selection for a new clone of tumor cells with increased proliferative capacity, invasiveness, and metastatic potential. For many cancers, it has been estimated that four to six such steps are required for the development of a malignant tumor.

Development of Colon Cancer

The pathogenesis of colon and rectum carcinomas is a good example of the multistep development of a common human malignancy (Fig. 2.3). The earliest stage in tumor development, prior to formation of an actual neoplasm, is increased proliferation of cells within the colonic epithelium. One of the cells within this hyperproliferative population is then thought to give rise, by clonal selection, to a small benign neoplasm, designated a small adenoma in Figure 2.3. Further rounds of clonal selection result in the formation of adenomas of increasing size and growth potential. Malignant carcinomas then arise from the benign adenomas. Further steps in tumor progression correspond to invasion of surrounding normal tissue by the tumor cells and to metastasis of the tumor cells to other body sites.

INVASION AND METASTASIS

Progression from Carcinoma *in Situ* to Invasive Carcinoma

The ability of malignant neoplasms to spread throughout the body rather than remaining confined to their site of origin is responsible for most cancer deaths. Benign neoplasms and **carcinomas *in situ,*** small tumors that have not yet invaded adjacent connective tissues, can be readily cured by localized surgical procedures. Once invasion of surrounding normal tissue has occurred, however, the effectiveness of surgery depends on removing all of the tissue that contains cancer cells. Finally, once the cancer has metastasized to distant body sites, surgery is no longer effective and must be combined with chemotherapy to treat the disseminated disease. The common skin cancers (basal and squamous cell

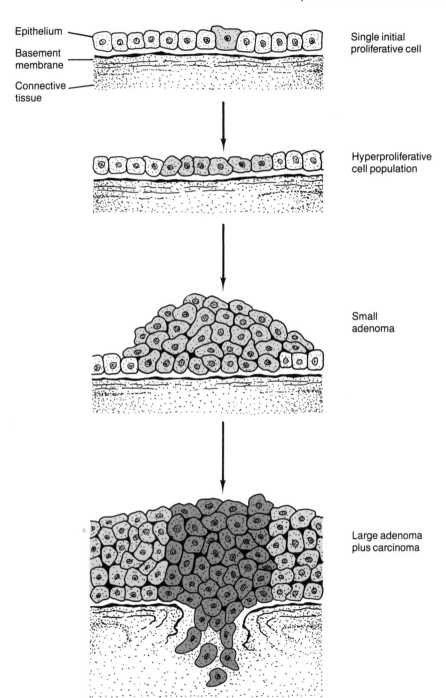

Epithelium

Basement membrane

Connective tissue

Single initial proliferative cell

Hyperproliferative cell population

Small adenoma

Large adenoma plus carcinoma

Figure 2.3   Development of colon/rectum carcinomas. A single initially altered cell gives rise to a hyperproliferative cell population, which progresses first to benign adenomas of increasing size and then to malignant carcinoma.

carcinomas) are highly curable because they rarely metastasize. Likewise, the Pap test has been effective at reducing mortality from cervical carcinoma because it allows detection of carcinoma of the uterine cervix *in situ,* at which stage simple curative treatment is possible. For other cancers, however, metastasis has occurred by the time of diagnosis in more than 50% of patients.

The first step in progression from carcinoma *in situ* to metastatic carcinoma is invasion of the tumor cells through the **basement membrane** into underlying connective tissue (Fig. 2.4). The cancer cells then continue to proliferate and spread through the normal tissue surrounding the site of the primary tumor. In some cases, the cancer cells can spread directly to adjacent organs. For example, colon carcinomas can penetrate through the wall of the colon and directly invade neighboring organs such as the bladder or small intestine. Most important, however, is entry of the tumor cells into the blood and lymphatic systems, since these are the major routes of tumor metastasis. Once a tumor has invaded normal tissue surrounding its site of origin, the cancer cells can penetrate blood and lymphatic vessels, allowing them to be carried throughout the body.

Metastasis via the Circulatory System

The circulatory system carries blood from the heart to all tissues of the body through the arteries and then returns it back to the heart through the veins. Tumor cells can enter the circulatory system by invading capillaries, the small vessels in tissues through which oxygen and nutrients in fresh blood are exchanged for carbon dioxide and waste products. Once in the circulatory system, the cancer cells can be carried to any site in the body. They can then initiate growth in a new organ by again penetrating through the capillaries and invading the adjacent tissue. Many tumors metastasize most frequently to the organ the tumor cells first reach via the circulatory system. For example, colon carcinomas often metastasize to the liver—the site to which the tumor cells are directly carried from the colon by the circulatory system. In the liver, the tumor cells reach a network of capillaries through which they can leave the circulatory system and establish a new metastatic growth.

The Lymphatic System and the Immune Response

The **lymphatic system** is a drainage system through which fluid from tissues is carried back to enter the circulation. During this process, the lymphatic fluid passes through a series of **lymph nodes,** which are aggregates of tissue containing lymphocytes. **Lymphocytes,** the principal cells of the immune system, are also carried throughout the body in the lymphatic fluid and blood. The lymphatic system thus plays a major role in the body's defense against infection. In addition, as discussed in subsequent chapters, it appears that lymphocytes can also recognize and eliminate at least some cancer cells, thereby providing a natural defense mechanism against cancer development.

Cancer cells can invade lymphatic vessels in tissues in a manner similar to their invasion of capillaries. Via the lymphatic system, they can spread through-

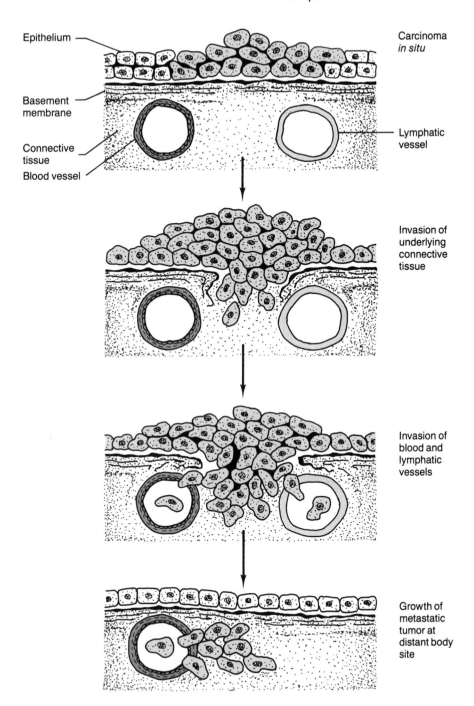

Epithelium

Basement
membrane

Connective
tissue

Blood vessel

Carcinoma
in situ

Lymphatic
vessel

Invasion of
underlying
connective
tissue

Invasion of
blood and
lymphatic
vessels

Growth of
metastatic
tumor at
distant body
site

Figure 2.4 *Invasion and metastasis.
Cancer cells first invade underlying normal
tissue and eventually reach and penetrate
blood and lymphatic vessels. The cancer cells
can then be carried throughout the body,
leading to the establishment of metastatic
tumors at distant body sites.*

out the body and enter the circulatory system as previously discussed. Since metastasis via the lymphatic system results in the deposition of cancer cells in the lymph nodes, the extent to which a tumor has spread is frequently estimated by examining lymph nodes in the area of the tumor for the presence of cancer cells.

Barriers to Metastasis

It is important to realize that entry of a cancer cell into the circulatory or lymphatic system is only the first step in metastasis of a tumor to a distant site. In order to establish a metastatic lesion successfully, the cancer cells must first survive a turbulent journey through the circulatory system and escape recognition and destruction by the immune system. They then must attach to and penetrate the walls of blood vessels to initiate a new growth at a distant site. These factors represent formidable barriers to metastasis, and the vast majority of tumor cells that enter the circulation are eliminated. In fact, it is estimated that less than one in ten thousand successfully establish a metastatic tumor. However, since rapidly growing tumors can shed millions of cells into the circulation daily, metastasis is the inevitable result of the progression of malignant neoplasms.

CLINICAL STAGING OF CANCER

Staging Systems

The extent of tumor progression at the time of diagnosis is of critical importance to prognosis and the determination of a plan of treatment. The degree to which cancers have progressed is generally described by **clinical staging,** with different systems frequently used to describe different kinds of neoplasms. However, the use of multiple staging systems, in which the same aspects of tumor progression are described in different ways, has created unnecessary complications of nomenclature. An alternative unified staging system, the **TNM system,** has been developed by the International Union Against Cancer and the American Joint Committee on Cancer. In this system, which is applicable to many different kinds of cancer, the extent of disease is described in terms of three considerations: T, the condition of the primary tumor; N, the extent of lymph node involvement; and M, the extent of distant metastases.

 As an example, staging of colon and rectum cancer by the TNM system is illustrated in Table 2.2. Tis indicates carcinoma *in situ,* and T1 through T4 indicate primary tumors characterized by increasing size and degrees of invasion of surrounding tissues. N0 indicates that the lymph nodes in the area of the tumor appear free of cancer cells, whereas N1, N2, and N3 indicate increasing extents of lymph node involvement. M0 indicates the absence of detectable metastases to distant organs, while M1 indicates their presence. Unfortunately, many M0 patients have small metastatic lesions that are not detectable at the time of diagnosis. The likelihood of such micrometastases is greater for patients with

Table 2.2   *Staging of colon and rectum cancer*

Classification	Description
Primary Tumor	
Tis	Carcinoma *in situ*
T1	Invasion into submucosa layer of colon wall
T2	Invasion through submucosa into underlying muscular layer
T3	Invasion through muscular layer
T4	Invasion into peritoneal cavity and adjacent organs
Regional Lymph Nodes	
N0	Regional lymph nodes free of tumor
N1	1–3 positive nodes
N2	4 or more positive nodes
N3	Positive nodes on named vascular trunk
Distant Metastases	
M0	No distant metastases
M1	Distant metastases present

larger or more invasive tumors (increasing grades of T) or with lymph node involvement (increasing N). Hence, T and N grading is important in predicting clinical course.

The Importance of Staging to Prognosis and Treatment

The significance of **tumor staging** to prognosis and treatment is also readily illustrated for colon and rectum cancer. The five-year survival rate after surgery is greater than 90% for patients with T1N0M0 or T2N0M0 stage colon cancer, about 80% for T3N0M0 stage, and 60–70% for T4N0M0 stage. Prognosis after surgery is significantly poorer if there is lymph node involvement. For example, five-year survival rates after surgery are approximately 50% for T3N1M0 stage colon cancer and approximately 40% for T4N1M0 stage. In addition to surgery, further treatment with radiation and/or chemotherapy appears advantageous for patients with more advanced primary tumors (T3 or T4) or with lymph node involvement (N1 or N2). Patients with detectable metastatic disease (M1 stage) are usually no longer curable, although appropriate treatment can significantly prolong life.

SUMMARY

The most important distinction in cancer pathology is that between benign neoplasms, which remain localized, and malignant neoplasms or cancers, which have the potential of invading normal tissue and spreading throughout the body.

The many different kinds of neoplasms are classified and named according to their site of origin and the type of cell within the tumor. Cancers are clonal in origin, since all cells in a tumor arise by the growth and division of a single initially altered progenitor cell. However, the pathogenesis of cancer is a complex multistep process in which malignant neoplasms develop through a series of progressive alterations. Such tumor progression is thought to occur by a series of clonal selections, involving genetic variation and selection for more rapidly growing cells within the tumor population. Cancers thus develop in a stepwise fashion from initially altered cells, which have begun to proliferate abnormally, to metastatic tumors, which have spread to distant body sites.

KEY TERMS

benign

malignant

neoplasm

tumor

cancer

carcinoma

sarcoma

leukemia

lymphoma

clonality

carcinogenesis

mutation

tumor initiation

tumor promotion

tumor progression

clonal selection

invasion

metastasis

carcinoma *in situ*

lymphatic system

lymph node

lymphocyte

immune system

clinical staging

tumor staging

TNM system

REFERENCES AND FURTHER READING

Classification of Neoplasms

Crowley, L.V. 1988. *Introduction to human disease.* 2nd ed. Jones and Bartlett, Boston.

Fawcett, D.W. 1986. *Bloom and Fawcett: a textbook of histology.* 11th ed. W.B. Saunders, Philadelphia.

Jandl, J.H. 1991. *Blood: pathophysiology.* Blackwell Scientific Publications, Boston.

Rubin, E., and Farber, J.L., eds. 1990. *Essential pathology.* J.B. Lippincott, Philadelphia.

Ruddon, R.W. 1987. *Cancer biology.* 2nd ed. Oxford University Press, New York.

Sadler, T.W. 1990. *Langman's medical embryology.* 6th ed. Williams and Wilkins, Baltimore.

Tumor Initiation and Progression

Fearon, E.R., and Vogelstein, B. 1990. A genetic model for colorectal tumorigenesis. *Cell* 61:759–767.

Fialkow, P.J. 1979. Clonal origin of human tumors. *Ann. Rev. Med.* 30:135–143.

Foulds, L. 1957. Tumor progression. *Cancer Res.* 17:355–356.

Morson, B.C. 1984. The evolution of colorectal carcinoma. *Clin. Radiol.* 35:425–431.

Muto, T., Bussey, H.J.R., and Morson, B.C. 1975. The evolution of cancer of the colon and rectum. *Cancer* 36:2251–2270.

Nowell, P.C. 1976. The clonal evolution of tumor cell populations. *Science* 194:23–28.

Nowell, P.C. 1986. Mechanisms of tumor progression. *Cancer Res.* 46:2203–2207.

Peto, R. 1977. Epidemiology, multistage models, and short-term mutagenicity tests. In *Origins of human cancer,* ed. Hiatt, H.W., Watson, J.D., and Winsten, J.A. Cold Spring Harbor Laboratory, New York. pp. 1403–1428.

Wainscot, J.S., and Fey, M.F. 1990. Assessment of clonality in human tumors: a review. *Cancer Res.* 50:1355–1360.

Invasion and Metastasis

Fidler, I.J. 1990. Critical factors in the biology of human cancer metastasis: Twenty-Eighth G.H.A. Clowes Memorial Award Lecture. *Cancer Res.* 50:6130–6138.

Liotta, L.A., and Stetler-Stevenson, W.G. 1989. Principles of molecular cell biology of cancer: cancer metastasis. In *Cancer: principles and practice of oncology,* ed. DeVita, V.T., Jr., Hellman, S., and Rosenberg, S.A. 3rd ed. J.B. Lippincott, Philadelphia. pp. 98–115.

Liotta, L.A., Steeg, P.S., and Stetler-Stevenson, W.G. 1991. Cancer metastasis and angiogenesis: an imbalance of positive and negative regulation. *Cell* 64:327–336.

Nicolson, G.L. 1987. Tumor cell instability, diversification, and progression to the metastatic phenotype: from oncogene to oncofetal expression. *Cancer Res.* 47:1473–1487.

Clinical Staging of Cancer

American Joint Committee on Cancer. 1987. *Manual for staging of cancer.* 3rd ed. J.B. Lippincott, Philadelphia.

Cohen, A.M., Shank, B., and Friedman, M.A. 1989. Colorectal cancer. In *Cancer: principles and practice of oncology,* ed. DeVita, V.T., Jr., Hellman, S., and Rosenberg, S.A. 3rd ed. J.B. Lippincott, Philadelphia. pp. 895–964.

International Union Against Cancer. 1990. *TNM atlas.* 3rd ed. Springer-Verlag, New York.

Chapter 3

Cancer and the Environment

THE PRECEDING CHAPTERS HAVE DESCRIBED CANCER as a family of diseases characterized by uncontrolled cell proliferation. What causes a normal cell to become cancerous? Since only limited success has been achieved in treating most cancers, the possibility of preventing cancer by identifying and eliminating the causative agents is an obviously important alternative.

As discussed in chapter 2, the development of cancer is a multistep process that involves a series of genetic alterations followed by selection for tumor cells with increased proliferative capacity. Since many steps are required for the development of malignancy, it is overly simplistic to speak of single agents as the cause of any given cancer. It is much more likely that multiple factors contribute to cancer development, potentially acting to increase the likelihood of any step in the series of events that culminates in malignancy. A number of different factors (called **risk factors**) may therefore affect the likelihood that any given individual will develop cancer. Such risk factors include the genetic makeup of each individual as well as the environmental agents to which he or she is exposed.

It is generally thought that the risk of developing many cancers is substantially affected by environmental agents, broadly defined as any external substance to which an individual is exposed. Environmental factors thus encompass all of the agents routinely encountered in the course of daily living, including substances present in food, air, and water. This chapter will consider chemicals, including dietary factors, and radiation as risk factors for human cancer. The role of viruses will be discussed in chapter 4 and genetic susceptibility to cancer in chapter 5. The mechanisms of action of some of these agents will be further considered in chapter 10, and the development of strategies for cancer prevention, including dietary modification, will be the subject of chapter 12.

THE IDENTIFICATION AND ACTION OF CARCINOGENS

Carcinogens Can Be Identified by Both Epidemiological and Experimental Approaches

Substances that can cause cancer, called **carcinogens,** have been identified by two general approaches. The first is **epidemiological analysis** of cancer frequencies in different human populations. In some cases, epidemiological studies have been able to establish specific causative factors related to increased cancer incidence. A striking example, discussed in detail below, is the increased frequency of lung cancer among cigarette smokers. The second general approach to identification of carcinogens is the testing of suspected agents for the ability to cause cancer in experimental animals. These two approaches are complementary. Epidemiological studies can suggest the carcinogenic potential of a suspected agent (e.g., cigarette smoke), but laboratory tests are then needed to establish whether the suspected substance really can cause cancer. Conversely, animal testing may establish the carcinogenic activity of a suspected chemical, but epidemiological studies are then needed to determine whether exposure to the suspected carcinogen is in fact associated with an increased frequency of human cancer.

Initiating Agents

Carcinogens can act in two general ways to increase the likelihood that cancer will develop (Fig. 3.1). As discussed in the preceding chapter, tumors develop as a consequence of genetic alterations that result in abnormal cell proliferation. Damage to the cell's genetic material (**DNA**) is thus a critical event in the development of cancer, so it might be expected that many carcinogens react with DNA to induce genetic alterations (**mutations**). These carcinogens are called **initiating agents,** since the induction of critical mutations is generally thought to be the initial event leading to cancer development. Such carcinogens include radiation

Figure 3.1 The action of carcinogens. Initiating agents react with DNA to induce mutations, whereas promoting agents act to stimulate cell proliferation.

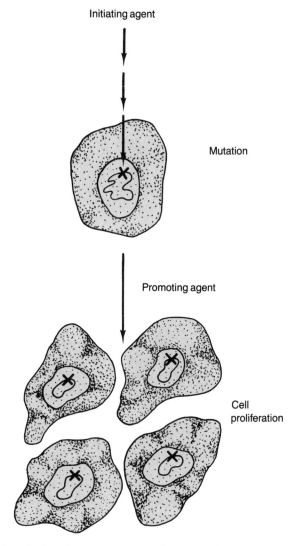

as well as a wide variety of chemicals, the structures of some of which are illustrated in Figure 3.2. Although these carcinogens clearly represent diverse types of chemicals, they share the common feature (sometimes after conversion to their active forms in the body) of reacting with DNA to cause mutations. When such mutations alter the function of critical cell regulatory genes (discussed in chapters 8–10), the outcome is abnormal cell growth, leading to cancer.

Promoting Agents

Other chemicals contribute to the development of cancer not by inducing mutations but by stimulating cell proliferation. The increased cell division resulting from exposure to these compounds, called **promoting agents,** enhances the

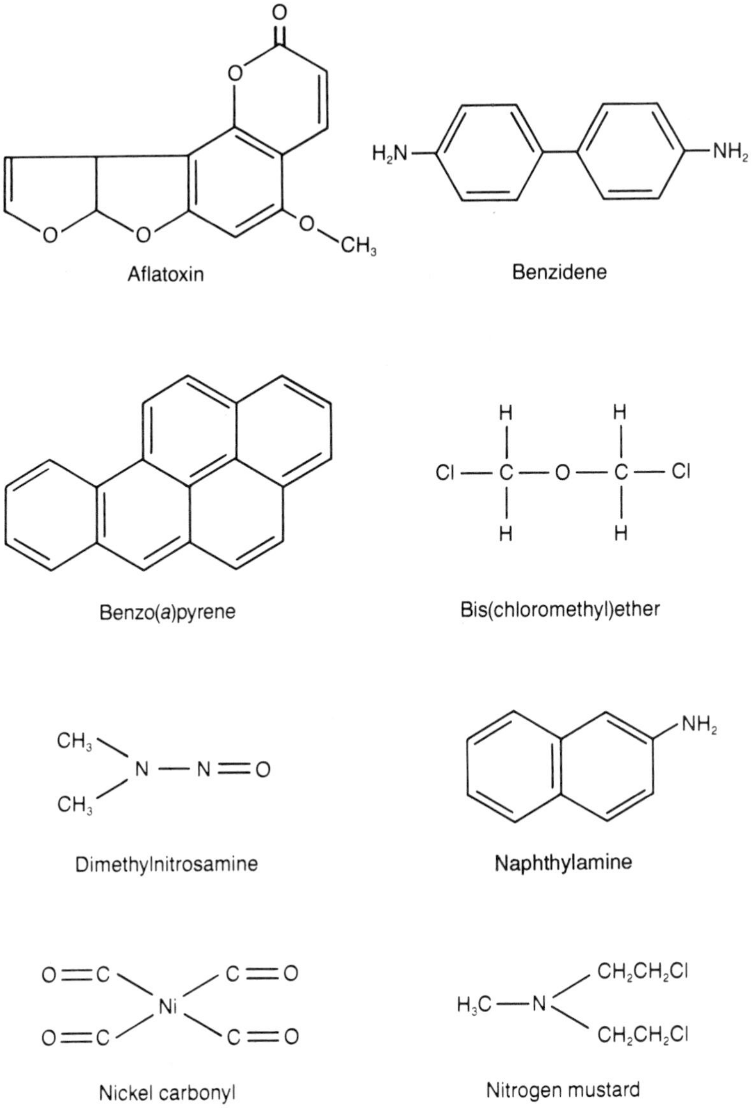

Aflatoxin

Benzidene

Benzo(a)pyrene

Bis(chloromethyl)ether

Dimethylnitrosamine

Naphthylamine

Nickel carbonyl

Nitrogen mustard

Figure 3.2   Structures of representative carcinogens.

process of carcinogenesis and is required for the initial development of a population of proliferating tumor cells. Many carcinogens act both to induce mutations and to stimulate cell proliferation, but some of the chemicals that are important contributors to the development of certain human cancers act solely by stimulating cell division. Particularly important among such promoting agents are **hormones,** especially estrogen. For example, as discussed further below, abnormal hormonal stimulation of cell proliferation by estrogen is a major factor in the development of endometrial carcinoma. Dietary fat is another example of a promoting agent that may contribute to the development of some human cancers, particularly colon carcinoma.

The Immune System

In addition to initiators and promoters, which act directly on the cancer cell, some agents increase the risk of cancer by inhibiting normal function of the immune system. The immune system, the body's natural defense against a variety of infections, also appears capable of acting against cancer cells, thereby preventing tumor development. Consequently, chemicals or other agents that interfere with normal immune function impart an increased risk that cancer will develop.

VARIATION IN CANCER INCIDENCE AMONG DIFFERENT COUNTRIES

Although many different agents can act as carcinogens in experimental studies, a much more limited number appear to be important contributors to the development of most human cancers. The remainder of this chapter will therefore focus on the major environmental sources of human exposure to chemical and radiation carcinogens.

The Frequencies of Most Cancers Vary Among Different National Populations

A principal argument linking cancer to environmental factors has come from comparisons of cancer incidence in different parts of the world. Since cancer incidence varies markedly with age (see chapter 1), such comparisons are usually based on age-adjusted rates, which normalize for differences in the age distributions of different populations. The important finding of such studies is that the frequency of specific kinds of cancer varies markedly, often more than tenfold, between different national populations. One illustration of these differences is provided by comparing those countries with the highest and lowest incidence of each kind of cancer (Table 3.1). For example, the age-adjusted incidence of colon cancer is highest in the United States (annual rate of approximately 34 per 100,000) and lowest in India (annual rate of approximately 1.8 per 100,000)—a difference of 19-fold. The cross-country variation in the incidence of different cancers ranges from approximately five-fold for leukemia to more than 150-fold for melanoma.

Studies of Migrant Populations Indicate the Importance of Environmental Factors

In principle, such variation in cancer incidence could be due either to genetic differences between the national populations or to variation in environmental factors to which the inhabitants of different countries are exposed. In some cases, these alternative possibilities have been distinguished by studies of mi-

Table 3.1   *Geographic Variation in Cancer Incidence*

Cancer	High-Incidence Area (rate per 100,000)	Low-Incidence Area (rate per 100,000)	Ratio
Melanoma	Australia (31)	Japan (0.2)	155
Prostate	United States (91)	China (1.3)	70
Liver	China (34)	Canada (0.7)	49
Cervix	Brazil (83)	Israel (3.0)	28
Stomach	Japan (82)	Kuwait (3.7)	22
Lung	United States (110)	India (5.8)	19
Colon	United States (34)	India (1.8)	19
Brain	New Zealand (9.7)	India (1.1)	9
Breast	Hawaii (94)	Israel (14)	7
Leukemia	Canada (11.6)	India (2.2)	5

Annual incidence rates for countries with the highest and lowest reported frequencies of the indicated kinds of cancer. (Data are from J.F. Fraumeni et al., 1989.)

grant populations. For example, an informative comparison can be made between the frequencies of some of the common cancers in the United States and Japan (Fig. 3.3). Breast and colon cancers are among the most common cancers in the United States but are rare in Japan. Conversely, stomach cancer, which is rare in the United States, is the most common cancer in Japan. The contribution of environmental versus genetic factors to these differences can be assessed by analysis of cancer incidence in the sizable populations of Japanese who have moved to Hawaii and California. Within one to two generations, the incidence of cancers in these Japanese-Americans shifts from the Japanese to the American pattern. The characteristic patterns of cancer incidence in Japan compared to the United States therefore appear to be determined primarily by environmental factors rather than by genetic differences.

Up to 80% of Cancers May Be Attributable to Environmental Risk Factors

Similar changes in the pattern of cancer incidence are observed among other migratory populations, suggesting that the worldwide variations in cancer frequencies are primarily due to environmental differences. On this basis, it has been estimated that environmental factors are responsible for up to 80% of all cancers. Such estimates assume that the lowest national frequency of any given type of cancer represents a minimum intrinsic level that cannot be reduced further. Frequencies above this minimum baseline level are then taken to represent the effect of environmental agents. For example, consider again the incidence of colon cancer in the United States (Table 3.1). The difference between the United States incidence (34 per 100,000) and the lowest national incidence (India, 1.8 per 100,000) would be attributed to environmental differences between the United States and India—a calculation suggesting that over 90% of colon cancer in the United States is attributable to environmental factors.

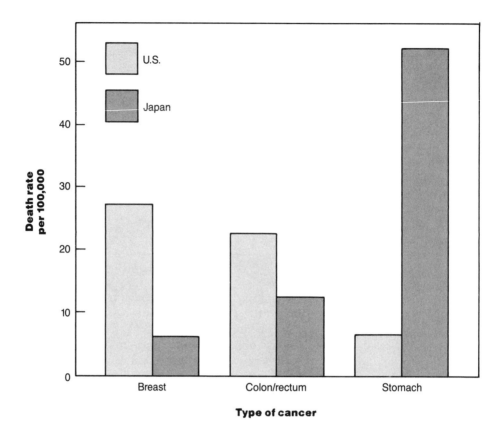

Figure 3.3   Comparison of cancer rates in the United States and Japan. Age-adjusted annual death rates from colon/rectum and stomach cancers are for both sexes. The death rate from breast cancer is for females only. (Data are plotted from H.S. Page and A.J. Asire, 1985.)

The notion that a large fraction of human cancers are associated with environmental risk factors suggests that many cancers could, in principle, be prevented by identifying and eliminating the causative agents. The identification of environmental carcinogens has therefore received a great deal of emphasis in cancer research.

SMOKING AND CANCER

Tobacco Causes About 30% of All Cancer Deaths

Cigarette smoking is unquestionably the major identified cause of human cancer, accounting for nearly one-third of all cancer deaths. Smoking is directly responsible for the vast majority (80–90%) of lung cancers. Since lung cancer is the most frequent lethal cancer in the United States, accounting for approximately 25% of all cancer deaths, it follows that a substantial fraction of total cancer mortality could be prevented by eliminating tobacco-induced lung cancer. As if this were not striking enough, smoking has also been implicated in the development of several other kinds of cancer, including cancer of the oral cavity, pharynx, larynx, esophagus, bladder, kidney, and pancreas. Combining the mortality from these cancers with that of lung cancer, it is estimated that smoking causes about 30% of all cancer deaths—clearly an impressive toll for a single environmental agent.

Relationships Between Cigarette Smoking and Lung Cancer

As noted in chapter 1, the incidence of lung cancer has increased more than tenfold since 1930. This is closely correlated with the increased use of tobacco in the early part of this century, as illustrated in Figure 3.4. A notable feature of this comparison is the lag time (or **latent period**), about 20 years, between the increase in smoking and the resultant increase in lung cancer incidence. This lag time reflects the gradual multistep development of cancer, as discussed in chapter 2; it is characteristic of carcinogen-induced human cancers. Generally, it takes 20–30 years (or more) for a cancer to develop following exposure to a carcinogen.

The cause-and-effect relationship between cigarette smoking and lung cancer is made clearer by considering differences between men and women with respect to smoking habits and lung cancer incidence. Cigarette smoking by young men in the United States began to increase around 1910, whereas ciga-

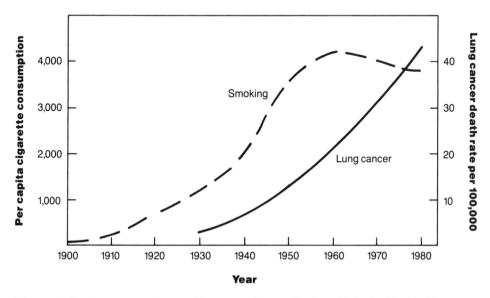

Figure 3.4   Cigarette smoking and lung cancer. Average annual per capita cigarette consumption and lung cancer death rates are shown for both sexes in the United States. (Data are from U.S. Dept. of Health and Human Services, *Reducing the Health Consequences of Smoking: 25 Years of Progress.* A report of the Surgeon General, 1989.)

rette smoking did not become popular among women until around 1940. This 30-year difference is directly reflected in lung cancer incidence (Fig. 3.5). Lung cancer in men began to increase about 1930, whereas lung cancer in women remained low until around 1960. In both men and women, increased incidence of lung cancer followed increased cigarette consumption with a characteristic 20-year lag time.

The risk of developing lung cancer depends on both the extent and the duration of smoking, as illustrated in Figure 3.6. The mortality rate from lung cancer among heavy smokers (two or more packs of cigarettes a day) is about 20 times greater than for nonsmokers. The risk for moderate smokers (one-half to one pack per day) is about half that of heavy smokers. The effect of duration of smoking on lung cancer incidence is even more dramatic. For example, the risk for an individual who began smoking at age 15 is nearly fivefold greater than that of someone who began smoking after age 25. Thus, prolonged exposure is a major factor in tobacco-induced lung cancer, suggesting that cigarette smoke may contribute to several stages of tumor development.

Figure 3.5   Annual lung cancer death rates for males and females in the United States. (Data are from U.S. Dept. of Health and Human Services, *Reducing the Health Consequences of Smoking: 25 Years of Progress.* A report of the Surgeon General, 1989.)

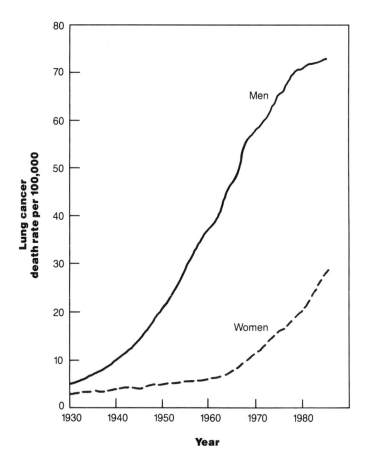

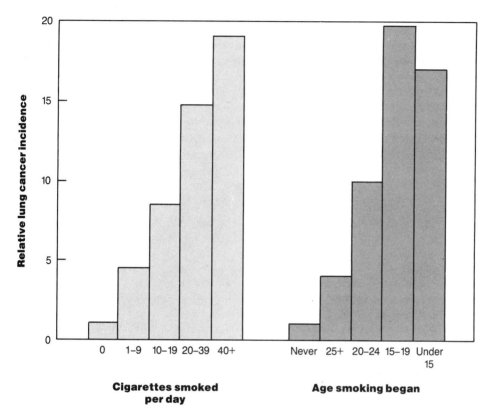

Figure 3.6 Relationship between lung cancer risk and the extent and duration of cigarette smoking. The lung cancer incidence for smokers is shown relative to that of nonsmokers. (Data are from the ACS 25-state study. U.S. Dept. of Health and Human Services, *The Health Consequences of Smoking: Cancer.* A report of the Surgeon General, 1982.)

Pipes, Cigars, Smokeless Tobacco, and Passive Smoking

Several other smoking variables also affect cancer incidence. The risk of lung cancer for smokers who inhale deeply is about twice that of smokers who inhale only slightly. Smoking filtered cigarettes with reduced tar and nicotine may also be associated with a lower cancer risk, but this is not a large difference. The risk of lung cancer for pipe or cigar smokers is less than that for cigarette smokers, although still greater than for nonsmokers. However, the effect of smoking pipes or cigars on the incidence of other cancers is similar to that of smoking cigarettes. The use of smokeless tobacco (for example, snuff) also increases cancer incidence, particularly that of oral cancer. In addition, exposure to the smoke of others (involuntary or passive smoking) may be associated with increased lung cancer risk, although the magnitude of this effect is much less than that of voluntary smoking.

The Effects of Quitting Smoking

Consistent with the role of prolonged exposure to cigarette smoke, the risk of lung cancer is substantially reduced by stopping smoking. For an ex-smoker, the risk of lung cancer remains about the same as it was at the time of quitting, rather than continuing to increase with age (Fig. 3.7). After about 20 years, the lung cancer risk for an ex-smoker becomes similar to that for a nonsmoker—approximately tenfold lower than if smoking had continued.

The epidemiological evidence implicating tobacco as a major cause of human cancer is abundantly supported by experimental studies in animals. Such studies have shown that tobacco smoke contains a variety of very potent carcinogenic chemicals, including benzo(a)pyrene, dimethylnitrosamine, and nickel compounds (see Fig. 3.2), which can act as both initiators and promoters of neoplasm development. There remains no question that smoking is the cause of a major fraction of human cancer mortality.

ALCOHOL

Cancers Associated with Excessive Alcohol Consumption

Excessive consumption of alcoholic beverages is clearly associated with an increased risk of some cancers, particularly those of the oral cavity, pharynx,

Figure 3.7 Lung cancer risk for ex-smokers. The relative incidence of lung cancer among ex-smokers, smokers, and nonsmokers is shown for 20 years after the ex-smokers quit smoking. (From U.S. Dept. of Health, Education, and Welfare, *Smoking and Health*. A report of the Surgeon General, 1979.)

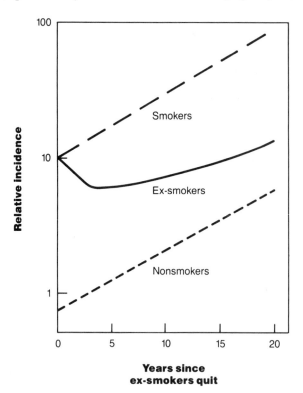

larynx, and esophagus. In addition, excess alcohol consumption can result in **cirrhosis,** leading to an increased incidence of liver cancer, probably as a consequence of excess cell proliferation resulting from chronic tissue damage.

Combined Effects of Alcohol and Smoking

The effect of alcohol on the development of oral, pharyngeal, laryngeal, and esophageal cancers seems to be exerted largely in combination with that of smoking (Fig. 3.8). For example, the risk of oral and pharyngeal cancer is increased less than twofold by either moderate smoking (1–2 packs per day) or moderate drinking (1–2 drinks per day) alone. In combination, however, moderate smoking and moderate drinking together result in more than a fourfold increased risk of developing these cancers. Heavy smoking (more than 2 packs per day) or heavy drinking (more than 4 drinks per day) each increases the risk of oral and pharyngeal cancers six- to sevenfold, while heavy drinking and heavy smoking together result in an increased risk of nearly 40-fold.

The combination of alcohol and smoking thus exerts a greater effect than either alone, suggesting that each enhances the carcinogenic activity of the

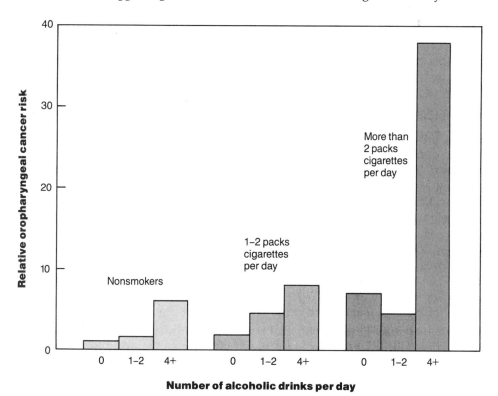

Figure 3.8   Combined effect of alcohol and smoking on oropharyngeal cancer. The risk of oropharyngeal cancer is shown relative to that of nonsmokers and nondrinkers. (From W.J. Blot et al., *Cancer Res.* 48:3282–3287, 1988.)

other. As will be noted in subsequent sections, many carcinogens behave in a similar cooperative fashion. Consequently, the risk associated with combined exposure to multiple carcinogenic substances is frequently much greater than that associated with each substance by itself.

Alcohol is only weakly carcinogenic in experimental animals, acting mainly to potentiate the action of other carcinogens, so the mechanism by which excessive consumption of alcoholic beverages increases human cancer risk is not known. In addition to alcohol itself, it is possible that other ingredients of alcoholic beverages are carcinogenic. In any event, the association between consumption of alcoholic beverages and human cancer is well established. For example, the combination of alcohol and tobacco accounts for about 75% of all oral and pharyngeal cancers, corresponding to over 6,000 deaths per year in the United States. Since most heavy drinkers are also heavy smokers, it is difficult to determine the number of cancers that can be attributed to alcohol alone. Overall, however, it has been estimated that alcohol may be a contributory factor in up to 3% of United States cancer mortality.

RADIATION

Solar Radiation and Skin Cancer

Solar radiation, in the form of ultraviolet light, is the major cause of human skin cancer. As discussed in chapter 1, skin cancer is extremely common but seldom lethal. The incidence of nonmelanoma skin cancers (basal and squamous cell carcinomas) is approximately 600,000 cases per year in the United States, almost all of which are thought to be induced by exposure to sunlight. For comparison, the incidence of lung cancer in the United States is approximately 160,000 cases per year, so solar radiation causes even more cancers than smoking. Fortunately, however, the nonmelanoma skin cancers metastasize very slowly and are consequently highly curable. This is reflected in the fact that these skin cancers result in only about 2,500 deaths per year in the United States. In contrast, lung cancer is highly lethal, resulting in over 140,000 deaths annually in the United States. Therefore, although nonmelanoma skin cancers caused by solar radiation are extremely frequent, they account for a comparatively small fraction of cancer mortality.

Exposure to excess sunlight is also associated with melanoma, which is a much more serious form of skin cancer, since it rapidly spreads to other parts of the body. The annual incidence of melanoma in the United States is about 27,000 cases, resulting in about 6,000 deaths per year. In addition, the incidence of melanoma is gradually increasing, both in the United States and throughout the rest of the world. Excessive exposure to sunlight thus appears to be a significant contributor to cancer mortality, associated with perhaps 1–2% of cancer deaths in the United States.

Ionizing Radiation

In addition to ultraviolet light, other forms of radiation can also cause cancer. In particular, the carcinogenic activity of higher-energy forms of radiation (**ionizing radiation**), including X-rays and radiation produced by the decay of radioactive particles, is well established. The induction of cancer by these forms of radiation has been demonstrated not only in laboratory animals, but also in humans exposed to excess radiation in a variety of unfortunate circumstances. For example, radiologists who used X-rays extensively in the early part of this century, before the danger was realized, suffered a three- to fourfold increased risk of leukemia. The carcinogenic effect of radiation produced by the decay of radioactive elements has likewise been demonstrated in several instances, including the increased rates of a number of cancers observed among the survivors of the World War II atomic bombings of Hiroshima and Nagasaki.

As with other carcinogens, the risk of developing cancer from exposure to ionizing radiation is related to the amount of radiation that an individual receives. In assessing the carcinogenic potential of radiation exposure, it is important to recognize that different kinds of radiation vary both in their ability to penetrate tissue and in the amount of biological damage they cause. Radiation exposure is usually assessed either as **rads** (**r**adiation **a**bsorbed **d**ose), the amount of radiation absorbed by tissue, or as **rems** (**r**adiation **e**quivalent **m**an), which are rads corrected for the biological effectiveness of the particular form of radiation being considered. For example, α-particles (positively charged particles emitted by some radioactive substances, such as uranium) are about 20-fold more damaging to biological tissue than X-rays. Therefore, while 1 rad of X-ray exposure is equal to 1 rem, 1 rad of α-particle exposure is equal to approximately 20 rems.

Medical X-Rays

The average annual radiation dose in the United States is 0.3–0.4 rem. About 80% of this radiation is from natural sources, including cosmic rays and radioactive substances in the earth's crust. Medical sources, particularly diagnostic X-irradiation, account for most of the rest of radiation exposure for the general population. About 250 million X-ray exams are performed each year in the United States, so diagnostic X-irradiation is clearly a significant source of radiation exposure. However, since the carcinogenic potential of X-irradiation has been recognized, appropriate precautions have effectively reduced the exposure and risk associated with medical X-rays for both the physician and the patient. The average radiation dose in diagnostic exams is about 0.05 rem. The risk associated with such exposure is small, estimated to be as low as one cancer induced per million X-ray examinations. According to this estimate, diagnostic X-irradiation would be responsible for approximately 250 cancers per year—less than 0.1% of total cancer mortality. This figure is imprecise, however, and may be in error by as much as several-fold. Nevertheless, although reducing exposure to medical X-rays is one effective way to avoid radiation, this must be weighed against the benefits derived from these procedures.

As discussed in chapter 14, X-rays and other forms of radiation are frequently used for treatment of cancer. Such radiation therapy involves the use of much higher doses of radiation, designed to kill the tumor cells. These higher doses carry a correspondingly higher probability of causing a second cancer to develop in the patient. However, it is again necessary to weigh risk against benefit. It is generally felt to be more urgent to treat the cancer that a patient already has than it is to worry about the possibility that a new cancer might later develop.

Radon Gas in the Home

A major source of radiation exposure for the general population is radon gas in the home, which accounts for three to four times more radiation than that received from medical X-rays. **Radon** is a natural source of α-particle radiation, formed as a decay product of uranium, which can seep into homes from underground. Radioactive decay products can then attach to aerosol particles, be inhaled, and become lodged in the lung. The carcinogenic effect of radon appears to combine with that of cigarette smoking, so the increased risk of lung cancer resulting from radon exposure is primarily observed among smokers. It has been estimated that radiation resulting from exposure to radon in United States homes may contribute to up to 10,000 lung cancer deaths per year—about 2% of total cancer mortality. The levels of indoor radon, and thus the associated risk, vary widely (over 1,000-fold) in homes throughout the United States. Many homes have much higher than average levels of radon, associated with substantially increased risks of lung cancer. Identification and modification of such homes to reduce indoor radon levels could be expected to reduce cancer risk significantly.

DIET

Variations in diet are an obvious possibility to account for the differences in cancer incidence between national populations. Many potential carcinogens are found in foods, whereas other dietary components may help to prevent the development of cancer. Moreover, a great deal of public and media attention has been focused on the role of diet in cancer. Indeed, it has been estimated that up to 30% of total cancer deaths in the United States are related to dietary factors. A number of dietary components have been suggested to increase or decrease cancer risk (Table 3.2). However, in contrast to the clear identification of tobacco, alcohol, and radiation as carcinogens, attempts to define specific dietary agents that affect cancer incidence have yielded controversial and contradictory results. Consequently, the role of potential dietary carcinogens as causes of human cancer has not yet been conclusively established.

Dietary Fat

Diets that are high in calories and fat have been repeatedly linked to increased cancer incidence. The association is strongest for dietary fat, which may contrib-

Table 3.2   *Dietary Factors and Cancer Risk*

Dietary Component	Effect on Cancer Risk
High fat	Increased risk of colon and possibly breast cancer
High calorie	Obesity resulting in increased risk of endometrial and possibly breast cancer
Cured, smoked, and pickled foods	Increased risk of stomach cancer
Aflatoxin	Increased risk of liver cancer
Vitamin A or β-carotene	Decreased risk of lung and other epithelial cancers
Vitamin C	Decreased risk of stomach cancer
Vitamin E and selenium	Deficiencies associated with increased cancer risk
Fiber	Decreased risk of colon cancer
Cruciferous vegetables	Decreased cancer risk

ute to development of breast and colon cancers. One indication of this relationship comes from comparing fat consumption with cancer incidence in different national populations. For example, there is clearly a strong cross-country correlation between dietary fat intake and breast cancer frequency (Fig. 3.9). The drawback of such correlations, however, is that there are many other differences between the countries under study besides fat intake. For example, with the exception of Japan and a few others, those countries with high rates of breast cancer also have higher levels of economic development. Consequently, there is also a good correlation between gross national product and breast cancer incidence, although this is not taken to indicate that economic development causes cancer. The question that must therefore be asked concerning studies such as that illustrated in Figure 3.9 is whether dietary fat is the real cause of high breast cancer rates. Alternatively, dietary fat might be only incidentally associated with some other unknown factor that represents the true cause of increased cancer frequencies.

The hypothesis that dietary fat is directly associated with increased cancer incidence has been supported by studies in experimental animals. For example, breast cancer incidence is substantially higher in mice that are fed a high-fat diet. On the other hand, a number of studies have failed to correlate fat intake with tumor incidence in individuals within the same national population. For example, one large study involved almost 90,000 United States women. During a four-year period, 601 cases of breast cancer were diagnosed within this group. Analysis of the dietary practices of these women failed to reveal any significant difference in fat intake between those women who developed breast cancer and those who did not. Although other studies have indicated that high fat intake may be associated with about a 1.5-fold increase in breast cancer risk, the potential association between high-fat diets and breast cancer incidence remains equivocal. On the other hand, the association between high-fat diets and colon cancer risk has been more reproducibly demonstrated. For example, analysis of the same group of United States women referred to above indicated that colon

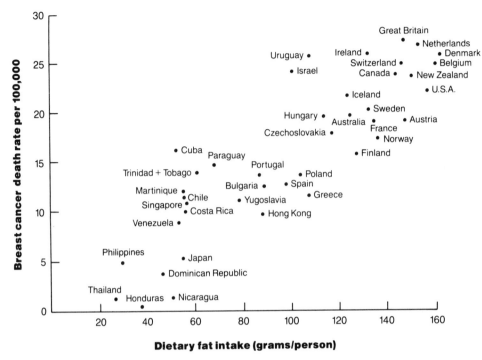

Figure 3.9   Cross-country correlation between dietary fat consumption and breast cancer incidence. The age-adjusted breast cancer death rate is shown in relation to average daily per capita consumption of dietary fat. (From K.K. Carroll. In *Carcinogens and Mutagens in the Environment*, ed. H.F. Stich, CRC Press, 1982.)

cancer was nearly twice as frequent among women whose diet contained approximately 44% of calories as fat than among women whose diet contained only 30% of calories as fat.

Although statistically significant in at least some studies, the associations between dietary fat and breast or colon cancer risk are clearly much less than the 20-fold increased risk of lung cancer that results from heavy cigarette smoking. On the other hand, since breast and colon cancer claim the lives of over 100,000 Americans per year, even a modest reduction in the risk of developing these cancers would result in a significant decrease in total cancer mortality. Unfortunately, there remain substantial inconsistencies between different studies. Although there is generally thought to be a correlation between high-fat diets and an increased risk of some cancers, particularly colon cancer, the extent to which high fat intake contributes to human cancer incidence remains unclear.

Obesity Increases Cancer Risk

Cancer of the uterine **endometrium** (the epithelial lining of the uterus) is associated with excess body weight, reflecting a high-calorie diet. For example, the risk of endometrial cancer has been estimated in different studies to be two- to fivefold higher for women weighing more than 165 pounds than for women weighing less than 125 pounds. The basis for this association may be hormone production by fat cells. Endometrial cancer is associated with increased levels of **estrogens**— hormones, produced by the ovaries, that act to stimulate proliferation of endometrial cells. Fat cells also produce estrogen and significantly contribute to estrogen levels in postmenopausal women. Consequently, production of this hormone by fat cells is thought to provide a link between obesity (defined as body weight 40% in excess of normal) and endometrial cancer. Although estrogen also stimulates the proliferation of breast epithelial cells, excess body weight is associated with only a modest increase (approximately 50%) in breast cancer risk, suggesting the importance of other factors as critical determinants of breast cancer development.

Some Dietary Factors May Decrease Cancer Risk

In contrast to dietary fat and high calorie intake, other dietary components, including dietary fiber, certain vitamins, selenium, and other compounds present in some vegetables, have been suggested to reduce cancer risk. In general, it appears that diets rich in fresh fruits and vegetables are associated with decreased cancer incidence. Such diets are high in fiber, **carotenoids** (a source of vitamin A), and vitamin C, as well as being low in fat and calories. However, studies evaluating the putative roles of individual dietary factors have generally been inconclusive.

Dietary Fiber

The possibility that dietary fiber protects against colon cancer has been investigated since the 1970s. A number of studies have suggested that the risk of colon cancer is reduced about twofold by consumption of foods that are rich in dietary fiber, including vegetables, fruits, and grains. Other studies, however, have failed to detect a protective effect of dietary fiber. Moreover, it is unclear whether the anticancer effects that have been associated with high-fiber diets are in fact due to fiber or to other vegetable components. Studies in experimental animals have been similarly inconclusive: Some experiments indicate that fiber exerts a protective effect, but others do not. Overall, it appears likely that diets rich in fiber are associated with a reduced risk of colon cancer, but this protective effect cannot unequivocally be attributed to fiber *per se*.

Vitamin A

Vitamin A and related compounds (**retinoids**) have been shown to block the development of a variety of epithelial cancers (carcinomas) in experimental animals. Diets that are rich in β-carotene, which is metabolized to form vitamin A, are associated with a decreased incidence of several cancers, including those of the lung, esophagus, stomach, bladder, and breast. The data are strongest for lung cancer, and several studies have indicated that diets deficient in green and yellow vegetables (which are rich in β-carotene) are associated with up to a twofold increase in lung cancer risk. However, it is not clear whether these protective effects are due to β-carotene itself, to vitamin A, or possibly to other vegetable components. It should be noted, however, that, as discussed in chapter 12, retinoids have been found in trial **chemoprevention** studies to reduce the incidence of secondary head and neck cancers (e.g., carcinomas of the oral cavity, pharynx, and larynx) among a population of patients who had been treated for one such cancer and were at high risk of developing a second. It thus appears that retinoids can act to inhibit the development of cancer in both experimental animals and humans, although the relevance of the high doses of retinoids used in these studies to normal dietary sources is unclear. This is particularly problematic, since the high doses of vitamin A used in both animal and human chemoprevention studies produce a number of toxic side effects.

Vitamin C

Possible anticancer effects of vitamin C have received a great deal of public attention, but there is only limited support for such claims. Some, but not all, studies have found a modest protective effect (less than twofold) of citrus fruit against stomach cancer, but it is unclear whether this is due to vitamin C or to other food components, possibly including carotenoids. Although vitamin C has been found to protect against the development of cancer in some laboratory studies, such protective effects have not been observed in other experiments.

Vitamin E and Selenium

Other vitamins have not been shown to reduce cancer incidence, although deficiencies of vitamin E combined with low levels of selenium may increase the risk of a variety of cancers. Selenium is a trace element derived from soil, and it has been observed that geographic areas with low selenium levels are associated with increased cancer incidence. Some studies have found that low blood levels of selenium in individual patients are also correlated with about a twofold increase in cancer risk, although these findings are not yet extensive enough to be conclusive. A possible anticancer role for selenium is further supported by laboratory experiments, in which high levels of dietary selenium protect against cancer development. At high doses, however, selenium is toxic, so caution must be employed in extrapolating such potential protective effects to human use.

Other Vegetable Components

In addition to fiber and vitamins, several other vegetable components may also protect against cancer. In particular, **cruciferous vegetables** (including broccoli, Brussels sprouts, cabbage, cauliflower, collards, kale, mustard greens, rutabagas, turnips, and turnip greens) contain several compounds (flavones, indoles, and isothiocyanates) that inhibit the action of carcinogens in experimental animal studies. These compounds may contribute to the protection against cancer conferred by vegetable-rich diets.

Cured, Smoked, and Pickled Foods

In addition to considerations of general dietary balance, a number of food additives have been considered potential carcinogens. An increased incidence of stomach cancer has been associated with consumption of cured, smoked, and pickled foods, which contain large amounts of salt, nitrates, and nitrites. The specific carcinogenic component associated with these foods is not known, but it is noteworthy that nitrites can be readily converted to a class of chemicals, **N-nitroso compounds** or **nitrosamines** (see Fig. 3.2), which are known to be potent carcinogens in animal studies. Vitamin C inhibits the formation of N-nitroso compounds, perhaps accounting for its suggested protective effect against stomach cancer.

Aflatoxin and Liver Cancer

Contaminants present in food can also be carcinogenic. A good example is provided by **aflatoxin** (see Fig. 3.2), a compound produced by some molds that can grow in improperly stored supplies of peanuts and grains. Aflatoxin is an extremely potent carcinogen in animals, and contaminated food supplies have been associated with liver cancer in humans. In particular, studies in Africa and Asia have shown that high rates of liver cancer in different geographic areas are specifically correlated with exposure to aflatoxin, such that the risk of liver cancer is about fivefold higher in areas with high levels of aflatoxin contamination in foods. In the United States, however, the levels of aflatoxin contamination are minimal and are not likely to contribute significantly to cancer incidence.

Other Potential Carcinogens in Food

A number of additional dietary components, both natural and synthetic, can act as carcinogens in animal tests. However, the role of these compounds in human cancer is not established. A good case in point is provided by the artificial sweetener, saccharin. Animal tests have clearly shown that high doses of saccharin cause bladder cancer in rats. However, the amount of saccharin required to induce these cancers is 100- to 1,000-fold greater than the amounts consumed by humans, and studies of possible correlations between saccharin use and human bladder cancer have been negative. It thus appears that, although saccharin is a potential dietary carcinogen, its normal use does not in fact confer increased cancer risk. Other potential carcinogens present in food include food colorings, pesticides, carcinogens produced during cooking (particularly during broiling of meat and fish), and a variety of natural carcinogens that are present in food plants. At present, there is no evidence that any of these substances makes a significant contribution to cancer incidence in the United States.

General Dietary Recommendations

Thus, although dietary factors are believed to contribute to a sizable proportion of cancers, attempts to identify specific dietary components that either increase or decrease human cancer incidence have largely been inconclusive. It is also unclear whether children may be more susceptible to some potential dietary carcinogens than adults, perhaps because of their lower body weight or the high rate of cell proliferation in growing tissues. At present, the clearest dietary risk factors are high-fat diets (for colon cancer), obesity (for endometrial cancer), and smoked, cured, and pickled foods (for stomach cancer). General dietary recommendations designed to reduce cancer risk have been made by a number of expert panels, such as the American Cancer Society. Such recommendations include reducing fat intake; eating fruits, vegetables, and high-fiber foods; and minimizing consumption of smoked, pickled, and cured foods. Such advice constitutes good general health practice and therefore represents a reasonable and prudent course of action, although the actual effects of such dietary recommendations on cancer incidence are unclear.

MEDICINES THAT CAUSE CANCER

The increased risk of cancer associated with exposure to radiation from medical X-rays was discussed earlier in this chapter. Likewise, some medications have been found to increase the risk of cancer as a side effect of their actions on cells of the patient receiving treatment. In total, it is estimated that such medications may account for up to 1% of the total cancer incidence in the United States. Some of these cancer-causing drugs have been eliminated from current practice, although other potential cancer-causing medicines remain in use because their therapeutic effects outweigh their possible dangers as carcinogens.

Estrogens

Hormones, particularly estrogens, have been a significant cause of some human cancers. An especially notable instance was the administration of **diethylstilbestrol** (DES), a synthetic estrogen, to pregnant women in the 1940s and 1950s. In the early 1970s, it was discovered that the daughters of women who had received diethylstilbestrol during pregnancy had an increased incidence of vaginal and cervical cancers. Fetal exposure to this hormone led to the development of cancer 10 to 20 years later.

In current practice, administration of diethylstilbestrol to pregnant women has, of course, been eliminated. However, estrogens are frequently used to alleviate symptoms of menopause and to prevent osteoporosis (bone thinning). Postmenopausal **estrogen replacement therapy,** particularly long-term treatment with high doses of estrogen alone, significantly increases the risk of endometrial cancer. Therefore, such therapy must be carefully considered from the standpoint of relative risk-benefit. This risk is reduced by treatment with lower doses of estrogen in combination with **progesterone,** which counteracts the stimulatory effect of estrogen on endometrial cell proliferation. Postmenopausal estrogen administration may also increase the risk of breast cancer, but this effect is more modest (less than 50%) and has not been conclusively established.

Estrogens are also the chief ingredient of birth control pills. Substantially increased rates of endometrial cancer were associated with an early form of birth control pills that contained only estrogen in relatively high doses. These pills were removed from the market in the 1970s. The birth control pills currently available contain lower doses of estrogen in combination with progesterone, and the use of these combination pills is not associated with increased endometrial cancer risk. Indeed, the incidence of endometrial cancer is actually decreased among users of these combination oral contraceptives, presumably due to the inhibition of endometrial cell proliferation by progesterone. Most studies have found no association between breast cancer incidence and use of oral contraceptives, although some studies have suggested a modest increase in risk associated with long-term use of birth control pills prior to first pregnancy.

Anticancer Drugs

Anticancer drugs are themselves frequently carcinogenic. As will be discussed in chapter 14, many of these drugs (e.g., **nitrogen mustard,** Fig. 3.2) damage the genetic material of the cell and, consequently, sometimes cause genetic changes that can convert a normal cell into a cancerous one. However, as noted above with regard to the use of radiation in cancer therapy, the benefits of anticancer drugs generally far outweigh the possibility that they will induce development of a second cancer.

Immunosuppressive Drugs

Drugs that suppress function of the immune system (**immunosuppressive drugs**) are used in organ transplant procedures to prevent rejection of the donor tissue. Studies of transplant patients have indicated that they suffer increased risk of some types of cancer, particularly lymphomas and **Kaposi's sarcoma** (a generally rare form of cancer that is also seen in AIDS patients). As will be discussed in chapter 4, these neoplasms are related to viral infections, which lead to cancer much more readily in the absence of normal immune function.

OCCUPATIONAL CARCINOGENS

Scrotal Cancer in Chimney Sweeps

Some of the clearest examples of environmental carcinogens are agents to which groups of workers are exposed in high doses. Indeed, the first observation of an **occupational carcinogen** was made in 1775 by Percival Pott, who noted a high incidence of scrotal cancer in young men who had been employed as chimney sweeps when they were children. Pott correctly attributed these cancers to the effect of soot that became lodged in the folds of the scrotum. Eventually, the identification of this occupational carcinogen led to preventive measures, and the incidence of scrotal cancer declined when chimney sweeps began to wear protective clothing and to bathe regularly.

Identified Occupational Carcinogens

Occupational carcinogens have been relatively easy to identify because, as in the case of scrotal cancer, a high incidence of a particular neoplasm becomes apparent in a specific group of workers. Consequently, studies of occupational carcinogens have identified more causes of human cancer than any other approach (Table 3.3). Once occupational carcinogens are recognized, appropriate actions can be taken to limit the exposure of affected workers. Unfortunately, as discussed in preceding sections, there is generally a long lag time between exposure to a carcinogen and development of a resulting cancer. Thus, a significant number of workers still suffer increased risks of cancer due to earlier exposure to occupational carcinogens that have since been recognized and controlled. In total, occupational exposure to carcinogens may account for up to 5% of cancer mortality.

Table 3.3 *Occupational Carcinogens*

Carcinogen	Occupational Exposure	Cancer Risk
4-aminobiphenyl	Chemical and dye workers	Bladder
Arsenic	Mining, pesticide workers	Lung, skin, and liver
Asbestos	Construction workers	Lung
Auramine	Dye workers	Bladder
Benzene	Leather, petroleum, rubber, and chemical workers	Leukemia
Benzidene	Chemical, dye, and rubber workers	Bladder
Bis (chloromethyl) ether	Chemical workers	Lung
Chromium	Metal workers, electroplaters	Lung
Isopropyl alcohol	Manufacturing by strong acid process	Nasal
Leather dust	Boot and shoe manufacturing and repair	Nasal and bladder
Mustard gas	Mustard gas workers	Lung, larynx, and nasal
Naphthylamine	Chemical, dye, and rubber workers	Bladder
Nickel dust	Nickel refining	Nasal and lung
Radon	Underground mining	Lung
Soots, tars, and oils	Coal, gas, and petroleum workers	Lung, skin, and bladder
Vinyl chloride	Rubber workers, polyvinyl chloride manufacturing	Liver
Wood dusts	Furniture manufacturing	Nasal

Asbestos and Lung Cancer

A good example of an industrial carcinogen is asbestos, which is widely used in the construction industry. The association between asbestos and lung cancer was first suggested in the 1930s and became clear by the 1950s, when it was recognized that factory workers who were heavily exposed to asbestos suffered as much as tenfold increased rates of lung cancer. Further studies have shown that asbestos exposure also causes mesothelioma, a comparatively rare cancer, and that the effect of asbestos on lung cancer appears to combine with that of smoking (Fig. 3.10). The first United States regulation limiting asbestos exposure in the workplace was put in place in the late 1960s, and such regulations have since been made increasingly stringent. However, since the lag time between exposure to asbestos and development of lung cancer can be 30 years or more, the effects of occupational exposure to high unregulated levels of asbestos are still being felt.

ENVIRONMENTAL POLLUTION

Low Levels of Carcinogenic Chemicals Are Released into the Environment

A large number of chemicals have been introduced into the environment as industrial pollutants. Many of these chemicals can induce cancer in experimental

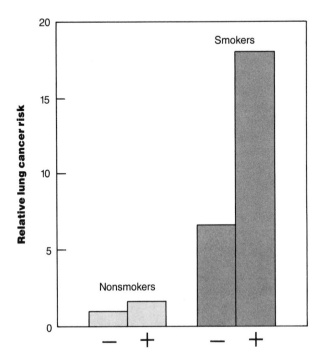

Figure 3.10 Combined effect of asbestos exposure and smoking on lung cancer. The risk of lung cancer is shown relative to that of nonsmokers who have not been exposed to asbestos. The individuals exposed to asbestos (indicated by a + sign) were shipyard workers during World War II. (From W.J. Blot et al., *N. Engl. J. Med.* 299:620–624, 1978.)

animals and must therefore be considered potential human carcinogens. In addition, most of the chemicals known to act as occupational carcinogens are also released into the environment as pollutants and might thereby confer increased cancer risk to the general population. Fortunately, however, these potential carcinogens are generally present at very low levels, and it does not seem likely that they are substantial contributors to total cancer incidence.

Industrial Pollution Has Not Had a Major Effect on Cancer Incidence

One line of argument suggesting that industrial pollution has not had a substantial impact derives from the analysis of cancer rates over the last 50 years. As noted in chapter 1, the incidence of most cancers has remained relatively constant since 1930. Lung cancer is an exception, having increased dramatically, but this is directly attributable to cigarette smoking. The absence of increasing rates for other major cancers suggests that industrial waste products introduced into the environment over this time period have not notably increased cancer incidence. However, given the usual lag time of 20 years or more between carcinogen exposure and cancer development, the effects of recently introduced pollutants might not yet be apparent.

Comparisons of cancer incidence in urban and rural areas similarly suggest that industrial pollution is not a major cancer risk factor. The overall incidence of lung cancer is higher in industrialized cities than in the country, initially suggesting the possibility that industrial pollution contributes to lung cancer risk. However, these differences appear instead to be primarily a result of an earlier increase in cigarette smoking among city dwellers. When populations with similar smoking habits are compared, the rates of lung cancer in urban and rural environments are not significantly different. A good example is provided by comparing lung cancer incidence in Finland and Britain. In spite of a much greater degree of industrial pollution in Britain, the rates of lung cancer in these two countries are similar, consistent with similar levels of cigarette consumption by the two national populations. It therefore appears that in spite of the many potential carcinogens introduced into the air by industrial processes, air pollution in cities has not in fact made a substantial contribution to overall cancer rates.

The Controversial Importance of Pollution in Cancer Risk

Consideration of the quantities of potential carcinogens released into the environment also suggests that pollution is not likely to be a major cause of cancer. An illustrative comparison is provided by noting that the amount of potentially carcinogenic burnt material inhaled per day by breathing Los Angeles smog is equivalent to smoking only one-tenth of a cigarette. As discussed earlier in this chapter, the effect of smoking on lung cancer is strongly dose dependent, and this level of smoking would not constitute a significant risk. Similar comparisons can be made between the amounts of industrial pollutants present in the general environment and those to which workers are exposed. For example, the level of asbestos generally present in city air is more than 1,000 times less than that currently permitted for occupational exposure.

Pollution has also introduced a number of carcinogens into drinking water, including known occupational carcinogens such as benzene and vinyl chloride. However, the amounts of these chemicals in drinking water are again very small compared to those in the workplace; they consequently do not seem likely to represent any significant carcinogenic risk.

All of these considerations lead to the conclusion that industrial pollution is not a major cause of human cancer. On the other hand, the number of potential carcinogens introduced into the environment by pollution is large, and entire populations are exposed to these agents throughout their lives. Moreover, the carcinogenic results of pollution are clearly evident in other animal species (particularly fish) that are exposed to higher amounts of industrial waste than humans. For example, flounder in contaminated sites of the Boston Harbor have a high incidence of liver cancer, apparently resulting from exposure to chemical pollutants. Therefore, it clearly seems prudent to control the release of carcinogens into the environment by reasonable regulations. However, the appropriate level of concern about environmental pollution in relation to cancer is a matter of some controversy, which will be further discussed in chapter 12.

SUMMARY

Comparisons of cancer incidence in different countries suggest that up to 80% of human cancers may be attributable to environmental risk factors. The major identified cause of human cancer is tobacco use, primarily cigarette smoking, which accounts for approximately 30% of total cancer mortality in the United States. Other identified risk factors for human cancer include radiation, excessive alcohol consumption, carcinogenic medicines, and occupational carcinogens, each of which accounts for only a few percent of total cancer deaths. Taken together, these known carcinogens account for about 35–40% of United States cancer mortality, which corresponds to about half of the total fraction of cancers (80%) estimated to be associated with environmental factors. A substantial portion of the remaining half may be related to diet, but specific dietary risk factors for most cancers have not been conclusively identified.

KEY TERMS

risk factor
carcinogen
epidemiology
initiating agent
promoting agent

hormone
immune system
tobacco use
latent period
alcohol
cirrhosis
ultraviolet light
ionizing radiation
X-rays
radon
dietary fat
obesity
estrogen
endometrial carcinoma
dietary fiber
carotenoids
vitamin A
retinoids
vitamin C
vitamin E
selenium
cruciferous vegetables
N-nitroso compounds
nitrosamines
aflatoxin
diethylstilbestrol
estrogen replacement therapy
progesterone
nitrogen mustard
immunosuppressive drugs
occupational carcinogens
environmental pollution

REFERENCES AND FURTHER READING

General References

Doll, R., and Peto, R. 1981. *The causes of cancer: quantitative estimates of avoidable risks of cancer in the United States today.* Oxford University Press, New York.

Hiatt, H.W., Watson, J.D., and Winsten, J.A., eds. 1977. *Origins of human cancer.* Cold Spring Harbor Laboratory, New York.

Page, H.S., and Asire, A.J. 1985. *Cancer rates and risks.* 3rd ed. National Institutes of Health, Bethesda.

The Identification and Action of Carcinogens

Ames, B.N. 1979. Identifying environmental chemicals causing mutations and cancer. *Science* 204:587–593.

Ames, B.N., and Gold, L.S. 1990. Too many rodent carcinogens: mitogenesis increases mutagenesis. *Science* 249:970–971.

Boutwell, R.K. 1974. The function and mechanism of promoters of carcinogenesis. *CRC Crit. Rev. Toxicol.* 2:419–431.

Cohen, S.M., and Ellwein, L.B. 1990. Cell proliferation in carcinogenesis. *Science* 249:1007–1011.

Friedberg, E.C. 1985. *DNA repair.* W.H. Freeman, New York.

Heidelberger, C. 1970. Chemical carcinogenesis, chemotherapy: cancer's continuing core challenges—G.H.A. Clowes Memorial Lecture. *Cancer Res.* 30:1549–1569.

IARC Working Group. 1980. An evaluation of chemicals and industrial processes associated with cancer in humans based on human and animal data: IARC monographs volumes 1 to 20. *Cancer Res.* 40:1–12.

Miller, J.A. 1970. Carcinogenesis by chemicals: an overview—G.H.A. Clowes Memorial Lecture. *Cancer Res.* 30:559–576.

Preston-Martin, S., Pike, M.C., Ross, R.K., Jones, P.A., and Henderson, B.E. 1990. Increased cell division as a cause of human cancer. *Cancer Res.* 50:7415–7421.

Variation in Cancer Incidence Among Different Countries

Armstrong, B., and Doll, R. 1975. Environmental factors and cancer incidence and mortality in different countries, with special reference to dietary practices. *Int. J. Cancer* 15:617–631.

Buell, P. 1973. Changing incidence of breast cancer in Japanese-American women. *J. Natl. Cancer Inst.* 51:1479–1483.

Doll, R. 1977. Strategy for detection of cancer hazards to man. *Nature* 265:589–596.

Fraumeni, J.F., Jr., Hoover, R.N., Devesa, S.S., and Kinlen, L.J. 1989. Epidemiology of cancer. In *Cancer: principles and practice of oncology,* ed. DeVita, V.T., Jr., Hellman, S., and Rosenberg, S.A. 3rd ed. J.B. Lippincott, Philadelphia. pp. 196–235.

Parkin, D.M., Laara, E., and Muir, C.S. 1988. Estimates of the worldwide frequency of sixteen major cancers in 1980. *Int. J. Cancer* 41:184–197.

Whittemore, A.S., Wu-Williams, A.H., Lee, M., Shu, Z., Gallagher, R.P., Deng-ao, J., Lun, Z., Xianghui, W., Kun, C., Jung, D., Teh, C.-Z., Chengde, L., Yao, X.J., Paffenbarger, R.S., Jr., and Henderson, B.E. 1990. Diet, physical activity, and colorectal cancer among Chinese in North America and China. *J. Natl. Cancer Inst.* 82:915–926.

Smoking and Cancer

Janerich, D.T., Thompson, W.D., Varela, L.R., Greenwald, P., Chorost, S., Tucci, C., Zaman, M.B., Melamed, M.R., Kiely, M., and McKneally, M.F. 1990. Lung cancer and exposure to tobacco smoke in the household. *N. Engl. J. Med.* 323:632–636.

U.S. Dept. of Health, Education, and Welfare. 1979. *Smoking and health.* A report of the Surgeon General.

U.S. Dept. of Health and Human Services. 1982. *The health consequences of smoking: cancer.* A report of the Surgeon General.

U.S. Dept. of Health and Human Services. 1989. *Reducing the health consequences of smoking: 25 years of progress.* A report of the Surgeon General.

Alcohol

Blot, W.J., McLaughlin, J.K., Winn, D.M., Austin, D.F., Greenberg, R.S., Preston-Martin, S., Bernstein, L., Schoenberg, J.B., Stemhagen, A., and Fraumeni, J.F., Jr. 1988. Smoking and drinking in relation to oral and pharyngeal cancer. *Cancer Res.* 48:3282–3287.

Radiation

Harwood, A.R., and Yaffe, M. 1982. Cancer in man after diagnostic or therapeutic irradiation. *Cancer Surveys* 1:703–731.

Kerr, R.A. 1988. Indoor radon: the deadliest pollutant. *Science* 240:606–608.

National Council on Radiation Protection and Measurements. 1987. *Ionizing radiation exposure of the population of the United States.* National Council on Radiation Protection and Measurements, Bethesda.

National Research Council. 1988. *Health risks of radon and other internally deposited alpha-emitters.* BEIR IV. Natl. Acad. Press, Washington.

Nero, A.V., Jr. 1988. Controlling indoor air pollution. *Sci. Amer.* 258:42–48.

Upton, A.C. 1977. Radiation effects. In *Origins of human cancer,* ed. Hiatt, H.W., Watson, J.D., and Winsten, J.A. Cold Spring Harbor Laboratory, New York. pp. 477–500.

Diet

Carroll, K.K. 1982. Dietary fat and its relationship to human cancer. In *Carcinogens and mutagens in the environment,* ed. Stich, H.F. CRC Press, Boca Raton, Fla. pp. 31–38.

Cohen, L.A. 1987. Diet and cancer. *Sci. Amer.* 257:42–48.

Graham, S., Dayal, H., Swanson, M., Mittelman, A., and Wilkinson, G. 1978. Diet in the epidemiology of cancer of the colon and rectum. *J. Natl. Cancer Inst.* 61:709–714.

Henderson, B.E., Ross, R., and Bernstein, L. 1988. Estrogens as a cause of human cancer: the Richard and Hinda Rosenthal Foundation Award Lecture. *Cancer Res.* 48:246–253.

Howe, G.R., Hirohata, T., Hislop, T.G., Iscovich, J.M., Yuan, J.-M., Katsouyanni, K., Lubin, F., Marubini, E., Modan, B., Rohan, T., Toniolo, P., and Shunzhang, Y. 1990. Dietary factors and risk of breast cancer: combined analysis of 12 case-control studies. *J. Natl. Cancer Inst.* 82:561–569.

Knekt, P., Aromaa, A., Maatela, J., Alfthan, G., Aaran, R.-K., Hakama, M., Hakulinen, T., Peto, R., and Teppo, L. 1990. Serum selenium and subsequent risk of cancer among Finnish men and women. *J. Natl. Cancer Inst.* 82:864–868.

McKeown-Eyssen, G.E. 1987. Fiber intake in different populations and colon cancer risk. *Prev. Med.* 16:532–539.

Moon, R.C., and Itri, L.M. 1984. Retinoids and cancer. In *The retinoids,* ed. Sporn, M.B., Roberts, A.B., and Goodman, D.S. Academic Press, New York. vol. 2, pp. 327–371.

National Research Council. 1982. *Diet, nutrition, and cancer.* Natl. Acad. Press, Washington.

National Research Council. 1989. *Diet and health: implications for reducing chronic disease risk.* Natl. Acad. Press, Washington.

Trock, B., Lanza, E., and Greenwald, P. 1990. Dietary fiber, vegetables, and colon cancer: critical review and meta-analyses of the epidemiologic evidence. *J. Natl. Cancer Inst.* 82:650–661.

Wattenberg, L.W. 1990. Inhibition of carcinogenesis by naturally-occurring and synthetic compounds. *Basic Life Sciences* 52:155–166.

Whittemore, A.S., Wu-Williams, A.H., Lee, M., Shu, Z., Gallagher, R.P., Deng-ao, J., Lun, Z., Xianghui, W., Kun, C., Jung, D., Teh, C.-Z., Chengde, L., Yao, X.J., Paffenbarger, R.S., Jr., and Henderson, B.E. 1990. Diet, physical activity, and colorectal cancer among Chinese in North America and China. *J. Natl. Cancer Inst.* 82:915–926.

Willett, W. 1989. The search for the causes of breast and colon cancer. *Nature* 338:389–394.

Willett, W.C., and MacMahon, B. 1984. Diet and cancer—an overview (two parts). *N. Engl. J. Med.* 310:633–638 and 697–703.

Willett, W.C., Stampfer, M.J., Colditz, G.A., Rosner, B.A., Hennekens, C.H., and Speizer, F.E. 1987. Dietary fat and the risk of breast cancer. *N. Engl. J. Med.* 316:22–28.

Willett, W.C., Stampfer, M.J., Colditz, G.A., Rosner, B.A., and Speizer, F.E. 1990. Relation of meat, fat, and fiber intake to the risk of colon cancer in a prospective study among women. *N. Engl. J. Med.* 323:1664–1672.

Medicines That Cause Cancer

Colditz, G.A., Stampfer, M.J., Willett, W.C., Hennekens, C.H., Rosner, B., and Speizer, F.E. 1990. Prospective study of estrogen replacement therapy and risk of breast cancer in postmenopausal women. *J. Amer. Med. Assoc.* 264:2648–2653.

Harwood, A.R., and Yaffe, M. 1982. Cancer in man after diagnostic or therapeutic irradiation. *Cancer Surveys* 1:703–731.

Henderson, B.E., Ross, R., and Bernstein, L. 1988. Estrogens as a cause of human cancer: the Richard and Hinda Rosenthal Foundation Award Lecture. *Cancer Res.* 48:246–253.

Hoover, R. 1977. Effects of drugs—immunosuppression. In *Origins of human cancer,* ed. Hiatt, H.W., Watson, J.D., and Winsten, J.A. Cold Spring Harbor Laboratory, New York. pp. 369–380.

Kelsey, J.L., and Berkowitz, G.S. 1988. Breast cancer epidemiology. *Cancer Res.* 48:5615–5623.

Miller, D.R., Rosenberg, L., Kaufman, D.W., Stolley, P., Warshauer, M.E., and Shapiro, S. 1989. Breast cancer before age 45 and oral contraceptive use: new findings. *Amer. J. Epidemiol.* 129:269–280.

Romieu, I., Berlin, J.A., and Colditz, G. 1990. Oral contraceptives and breast cancer: review and meta-analysis. *Cancer* 66:2253–2263.

Romieu, I., Willett, W.C., Colditz, G.A., Stampfer, M.J., Rosner, B., Hennekens, C.H., and Speizer, F.E. 1989. Prospective study of oral contraceptive use and risk of breast cancer in women. *J. Natl. Cancer Inst.* 81:1313–1321.

Schilsky, R.L., and Erlichman, C. 1990. Infertility and carcinogenesis: late complications of chemotherapy. In *Cancer chemotherapy: principles and practice,* ed. Chabner, B.A., and Collins, J.M. J.B. Lippincott, Philadelphia. pp. 32–58.

Occupational Carcinogens

Blot, W.J., Harrington, J.M., Toledo, A., Hoover, R., Heath, C.W., Jr., and Fraumeni, J.F., Jr. 1978. Lung cancer after employment in shipyards during World War II. *N. Engl. J. Med.* 299:620–624.

IARC Working Group. 1980. An evaluation of chemicals and industrial processes associated with cancer in humans based on human and animal data: IARC monographs volumes 1 to 20. *Cancer Res.* 40:1–12.

Selikoff, I.J. 1977. Cancer risk of asbestos exposure. In *Origins of human cancer,* ed. Hiatt, H.W., Watson, J.D., and Winsten, J.A. Cold Spring Harbor Laboratory, New York. pp. 1765–1784.

Environmental Pollution

Ames, B.N. 1989. What are the major carcinogens in the etiology of human cancer? Environmental pollution, natural carcinogens, and the causes of human cancer: six errors. In *Important advances in oncology,* ed. DeVita, V.T., Jr., Hellman, S., and Rosenberg, S.A. J.B. Lippincott, Philadelphia. pp. 237–247.

Ames, B.N., Magaw, R., and Gold, L.S. 1987. Ranking possible carcinogenic hazards. *Science* 236:271–280.

McMahon, G., Huber, L.J., Moore, M.J., Stegeman, J.J., and Wogan, G.N. 1990. Mutations in c-Ki-*ras* oncogenes in diseased livers of winter flounders from Boston Harbor. *Proc. Natl. Acad. Sci. USA* 87:841–845.

Perera, F.P., Boffetta, P., and Nisbet, I.C.T. 1989. What are the major carcinogens in the etiology of human cancer? Industrial carcinogens. In *Important advances in oncology,* ed. DeVita, V.T., Jr., Hellman, S., and Rosenberg, S.A. J.B. Lippincott, Philadelphia. pp. 249–265.

Chapter 4
Viruses, Cancer, and AIDS

THE ENVIRONMENTAL RISK FACTORS for human cancer include viruses in addition to chemicals and radiation. **Viruses** are the simplest and smallest forms of life. They are not cells and cannot multiply on their own. Instead, they replicate by infecting cells of a host organism and taking over the cellular machinery to propagate more virus particles. Viruses can therefore be considered intracellular parasites that reproduce inside cells by subverting normal cellular functions. In many cases, viruses kill the cells in which they replicate. Sometimes, however, cells are not killed by virus infection and can instead be altered to become cancer cells. Most viruses, including those that cause the common cold, polio, and influenza, do not cause cancer. However, a number of viruses (called **oncogenic viruses** or **tumor viruses**) are known to induce cancer in experimental animals, and several kinds of viruses are associated with human cancers (Table 4.1).

The role of viruses in human cancer will be the subject of this chapter. The mechanisms by which viruses act to convert normal cells to cancer cells will be discussed in Part II.

HEPATITIS B VIRUS

Hepatitis B virus, the major risk factor for **hepatocellular carcinoma** (liver cancer), is responsible for a substantial fraction of worldwide human cancer. Hepatitis B virus specifically infects liver cells and can cause acute liver damage. In 5–10% of cases, infection with hepatitis B virus is not resolved, and a persistent, usually lifelong, chronic infection of the liver can develop. Such persistent infection imparts a high risk of hepatocellular carcinoma to chronic hepatitis B virus carriers.

Correlation Between Hepatitis B Virus Infection and Hepatocellular Carcinoma

Although rare in the United States and Europe, hepatocellular carcinoma is extremely common in other parts of the world, particularly in parts of Asia and Africa. For example, the annual incidence of hepatocellular carcinoma in the United States is less than 4 per 100,000 people, whereas its incidence in China, Korea, and parts of Africa is as high as 150 per 100,000—a difference of more than 30-fold. Throughout the world, it has been estimated that there are 250,000 to 1,000,000 cases of hepatocellular carcinoma annually. Hepatocellular carcinoma is thus one of the most frequent human cancers, accounting for approximately 5–15% of worldwide cancer incidence.

The first indication of the causal link between hepatitis B virus and hepatocellular carcinoma came from epidemiological studies, which established a strong correlation between the frequency of hepatitis B virus infection and the incidence of hepatocellular carcinoma in different countries (Fig. 4.1). For example, about 10% of adults in the United States have been infected with hepatitis B virus, and less than 1% are chronic virus carriers. In contrast, virtually 100% of

Table 4.1   Viruses Associated with Human Cancers

Virus	Type of Cancer
Hepatitis B virus (HBV)	Hepatocellular carcinoma
Human papillomaviruses (HPV)	Cervical and other anogenital carcinomas, squamous cell skin carcinomas
Epstein-Barr virus (EBV)	Burkitt's and other B-cell lymphomas, nasopharyngeal carcinoma
Human T-cell lymphotropic virus (HTLV-1)	Adult T-cell leukemia
Human immunodeficiency virus (HIV)	Lymphomas, Kaposi's sarcoma, anogenital carcinomas

Figure 4.1   Worldwide distribution of hepatitis B virus and hepatocellular carcinoma. Areas of the world with a high frequency of hepatitis B virus infection also have a high incidence of hepatocellular carcinoma.

adults in China have been infected, and 10–15% are virus carriers. These differences in the frequency of chronic infection clearly correlate with the much higher incidence of hepatocellular carcinoma in China.

Comparisons of the frequency of hepatocellular carcinoma in chronically infected and uninfected individuals in the same national population further substantiate the role of hepatitis B virus inferred from cross-country studies. For example, a study of over 20,000 Chinese showed that hepatocellular carcinoma was more than 200-fold more frequent among chronic hepatitis B virus carriers. In addition, hepatitis B virus is regularly found in hepatocellular carcinoma tissues. This is consistent with the idea that infection with this virus contributes directly to changing a normal liver cell into a neoplastic one.

Hepatitis B Viruses and Liver Cancer in Experimental Animals

The role of hepatitis B virus in human hepatocellular carcinoma is also supported by studies in experimental animals. A related member of the hepatitis B virus family has been shown to cause hepatocellular carcinoma in woodchucks, and it has been demonstrated that introduction of human hepatitis B virus into mice can lead to hepatocellular carcinoma development.

Worldwide Prevalence of Hepatitis B Virus Infection

Thus, both epidemiological and experimental studies convincingly demonstrate the role of hepatitis B virus as a causative agent of hepatocellular carcinoma. In areas with high frequencies of hepatitis B virus infection, chronic carriers usually become infected early in life, although hepatocellular carcinoma does not develop until around 40 years of age. Thus, as in the case of chemical and radiation carcinogens (discussed in the preceding chapter), there is a lag time of several decades between initial carcinogenic exposure (hepatitis B virus infection) and development of a resulting cancer. Other carcinogens, such as alcohol and aflatoxin, may also contribute to hepatocellular carcinoma development. However, the role of hepatitis B virus as a major risk factor is clear. Worldwide, more than 250 million people are chronic hepatitis B virus carriers and consequently suffer more than a 100-fold increased risk of hepatocellular carcinoma.

PAPILLOMAVIRUSES

Different Papillomaviruses Cause Different Types of Tumors

The **papillomaviruses** are a widespread group of viruses that cause both benign and malignant neoplasms. The first of these viruses to be isolated caused benign skin tumors (**papillomas**) in rabbits. In humans, approximately 60 different types of papillomaviruses, which infect cells of several tissues, have been identified (Table 4.2). Some of these viruses cause only benign tumors (such as common skin warts), whereas others are important causative agents of malignant neoplasms, particularly cervical carcinoma and other genital cancers.

Many different types of human papillomaviruses infect skin cells and cause warts (Table 4.2). Generally, skin warts are harmless benign growths that do not become malignant. However, in patients with a rare skin disease (epidermodysplasia verruciformis), squamous cell carcinomas frequently develop from the warts induced by some types of human papillomaviruses. Thus, in this unusual circumstance, infection with some human papillomaviruses leads to skin cancer development.

Human Papillomaviruses and Cervical Carcinoma

The critical link between papillomaviruses and common human cancers, however, was the identification of other types of human papillomaviruses as caus-

Table 4.2   *Examples of Human Papillomaviruses*

Virus Type	Associated Neoplasm
HPV-1 and 4	Plantar warts
HPV-2, 7, 27, and 29	Common warts
HPV-3, 10, 26, and 28	Flat warts
HPV-5 and 8	Skin cancer in patients with epidermodysplasia verruciformis
HPV-6 and 11	Genital warts
HPV-16, 18, and 33	Anogenital cancers

ative agents of cervical carcinoma. Epidemiological considerations had indicated for many years that cervical carcinoma was a sexually transmitted disease. For example, cervical carcinoma is extremely rare among nuns, and most frequent among women who have had multiple sexual partners. In addition, the risk of cervical carcinoma is increased for women married to men whose former wives had the disease. Such correlations strongly suggest that cervical carcinoma is caused by a venereally transmitted infectious agent, such as a virus.

However, the causative agent resisted identification until 1983, when a distinct papillomavirus, human papillomavirus type 16 (HPV-16), was first isolated from a cervical carcinoma specimen. Subsequent studies then detected specific types of human papillomaviruses in several anogenital cancers, including cervical, vulval, penile, and perianal carcinomas. HPV-16 is the type of virus most frequently found, being present in approximately 50% of these neoplasms. Another 20% of these cancers contain HPV-18, 10% contain HPV-33, and 10% contain other human papillomavirus types. Thus, all told, about 90% of anogenital cancers are associated with papillomavirus infection. Interestingly, genital infection with different types of human papillomaviruses has quite distinct pathological consequences. For example, HPV-6 and HPV-11 cause genital warts that almost always remain benign, whereas infection with HPV-16 and HPV-18 appears to confer a high risk of progression to malignancy.

The association of human papillomaviruses with genital carcinomas is supported by a variety of experimental studies that clearly demonstrate the oncogenic potential of these viruses. It thus appears that some types of human papillomaviruses are causative agents of lethal malignancies, the most frequent of which is cervical carcinoma. As in the case of hepatitis B virus and other carcinogens, there is a lag time of several decades between primary papillomavirus infection and cancer development.

As noted in chapter 1, the mortality from cervical carcinoma is declining in the United States, where it currently accounts for only about 1% of total cancer deaths. This decline is due in large part to early diagnosis and treatment made possible by the Pap smear. Worldwide, however, cervical carcinoma is much more common, accounting for approximately 460,000 cases per year, or about 7% of total cancer incidence.

EPSTEIN-BARR VIRUS

Epstein-Barr Virus Association with Cancer Risk Is Restricted to Specific Geographic Regions

Epstein-Barr virus was the first virus recognized as a cause of human cancer. It is associated with Burkitt's lymphoma in regions of Africa, with nasopharyngeal carcinoma in parts of China, and with lymphomas in immunosuppressed individuals, including patients suffering from AIDS. Although most of the neoplasms caused by Epstein-Barr virus are confined to highly restricted geographic areas, infection with the virus is common throughout the world. In most areas, including the United States, Epstein-Barr virus infection of normal individuals does not lead to cancer. The fact that this virus is a major risk factor for neoplasms in specific geographic regions thus indicates that other factors must also contribute to the carcinogenic process.

Epstein-Barr Virus, Burkitt's Lymphoma, and the Immune System

Burkitt's lymphoma is a childhood cancer. Although rare throughout most of the world, it is the most frequent cancer in African children, occurring with an annual incidence of up to 10 cases per 100,000 children in some areas (Fig. 4.2). In these high-incidence areas, infection with Epstein-Barr virus is clearly

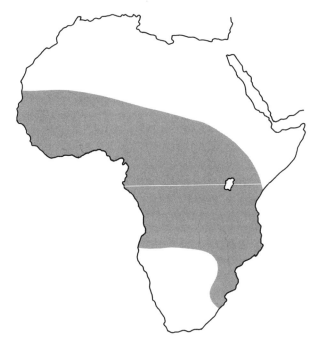

Figure 4.2   Regions of Africa with a high incidence of Burkitt's lymphoma caused by Epstein-Barr virus.

associated with lymphoma development. First, virtually all Burkitt's lympho-mas in African children contain Epstein-Barr virus. In addition, an extensive study of over 40,000 Ugandan children has demonstrated a close correlation between infection with Epstein-Barr virus and subsequent development of Burkitt's lymphoma, after a lag time of several years. These epidemiological associations are further supported by experimental studies, which demon-strate the oncogenic potential of Epstein-Barr virus in a variety of laboratory tests.

Although Burkitt's lymphoma is common only in regions of Africa, over 90% of people throughout the world have been infected with Epstein-Barr virus. Outside of Africa, however, infection with this virus generally has either no pathologic consequence or causes mononucleosis, not cancer. In mononucleosis, the Epstein-Barr virus infects and stimulates proliferation of the same kind of cells (**B lymphocytes**) that give rise to Burkitt's lymphomas. However, proliferation of these cells in mononucleosis is limited, and they do not become malig-nant. Thus, Epstein-Barr virus infection is clearly not sufficient to cause Burkitt's lymphoma, and some other factor must contribute to the high incidence of the disease in regions of Africa.

One possibility is that malaria, which is widespread in those regions of Africa with a high Burkitt's lymphoma incidence, contributes to development of this cancer by interfering with the normal functioning of the immune system. This possibility is consistent with the increased frequency of lymphomas in immunodeficient patients in non-African countries. Individuals who lack a nor-mally functioning immune system, either because of genetic abnormalities (see chapter 5), medical treatments (e.g., suppression of immune function to prevent rejection of organ transplants), or infection (e.g., AIDS), have a high risk of developing lymphomas that are associated with Epstein-Barr virus infection. It is thus possible that Epstein-Barr virus infection of a normal individual results in mononucleosis, a disease in which the proliferation of infected lymphocytes is limited by normal immune function. In the absence of a normal immune system, however, infection with the same virus may lead to continual lymphocyte prolif-eration, culminating in malignancy.

Nasopharyngeal Carcinoma

A distinct kind of cancer, nasopharyngeal carcinoma, is also caused by Epstein-Barr virus. This carcinoma is common in China, where it occurs with an annual incidence of about 10 per 100,000—a frequency about 100 times higher than in the United States. Epstein-Barr virus is found in nearly all nasopharyngeal carci-noma specimens, and epidemiological studies in China indicate that develop-ment of the disease is closely associated with Epstein-Barr virus infection. As in the case of Burkitt's lymphoma in Africa, however, other factors must be in-voked to explain why infection with this virus results in a high incidence of naso-pharyngeal carcinoma specifically in the Chinese population.

HUMAN T-CELL LYMPHOTROPIC VIRUSES

Human Retroviruses

The **human T-cell lymphotropic viruses** are members of the **retrovirus** family, a large group of viruses that cause a variety of cancers in several different animal species. The oncogenic potential of retroviruses was first discovered in experimental animals (chickens) in 1908, and members of this virus family have been studied for many years because of their ability to induce cancer in laboratory experiments. Only within the last decade, however, have retroviruses that cause human cancer been identified.

HTLV-1 and Adult T-Cell Leukemia

Human T-cell lymphotropic virus type 1 (HTLV-1) has been identified as the causative agent of adult T-cell leukemia, a disease that is rare in the United States and Europe but frequent in parts of Japan, the Caribbean, and Africa. The virus was initially isolated from patients with this disease, and epidemiological studies have provided strong evidence for its causal role. These studies have shown that HTLV-1 is prevalent in those geographic areas that have a high incidence of adult T-cell leukemia, and have established that infection with the virus is closely correlated with development of the disease. For example, adult T-cell leukemia occurs with high incidence in the two southwestern islands of Japan, Shikoku and Kyushu, but not in the rest of the country (Fig. 4.3). Virtually all individuals with the disease are infected with HTLV-1, as are approximately 20% of healthy individuals on these two islands. In contrast, infection with HTLV-1 is rare in the rest of Japan (only 1–2% of healthy individuals infected), as it is in the United States and Europe. There is thus a clear relationship between infection with HTLV-1 and development of adult T-cell leukemia, which occurs with a characteristic lag time of many years after initial infection with the virus. In addition to this strong epidemiological evidence, HTLV-1 is reproducibly found in tumor cells of adult T-cell leukemias and has readily demonstrable oncogenic potential in laboratory studies.

HTLV-2

A related virus, HTLV-2, may also be associated with human T-cell leukemia, but the evidence for its role as a cause of human cancer is much less clear. HTLV-2 has been isolated from two unusual cases of a particular kind of leukemia called hairy T-cell leukemia. Like HTLV-1, HTLV-2 has been shown to possess oncogenic potential in laboratory experiments. In contrast to HTLV-1, however, HTLV-2 has not been epidemiologically associated with the occurrence of leukemia. Its role in human cancer is therefore uncertain.

Hokkaido

Honshu

Shikoku

Kyushu

Figure 4.3 Islands of Japan with a high frequency of HTLV-1 infection and a high incidence of adult T-cell leukemia.

HUMAN IMMUNODEFICIENCY VIRUS

AIDS Patients Suffer a High Incidence of Some Neoplasms

Acquired immune deficiency syndrome (AIDS) is caused by another human retrovirus, **human immunodeficiency virus** (HIV). AIDS is not itself a cancer, nor does HIV infection appear to cause cancer directly by converting a normal cell into a neoplastic one. However, AIDS patients suffer an extremely high incidence of certain neoplasms, particularly Kaposi's sarcoma and lymphomas. These cancers appear to develop as a consequence of immunodeficiency and therefore represent a secondary effect of HIV infection.

HIV, like HTLV, infects a type of **T lymphocyte** (the T4 cell), which is an important component of a functional immune system. In contrast to HTLV, however, HIV does not cause these cells to proliferate and become cancerous. Instead, HIV kills the cells in which it replicates, eventually resulting in depletion of the population of T4 lymphocytes in an infected individual. The result of the depletion of these lymphocytes is immune deficiency. In the absence of a normally functioning immune system, AIDS sufferers are sensitive to infection by a variety of agents to which a healthy individual would be resistant. Moreover, AIDS victims suffer a high frequency of the same types of cancers that are also seen in other immunosuppressed individuals, such as patients who have undergone organ transplants (see chapter 3).

Kaposi's Sarcoma

The neoplasm most frequently encountered in AIDS patients is Kaposi's sarcoma, a tumor that is extremely rare among the general population. Kaposi's sarcoma develops in approximately 15% of AIDS patients—a frequency about 20,000 times higher than in normal individuals. Notably, the incidence of Kaposi's sarcoma in AIDS patients is also significantly higher (about 100-fold) than in other immunosuppressed individuals. The particularly high incidence of this neoplasm in AIDS patients may be due both to the effects of immunosuppression and to the fact that HIV-infected lymphocytes appear to produce factors that stimulate the growth of Kaposi's sarcoma cells.

Lymphomas and Other Tumors

Lymphomas are the second most common cancer associated with HIV, occurring in up to 10% of AIDS patients. In this case, the incidence of lymphoma in AIDS patients is similar to that in other immunosuppressed individuals. The majority of lymphomas in AIDS patients are caused by infection with Epstein-Barr virus, which (as discussed earlier in this chapter) leads to neoplasm development in the absence of normal immune function. Anogenital carcinomas associated with human papillomavirus infection also occur frequently in patients with AIDS.

A large fraction of HIV-infected individuals thus develop cancer, particularly those types of cancer associated with infection by other viruses. In contrast to the other viruses discussed in this chapter, however, these neoplasms in AIDS patients are not induced by direct action of HIV in the cancer cell. Rather, they develop as a secondary consequence of HIV infection and resulting immunodeficiency.

SUMMARY

Four kinds of viruses act directly to cause human cancers. Two of these, hepatitis B virus and the papillomaviruses, cause cancers that are common worldwide. In particular, hepatitis B virus-induced hepatocellular carcinoma and papillomavirus-induced cervical carcinoma may account for 10–20% of overall cancer incidence. Cancers induced by Epstein-Barr virus and human T-cell lymphotropic virus (HTLV) are not common worldwide, but they occur with high frequency in limited geographic areas. In addition to these viruses, human immunodeficiency virus (HIV) indirectly causes a high frequency of Kaposi's sarcoma and lymphomas, which develop as a result of immunodeficiency, in AIDS patients. A significant fraction of human cancers are thus caused by viruses. Worldwide, virus-induced cancers may account for up to one-fourth of the 80% of total cancers attributed to environmental factors.

KEY TERMS

virus
oncogenic virus

tumor virus

hepatitis B virus

hepatocellular carcinoma

papillomaviruses

warts

cervical carcinoma

Epstein-Barr virus

Burkitt's lymphoma

B lymphocyte

nasopharyngeal carcinoma

human T-cell lymphotropic viruses

retrovirus

adult T-cell leukemia

human immunodeficiency virus

AIDS

T lymphocyte

Kaposi's sarcoma

REFERENCES AND FURTHER READING

General References

Gross, L. 1983. *Oncogenic viruses.* 3rd ed. Pergamon Press, New York.

Tooze, J., ed. 1981. *Molecular biology of tumor viruses: DNA tumor viruses.* 2nd ed. Cold Spring Harbor Laboratory, New York.

Weiss, R., Teich, N., Varmus, H., and Coffin, J., eds. 1985. *Molecular biology of tumor viruses: RNA tumor viruses.* 2nd ed. Cold Spring Harbor Laboratory, New York.

Hepatitis B Virus

Beasley, R.P., Lin, C.-C., Hwang, L.-Y., and Chien, C.-S. 1981. Hepatocellular carcinoma and hepatitis B virus. *Lancet* 2:1129–1133.

Chisari, F.V., Klopchin, K., Moriyama, T., Pasquinelli, C., Dunsford, H.A., Sell, S., Pinkert, C.A., Brinster, R.L., and Palmiter, R.D. 1989. Molecular pathogenesis of hepatocellular carcinoma in hepatitis B virus transgenic mice. *Cell* 59:1145–1156.

Di Bisceglie, A.M., Rustgi, V.K., Hoofnagle, J.H., Dusheiko, G.M., and Lotze, M.T. 1988. Hepatocellular carcinoma. *Ann. Int. Med.* 108:390–401.

Ganem, D., and Varmus, H.E. 1987. The molecular biology of the hepatitis B viruses. *Ann. Rev. Biochem.* 56:651–693.

Harris, C.C. 1990. Hepatocellular carcinogenesis: recent advances and speculations. *Cancer Cells* 2:146–148.

Szmuness, W. 1978. Hepatocellular carcinoma and the hepatitis B virus: evidence for a causal association. *Prog. Med. Virol.* 24:40–69.

Papillomaviruses

Broker, T.R., and Botchan, M. 1986. Papillomaviruses: retrospectives and prospectives. In *Cancer cells*, Vol. 4, *DNA tumor viruses*, ed. Botchan, M., Grodzicker, T., and Sharp, P.A. Cold Spring Harbor Laboratory, New York. pp. 17–36.

Dürst, M., Gissmann, L., Ikenberg, H., and Zur Hausen, H. 1983. A papillomavirus DNA from a cervical carcinoma and its prevalence in cancer biopsy samples from different geographic regions. *Proc. Natl. Acad. Sci. USA* 80:3812–3815.

Kessler, I.I. 1976. Human cervical cancer as a venereal disease. *Cancer Res.* 36:783–791.

Zur Hausen, H. 1989. Papillomaviruses in anogenital cancer as a model to understand the role of viruses in human cancers. *Cancer Res.* 49:4677–4681.

Epstein-Barr Virus

De-Thé, G., Geser, A., Day, N.E., Tukei, P.M., Williams, E.H., Beri, D.P., Smith, P.G., Dean, A.G., Bornkamm, G.W., Feorino, P., and Henle, W. 1978. Epidemiological evidence for causal relationship between Epstein-Barr virus and Burkitt's lymphoma from Ugandan prospective study. *Nature* 274:756–761.

Epstein, M.A., and Achong, B.G., eds. 1986. *The Epstein-Barr virus: recent advances.* J. Wiley, New York.

Epstein, M.A., Achong. B.G., and Barr, Y.M. 1964. Virus particles in cultured lymphoblasts from Burkitt's lyphoma. Lancet 1:702–703.

Henle, W., and Henle, G. 1985. Epstein-Barr virus and human malignancies. *Adv. Viral Oncol.* 5:201–238.

Human T-Cell Lymphotropic Viruses

Blattner, W.A., Blayney, D.W., Robert-Guroff, M., Sarngadharan, M.G., Kalyanaraman, V.S., Sarin, P.S., Jaffe, E.S., and Gallo, R.C. 1983. Epidemiology of human T-cell leukemia/lymphoma virus. *J. Inf. Dis.* 147:406–416.

Gallo, R.C., Essex, M.E., and Gross, L., eds. 1984. *Human T-cell leukemia/lymphoma virus.* Cold Spring Harbor Laboratory, New York.

Hinuma, Y., Komoda, H., Chosa, T., Kondo, T., Kohakura, M., Tanenaka, T., Kikuchi, M., Ichimaru, M., Yunoki, K., Sato, I., Matsuo, R., Takiuchi, Y., Uchino, H., and Hanaoka, M. 1982. Antibodies to adult T-cell leukemia-virus-associated antigen (ATLA) in sera from patients with ATL and controls in Japan: a nation-wide sero-epidemiologic study. *Int. J. Cancer* 29:631–635.

Hinuma, Y., Nagata, K., Hanaoka, M., Nakai, M., Matsumoto, T., Kinoshita, K.-I., Shirakawa, S., and Miyoshi, I. 1981. Adult T-cell leukemia: antigen in an ATL cell line and detection of antibodies to the antigen in human sera. *Proc. Natl. Acad. Sci. USA* 78:6476–6480.

Poiesz, B.J., Ruscetti, F.W., Gazdar, A.F., Bunn, P.A., Minna, J.D., and Gallo, R.C. 1980. Detection and isolation of type C retrovirus particles from fresh and cultured lymphocytes of a patient with cutaneous T-cell lymphoma. *Proc. Natl. Acad. Sci. USA* 77:7415–7419.

Robert-Guroff, M., Nakao, Y., Notake, K., Ito, Y., Sliski, A., and Gallo, R.C. 1982. Natural antibodies to human retrovirus HTLV in a cluster of Japanese patients with adult T-cell leukemia. *Science* 215:975–978.

Rosenblatt, J.D., Golde, D.W., Wachsman, W., Giorgi, J.V., Jacobs, A., Schmidt, G.M., Quan, S., Gasson, J.C., and Chen, I.S.Y. 1986. A second isolate of HTLV-II associated with atypical hairy-cell leukemia. *N. Engl. J. Med.* 315:372–377.

Human Immunodeficiency Virus

Beral, V., Peterman, T.A., Berkelman, R.L., and Jaffe, H.W. 1990. Kaposi's sarcoma among persons with AIDS: a sexually transmitted infection? *Lancet* 335:123–128.

Ensoli, B., Barillari, G., Salahuddin, S.Z., Gallo, R.C., and Wong-Staal, F. 1990. Tat protein of HIV-1 stimulates growth of cells derived from Kaposi's sarcoma lesions of AIDS patients. *Nature* 345:84–86.

Fauci, A.S. 1988. The human immunodeficiency virus: infectivity and mechanisms of pathogenesis. *Science* 239:617–622.

Gallo, R.C. 1990. Mechanism of disease induction by HIV. *J. Acq. Immun. Def. Synd.* 3:380–389.

Pinching, A.J., and Weiss, R.A. 1986. AIDS and the spectrum of HTLV-III/LAV infection. *Int. Rev. Exptl. Path.* 28:1–44.

Chapter 5

Heredity and Cancer

THE ENVIRONMENTAL AGENTS discussed in preceding chapters (chemicals, radiation, and viruses) constitute major risk factors for many cancers. In most cases, the action of these or other carcinogens leads to the development of cancer in otherwise normal individuals. Moreover, the vast majority of patients with cancer have not inherited the disease and do not pass it on to their children. Cancer in general is therefore not considered a hereditary disease.

In spite of this generalization, there are a number of instances in which an individual's susceptibility to cancer is affected by heredity. These include rare forms of cancer that are directly inherited. Several very rare genetic diseases, such as inherited immunodeficiencies, are also associated with striking predispositions to development of cancer. In addition, less well-characterized hereditary factors appear to affect susceptibility to many of the common cancers, including breast, lung, and colon carcinomas. An individual's risk of developing cancer may therefore be determined by genetic susceptibility as well as exposure to environmental carcinogens.

INHERITED CANCERS

Hereditary Forms of Cancer

Although directly inherited cancers constitute only a small percentage of the total cancer incidence, there are rare hereditary forms of many different kinds of cancer (Table 5.1 is a representative listing). In these cases, a strong predisposition to cancer is transmitted directly from parent to child, and development of cancer is inherited like any other genetic trait, such as hair or eye color. Most of these inherited predispositions lead to the development of only one or a few specific types of cancer, not to all cancers in general. The mode of inheritance of these cancers suggests that susceptibility is determined by single genes, which are transmitted in a genetically dominant fashion (Fig. 5.1). Thus, one-half of the children of an affected parent will inherit the cancer susceptibility gene from that parent. Since cancer susceptibility is dominant, the children who have inherited this gene will almost always develop cancer, even in the presence of a normal gene copy from the other parent. Such inherited cancers generally occur early in life, and affected individuals frequently develop multiple independent tumors.

Inheritance of Retinoblastoma

Many of the inherited cancers are rare diseases of childhood. A good example is **retinoblastoma,** an eye tumor that usually develops in children by age 3. Retinoblastoma is a neoplasm of embryonic retinal cells ("retino" = retinal, "blast" = embryonic cell, and "oma" = tumor). Provided that the disease is detected early, retinoblastoma can be successfully treated by surgery and radiotherapy, so most children with retinoblastoma now survive to have families. This has allowed inheritance of the disease to be studied by following the family history and offspring of retinoblastoma patients. Such studies have shown that

Table 5.1 Examples of Inherited Cancers

Genetic Disease	Types of Cancer
Beckwith-Wiedemann syndrome	Wilms' tumor, hepatoblastoma, rhabdomyosarcoma, adrenal carcinoma
Dysplastic nevus syndrome	Melanoma
Familial adenomatous polyposis	Colon carcinoma
Li-Fraumeni cancer family syndrome	Sarcomas, breast carcinoma, brain tumors, leukemia, adrenocortical tumors
Lynch cancer family syndrome	Breast and ovarian carcinoma
Multiple endocrine neoplasia–1	Pituitary, parathyroid, adrenal, and pancreatic tumors
Multiple endocrine neoplasia–2a	Pheochromocytoma, medullary thyroid carcinoma
Multiple endocrine neoplasia–2b	Pheochromocytoma, medullary thyroid carcinoma, mucosal neuroma
Neuroblastoma	Neuroblastoma
Neurofibromatosis type 1 (von Recklinghausen's disease)	Neurofibrosarcoma, malignant Schwannoma
Neurofibromatosis type 2	Acoustic neuroma, meningioma
Nevoid basal cell carcinoma	Basal cell skin carcinoma
Retinoblastoma	Retinoblastoma, osteosarcoma
Von Hippel-Lindau syndrome	Retinal and cerebellar angioma, renal carcinoma
Warthin cancer family syndrome	Colon and endometrial carcinomas
Wilms' tumor (WAGR syndrome)	Wilms' tumor

retinoblastoma can occur in two forms: either as an inherited disease or in a sporadic, noninherited manner (Fig. 5.2). Individuals with the inherited form of the disease transmit retinoblastoma to approximately half of their offspring, consistent with the expected pattern of inheritance of a single dominant gene. In contrast, sporadic retinoblastoma occurs without prior family history and is not transmitted to the patients' offspring. As is typical of the hereditary cancers, most children with inherited retinoblastoma develop multiple tumors in both eyes, whereas children with the sporadic form of the disease develop only a single tumor in one eye. In addition, children who have inherited the disease usually develop tumors at a younger age than children with the sporadic form.

Retinoblastoma is an infrequent disease, affecting about 1 in 20,000 children, with inherited disease accounting for about 40% of the total incidence. Other childhood cancers for which hereditary forms are known, such as **Wilms' tumor** (a kidney cancer), also occur infrequently, affecting about 1 in 10,000 children. As in the case of retinoblastoma, patients with inherited Wilms' tumor usually develop multiple neoplasms in both kidneys. In contrast to retinoblastoma, however, less than 10% of Wilms' tumors (and other childhood cancers) are hereditary. Inherited cancers thus constitute a small fraction of total childhood cancer incidence.

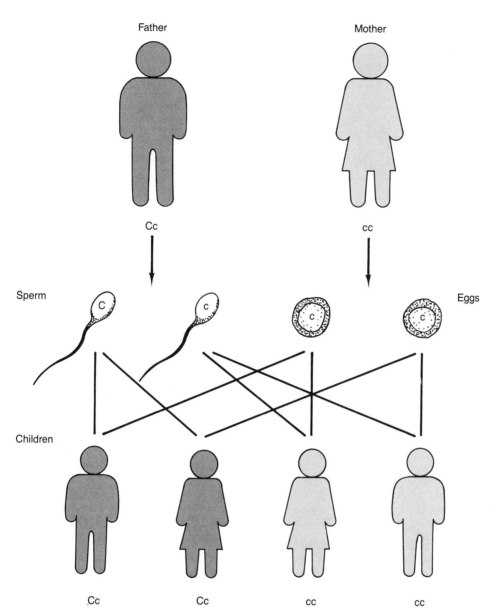

Figure 5.1   Inheritance of a dominant cancer susceptibility gene. The gene for cancer susceptibility is designated C, and its normal equivalent is designated c. In this example, one parent (the father) has one copy of C and one copy of c. Since C is dominant, he develops cancer. The mother is normal and has two copies of c. The father transmits the C gene to one-half of his children, resulting in cancer development.

Figure 5.2 A family pedigree illustrating the inheritance of retinoblastoma. Affected individuals are indicated by filled symbols. In inherited retinoblastoma, the retinoblastoma susceptibility gene is transmitted to approximately one-half of offspring.

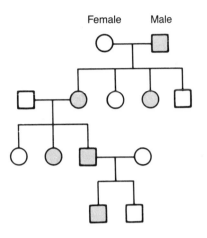

Inheritance of Rare Adult Cancers

Hereditary cancers are not limited to rare cancers of childhood. There are also inherited forms of many common adult cancers, including colon and breast carcinomas (Table 5.1). In these cases, however, the inherited forms account for no more than a few percent of total disease incidence.

Colon carcinoma is a good example of a common cancer with both inherited and sporadic forms. About 1 in 20 Americans are affected by colon cancer, which occurs nearly 1,000 times more frequently than the rare childhood cancers discussed above. The majority of colon cancers occur sporadically, but two inherited forms of the disease have been identified. The most frequently recognized is known as **familial adenomatous polyposis (adenoma** = a benign tumor of glandular epithelium and **polyp** = a benign tumor projecting from an epithelial surface). This disease, like retinoblastoma, is inherited as a single dominant genetic trait. During the first 20 years of life, affected individuals develop hundreds of colon adenomas (polyps). The likelihood that one or more of these multiple benign adenomas will progress to malignancy is extremely high, so most affected individuals (more than 75%) will develop colon carcinoma by age 40 if the disease is not treated. The colons of these patients are therefore usually removed before cancer has a chance to develop. The frequency of familial adenomatous polyposis is about 1 in 10,000, so this inherited form accounts for less than 0.5% of total colon cancer incidence. The second inherited form, hereditary nonpolyposis colon cancer, in which affected individuals develop colon carcinoma without the large number of polyps characteristic of familial adenomatous polyposis, is similarly infrequent. Thus, in spite of the existence of at least two inherited forms, over 95% of colon cancer appears to represent noninherited sporadic disease.

Cancer Family Syndromes

Rare hereditary forms of most of the other common cancers (including leukemias and lymphomas, sarcomas, melanoma, brain tumors, and carcinomas of a variety of sites) are similarly transmitted as dominant genetic traits. Usually, a propensity to develop only one or a few kinds of cancer is inherited, but some hereditary predispositions lead to the development of several different neoplasms. An example of such multiple neoplasm inheritance is the **Li-Fraumeni cancer family syndrome,** which involves dominant inheritance of several types of tumors—primarily sarcomas and breast carcinomas, but also leukemias, brain tumors, adrenocortical carcinomas, and other neoplasms (Fig. 5.3). Other such cancer family syndromes include inherited predispositions to development of both breast and ovarian carcinomas (**Lynch cancer family syndrome**) and to both nonpolyposis colon and endometrial carcinomas (Lynch II or **Warthin cancer family syndrome**).

A wide variety of both childhood and adult cancers can thus be inherited. In each case, these inherited cancers are transmitted as single dominant genes that impart a very high risk of tumor development. All of these hereditary forms of cancer are quite rare, however; such directly inherited cancers constitute only a small fraction of the total disease incidence.

GENETIC DISEASES THAT PREDISPOSE TO CANCER

Indirect Predispositions to the Development of Cancer

The hereditary cancers discussed above represent diseases in which the inherited genetic defect exerts a direct effect on the behavior of the cells that become neoplastic. For example, as will be discussed in detail in chapter 9, the mutant gene whose inheritance leads to retinoblastoma directly affects proliferation of the retinal cells from which the tumor will develop. In contrast, other genetic diseases confer an indirect predisposition to increased cancer incidence. The primary disorders in these diseases affect either the stability of the cellular genetic material or the function of the immune system. A secondary consequence of such defects is a high frequency of tumor development in affected individuals. The diseases of this group also differ from the inherited cancers in their mode of genetic transmission. They are transmitted as recessive rather than dominant traits, so that development of disease requires the inheritance of two abnormal gene copies—one from each parent.

Diseases Associated with Genetic Instability

Xeroderma pigmentosum is a good example of a disease in which defective maintenance of the genetic material confers increased cancer susceptibility, in this case to skin cancer (Table 5.2). Individuals with this disease suffer from several skin ("derma") disorders, particularly extreme dryness ("xerosis") and areas of nonuniform pigmentation ("pigmentosum"). The disease is extremely rare, with an incidence of about 1 in 250,000. The basic defect in xeroderma pigmentosum is an inability to repair the genetic damage caused by ultraviolet light, which, as discussed in chapter 3, is the major environmental risk factor for

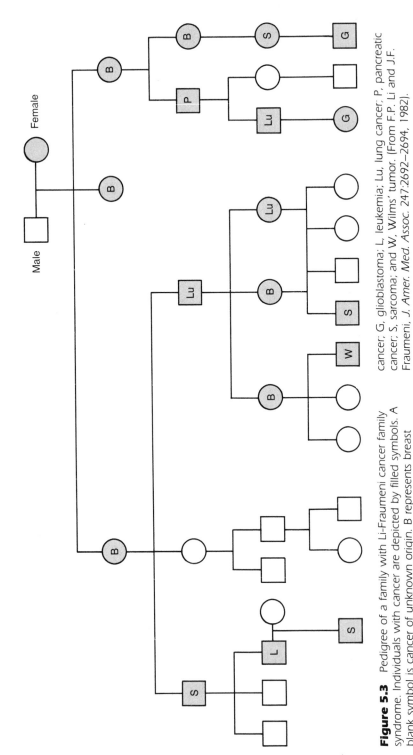

Figure 5.3 Pedigree of a family with Li-Fraumeni cancer family syndrome. Individuals with cancer are depicted by filled symbols. A blank symbol is cancer of unknown origin. B represents breast cancer; G, glioblastoma; L, leukemia; Lu, lung cancer; P, pancreatic cancer; S, sarcoma; and W, Wilms' tumor. (From F.P. Li and J.F. Fraumeni, *J. Amer. Med. Assoc.* 247:2692–2694, 1982).

Table 5.2   Inherited Predispositions to Cancer Associated with Genetic Instability

Genetic Disease	Types of Cancer
Ataxia telangiectasia	Leukemias and lymphomas
Bloom's syndrome	Leukemias and lymphomas
Fanconi's anemia	Leukemias and squamous cell carcinomas
Xeroderma pigmentosum	Skin cancers

skin cancer. Because of their inability to repair such damage, the skin of xeroderma pigmentosum patients is much more sensitive to solar radiation than that of normal individuals. As a result, individuals with this disease characteristically develop multiple skin tumors with high frequency. Other diseases of this general type, in which abnormalities in genetic maintenance lead to increased cancer incidence, include **ataxia telangiectasia, Bloom's syndrome, and Fanconi's anemia.** The central feature of such diseases is that genetic damage occurs with an abnormally high frequency in affected individuals. As discussed in detail in subsequent chapters, the conversion of a normal cell to a cancer cell results from damage to certain critical genes that regulate cell growth. Consequently, an increased frequency of genetic damage in the cells of patients suffering from diseases of this type leads to a high likelihood that they will develop cancer.

Inherited Immunodeficiencies

The increased incidence of certain cancers resulting from a lack of normal immune function, due either to immunosuppressive drugs or to AIDS, was discussed in chapters 3 and 4. In addition to these acquired immunodeficiencies, there are a number of diseases in which immunodeficiency is inherited (Table 5.3). Patients with such inherited immunodeficiencies, like those with acquired immunodeficiencies, suffer an increased risk of developing cancer. The inherited immunodeficiency diseases include ataxia telangiectasia; the increased cancer incidence seen in these patients may be due to both genetic instability and abnormal immune function.

Lymphomas, in particular, occur about 100 times more frequently in immunosuppressed individuals. As discussed in chapter 4, these lymphomas are associated with Epstein-Barr virus infection, which regularly leads to lymphoma development in immunodeficient patients. In normal individuals, the immune system effectively limits the proliferation of Epstein-Barr virus-infected cells, preventing lymphoma development. However, in immunodeficient individuals, Epstein-Barr virus infection leads to unlimited lymphocyte proliferation, eventually resulting in malignancy.

Table 5.3 Inherited Immunodeficiency Syndromes

Genetic Disease	Types of Cancer
Ataxia telangiectasia	Leukemias and lymphomas
Common variable immunodeficiency	Lymphomas and gastric carcinomas
Severe combined immunodeficiency	Leukemias and lymphomas
Wiskott-Aldrich syndrome	Leukemias and lymphomas
X-linked agammaglobulinemia	Leukemias and lymphomas
X-linked lymphoproliferative syndrome	Lymphomas

RACIAL AND FAMILIAL RISK FACTORS

Inherited Susceptibilities to Common Cancers

Both types of disorders discussed above—directly inherited cancers and genetic diseases that predispose to cancer development—are transmitted as single genes with classical Mendelian patterns of dominant or recessive inheritance. Inheritance of these genes results, either directly or indirectly, in a very high risk of cancer development in affected individuals. In addition to these clearly inherited predispositions to cancer, other genetic determinants appear to exert weaker, but still significant, effects on cancer susceptibility. These inherited susceptibilities constitute racial and familial risk factors for some of the common cancers of adults.

Racial Differences in Susceptibility to Melanoma

Melanoma is a good example of a neoplasm for which there are hereditary differences in racial susceptibility. The incidence of melanoma is about ten times higher among whites than blacks. This difference probably reflects the greater degree of pigmentation of black skin, which provides substantial protection against the carcinogenic effect of solar ultraviolet radiation. An individual's risk of developing melanoma is thus determined by the combination of genetic susceptibility (skin pigmentation) and exposure to an environmental carcinogen (sunlight). Genetic factors may also contribute to some of the other variations in cancer incidence among ethnic groups, including the high incidence of Epstein-Barr virus-induced nasopharyngeal carcinoma among the Chinese (see chapter 4), and the low incidence of Ewing's sarcoma and testicular cancer among blacks. As discussed in chapter 3, however, studies of migrant populations clearly indicate that most of the worldwide variation in cancer incidence is due to environmental rather than hereditary differences in national populations.

Familial Risk Factors for Breast, Lung, and Colon Cancers

In addition to heritable racial differences in the incidence of certain kinds of cancer, there also appear to be familial risk factors for a number of common cancers.

As discussed above, only rare cases of these cancers are directly inherited by transmission of single dominant genes. There are, however, less well-understood genetic determinants that affect susceptibility to the common sporadic forms of several frequently occurring cancers, including breast, lung, and colon carcinomas. In these cases, the risk of developing cancer is generally increased two- to threefold for individuals with first-degree relatives (parents or siblings) who have had the disease. This increased familial risk is much less than that associated with the directly inherited forms of cancer. For example, the age-specific risk of developing colon cancer is increased more than 1,000-fold by inheritance of familial adenomatous polyposis. The familial risk factors for common cancers thus appear to represent relatively small differences in cancer susceptibility. On the other hand, they represent a significant determinant of cancer risk for a much larger number of individuals than those affected by the rare cancers that are directly inherited.

Neither the genetic basis nor the mode of inheritance of these familial risk factors has yet been defined. Some risk factors of this type may represent inherited differences in an individual's sensitivity to carcinogens. For example, recent studies suggest that genetic differences in the ability to metabolize some of the carcinogenic chemicals in cigarette smoke may affect the risk of lung cancer by five- to tenfold. In addition, it is estimated that such inherited susceptibility may contribute to about 20% of all lung cancer cases. Genes that confer increased susceptibility to breast and colon carcinomas have also been estimated to be inherited by 10–20% of the population, and such inherited susceptibilities may play a role in the development of a substantial fraction, perhaps most, of these common adult neoplasms.

SUMMARY

Although the vast majority of cancers are not directly inherited, there are a number of ways in which susceptibility to cancer can be genetically transmitted. Rare inherited forms of many childhood and adult cancers are transmitted as dominant genes, which directly confer a high likelihood (virtually 100%) of development of particular neoplasms. Other rare hereditary diseases indirectly lead to the development of cancer by affecting either the stability of cellular genetic material or the function of the immune system. Both of these types of inherited cancer susceptibilities are extremely rare and account for only a small fraction of total cancer incidence. In addition, however, heredity appears to govern some racial and familial differences in susceptibility to more frequently occurring cancers. Compared to the rare inherited cancers, such familial factors impart smaller increases in risk, but they may contribute to a significant fraction of common adult cancers.

KEY TERMS

inherited cancer
cancer susceptibility

retinoblastoma
Wilms' tumor
familial adenomatous polyposis
Li-Fraumeni cancer family syndrome
Lynch cancer family syndrome
Warthin cancer family syndrome
xeroderma pigmentosum
ataxia telangiectasia
Bloom's syndrome
Fanconi's anemia
inherited immunodeficiency
familial risk factor

REFERENCES AND FURTHER READING

General References

Knudson, A.G., Jr. 1977. Genetics and etiology of human cancer. *Adv. Human Genet.* 8:1–66.

Knudson, A.G., Jr. 1986. Genetics of human cancer. *Ann. Rev. Genet.* 20:231–251.

Inherited Cancers

Bodmer, W.F., Bailey, C.J., Bodmer, J., Bussey, H.J.R., Ellis, A., Gorman, P., Lucibello, F.C., Murday, V.A., Rider, S.H., Scrambler, P., Sheer, D., Solomon, E., and Spurr, N.K. 1987. Localization of the gene for familial adenomatous polyposis on chromosome 5. *Nature* 328:614–616.

Go, R.C.P., King, M.-C., Bailey-Wilson, J., Elston, R.C., and Lynch, H.T. 1983. Genetic epidemiology of breast cancer and associated cancers in high-risk families. I. Segregation analysis. *J. Natl. Cancer Inst.* 71:455–461.

Hall, J.M., Lee, M.K., Newman, B., Morrow, J.E., Anderson, L.A., Huey, B., and King, M.-C. 1990. Linkage of early-onset familial breast cancer to chromosome 17q21. *Science* 250:1684–1689.

Hansen, M.F., and Cavenee, W.K. 1987. Genetics of cancer predisposition. *Cancer Res.* 47:5518–5527.

Knudson, A.G., Jr. 1971. Mutation and cancer: statistical study of retinoblastoma. *Proc. Natl. Acad. Sci. USA* 68:820–823.

Knudson, A.G., Jr. 1985. Hereditary cancer, oncogenes, and antioncogenes. *Cancer Res.* 45:1437–1443.

Li, F.P., and Fraumeni, J.F., Jr. 1982. Prospective study of a family cancer syndrome. *J. Amer. Med. Assoc.* 247:2692–2694.

Li, F.P., Fraumeni, J.F., Jr., Mulvihill, J.J., Blattner, W.A., Dreyfus, M.G., Tucker, M.A., and Miller, R.W. 1988. A cancer family syndrome in twenty-four kindreds. *Cancer Res.* 48:5358–5362.

Lynch, H.T., Bronson, E.K., Strayhorn, P.C., Smyrk, T.C., Lynch, J.F., and Ploetner, E.J. 1990. Genetic diagnosis of Lynch syndrome II in an extended colorectal cancer-prone family. *Cancer* 66:2233–2238.

Matsunaga, E. 1981. Genetics of Wilms' tumor. *Hum. Genet.* 57:231–246.

Genetic Diseases That Predispose to Cancer

Abbas, A.K., Lichtman, A.H., and Pober, J.S. 1991. *Cellular and molecular immunology.* W.B. Saunders, Philadelphia.

Friedberg, E.C. 1985. *DNA repair.* W.H. Freeman, New York.

Knudson, A.G., Jr., Strong, L.C., and Anderson, D.E. 1973. Heredity and cancer in man. *Prog. Med. Genet.* 9:113–158.

Racial and Familial Risk Factors

Adami, H.-O., Hansen, J., Jung, B., and Rimsten, A. 1981. Characteristics of familial breast cancer in Sweden: absence of relation to age and unilateral versus bilateral disease. *Cancer* 48:1688–1695.

Anderson, D.E. 1974. Genetic study of breast cancer: identification of a high risk group. *Cancer* 34:1090–1097.

Anderson, D.E., and Badzioch, M.D. 1985. Risk of familial breast cancer. *Cancer* 56:383–387.

Cannon-Albright, L.A., Skolnick, M.H., Bishop, D.T., Lee, R.G., and Burt, R.W. 1988. Common inheritance of susceptibility to colonic adenomatous polyps and associated colorectal cancers. *N. Engl. J. Med.* 319:533–537.

Caporaso, N.E., Tucker, M.A., Hoover, R.N., Hayes, R.B., Pickle, L.W., Issaq, H.J., Muschik, G.M., Green-Gallo, L., Buivys, D., Aisner, S., Resau, J.H., Trump, B.F., Tollerud, D., Weston, A., and Harris, C.C. 1990. Lung cancer and the debrisoquine metabolic phenotype. *J. Natl. Cancer Inst.* 82:1264–1272.

McLemore, T.L., Adelberg, S., Liu, M.C., McMahon, N.A., Yu, S.J., Hubbard, W.C., Czerwinski, M., Wood, T.G., Storeng, R., Lubet, R.A., Eggleston, J.C., Boyd, M.R., and Hines, R.N. 1990. Expression of CYP1A1 gene in patients with lung cancer: evidence for cigarette smoke-induced gene expression in normal lung tissue and for altered gene regulation in primary pulmonary carcinomas. *J. Natl. Cancer Inst.* 82:1333–1339.

Mulvihill, J.J. 1985. Clinical ecogenetics: cancer in families. *N. Engl. J. Med.* 312:1569–1570.

Ooi, W.L., Elston, R.C., Chen, V.W., Bailey-Wilson, J.E., and Rothschild, H. 1986. Increased familial risk for lung cancer. *J. Natl. Cancer Inst.* 76:217–222.

Page, H.S., and Asire, A.J. 1985. *Cancer rates and risks.* 3rd ed. National Institutes of Health, Bethesda.

Sattin, R.W., Rubin, G.L., Webster, L.A., Huezo, C.M., Wingo, P.A., Ory, H.W., Layde, P.M., and the Cancer and Steroid Hormone Study. 1985. Family history and the risk of breast cancer. *J. Amer. Med. Assoc.* 253:1908–1913.

Sellers, T.A., Bailey-Wilson, J.E., Elston, R.C., Wilson, A.F., Elston, G.Z., Ooi, W.L., and Rothschild, H. 1990. Evidence for Mendelian inheritance in the pathogenesis of lung cancer. *J. Natl. Cancer Inst.* 82:1272–1279.

Skolnick, M.H., Cannon-Albright, L.A., Goldgar, D.E., Ward, J.H., Marshall, C.J., Schumann, G.B., Hogle, H., McWhorter, W.P., Wright, E.C., Tran, T.D., Bishop, D.T., Kushner, J.P., and Eyre, H.J. 1990. Inheritance of proliferative breast disease in breast cancer. *Science* 250:1715–1720.

PART II

CANCER AT THE CELLULAR AND MOLECULAR LEVELS

THE FIRST PART OF THIS BOOK discussed the varieties, development, and causes of cancer from the standpoint of the intact organism—the human being. We now turn to a discussion of cancer at a more basic level: that of the individual cell. As already noted, cancer is fundamentally a disease at the cellular level, in which tumor cells fail to respond to the controls that regulate normal cell growth. The second part of this book will thus focus on the mechanisms that control normal cell growth and differentiation and on the defects in those mechanisms that cause normal cells to become neoplastic.

Chapter 6

Cell Growth and Differentiation

KEY TERMS

REFERENCES AND FURTHER READING

THE FUNDAMENTAL UNIT OF BIOLOGY is the cell—the smallest living structure capable of independent self-replication. Many organisms, such as bacteria and yeast, consist of only single cells. More complex organisms are composed of collections of cells that function in a coordinated manner, with different cell types specialized to perform particular tasks required by the organism as a whole. The human body, for example, consists of about 50 trillion (5×10^{13}) cells of approximately 100 different kinds, including cells that are responsible for such diverse activities as digestion, movement, sight, and thought. The organization of such a complex system, which arises during the development of each individual, is clearly a marvel of biology. Each cell must be regulated so that it functions to meet the needs of the organism as a whole, rather than replicating as an autonomous unit. The breakdown of this regulatory system can lead to uncontrolled cell division and the development of cancer.

THE ORGANIZATION OF CELLS

Cell Structures

The structures of three representative types of human cells, which are on the order of 10 microns (about 1/2,500 inch) in diameter, are illustrated in Figure 6.1. **Epithelial cells,** from which carcinomas arise, form continuous sheets covering the surface of the body and lining the internal organs. **Fibroblasts** are a common type of connective tissue cell, the malignant derivatives of which are fibrosarcomas. **Lymphocytes,** which give rise to leukemias and lymphomas, are cells of the blood and lymph system that function in the immune response.

All cells consist of two major compartments: the nucleus and the cytoplasm. The nucleus contains the genetic material of the cell and thus serves to direct all of the cell's activities. Most cellular functions are carried out in the cytoplasm, which contains a variety of subcellular structures specialized to perform different tasks, such as the synthesis of proteins (**ribosomes**), the generation of energy within the cell (**mitochondria**), the digestion of materials taken up by the cell from its external environment (**lysosomes**), and the secretion of material out of the cell (**Golgi complexes**).

DNA and Chromosomes

The genetic material that directs the development and function of an organism consists of **deoxyribonucleic acid (DNA)**, which is packaged into **chromosomes.** Each cell contains a copy of this genetic material within its nucleus. Normal cells of each species contain a characteristic number of chromosomes, which for hu-

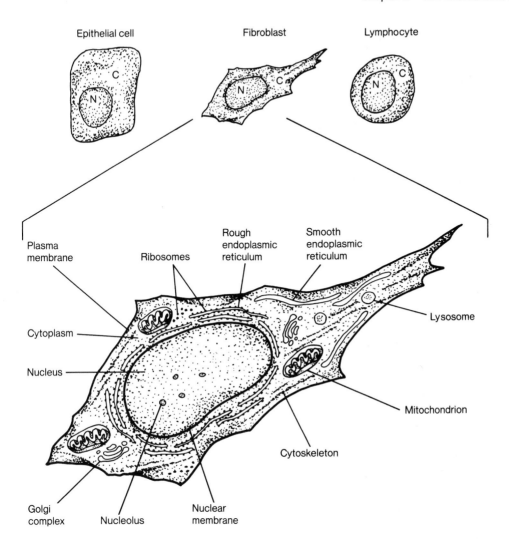

Epithelial cell   Fibroblast   Lymphocyte

Plasma
membrane   Ribosomes

Rough
endoplasmic
reticulum

Smooth
endoplasmic
reticulum

Lysosome

Cytoplasm

Nucleus

Mitochondrion

Cytoskeleton

Golgi
complex   Nucleolus

Nuclear
membrane

Figure 6.1   The structure of cells. The top part of the figure illustrates the morphology of three basic cell types: epithelial cells, fibroblasts, and lymphocytes. Each consists of two major compartments, the cytoplasm (C) and the nucleus (N). The bottom part of the figure is a more detailed view of a fibroblast, indicating several subcellular structures. These include the **cytoskeleton**, which provides the structural framework of the cell; the **endoplasmic reticulum**, an array of membranes within the cell; the **Golgi complex**, which functions in the transport of material out of the cell; **lysosomes**, which contain digestive enzymes; **mitochondria**, which generate sources of energy (e.g., ATP) within the cell; **nucleoli**, the sites of ribosome synthesis; the **plasma membrane**, a bilayer that separates the cell from its environment; and **ribosomes**, the sites of protein synthesis. Ribosomes can be either free in the cytoplasm or membrane-associated, forming rough endoplasmic reticulum.

man cells is 46. These 46 human chromosomes consist of 23 pairs, one member of which is inherited from each parent. The sperm and the egg each contain only one member of each chromosome pair (23 chromosomes total), so their union results in the assembly of a new set of 46 chromosomes at the time of fertilization. The information contained in these 46 chromosomes (the human genome) directs the development of the fertilized egg, a single cell, into a complete human being.

NORMAL DEVELOPMENT: FROM EGG TO ORGANISM

At the cellular level, two distinct kinds of processes must take place during development of an organism: cell division and cell differentiation. Cell division generates the 5×10^{13} cells that make up an adult human. Cell differentiation is the specialization of these cells: Some function as nerve cells, some as liver cells, some as muscle cells, and so forth. Both cell division and cell differentiation must be carefully regulated and coordinated in order for normal growth and development to take place.

Early Development

The first cell division of the new organism results in the formation of two cells from the fertilized egg (Fig. 6.2). This is followed by a series of divisions that give rise to the collection of undifferentiated cells called the morula. The first differentiated cell types arise at the blastocyst stage of embryo development, about a week after fertilization, when the human embryo is composed of a few hundred cells. Two distinct cell types are formed at this point in development: the inner cell mass and the trophoblast, which give rise to the embryo proper and to the extraembryonic membranes (e.g., the placenta), respectively. During the next stages of development, the inner cell mass gives rise to three distinct cell layers known as the germ layers. The cells of each of these three germ layers—the **ectoderm, mesoderm,** and **endoderm**—are destined to form specific tissues and organs in the individual. The ectoderm gives rise to skin (the epidermis), the nervous system, pigment cells, hair, nails, enamel of the teeth, the pituitary gland, the adrenal medulla, and the mammary glands. The mesoderm gives rise to muscle, bone, connective tissue (e.g., fibroblasts and fat cells), the cells of the blood and lymph systems, the spleen, the adrenal cortex, the urogenital system, and the dermis. The endoderm gives rise to the epithelial linings of the gastrointestinal and respiratory tracts, the tympanic cavity and Eustachian tube, and part of the urinary bladder, in addition to the epithelial portions of the liver, pancreas, tonsils, thyroid and parathyroid glands, and thymus.

Cell Differentiation and Gene Expression

What causes cell differentiation, driving one cell to become a liver cell and another to become a muscle cell? This is one of the fundamental unanswered

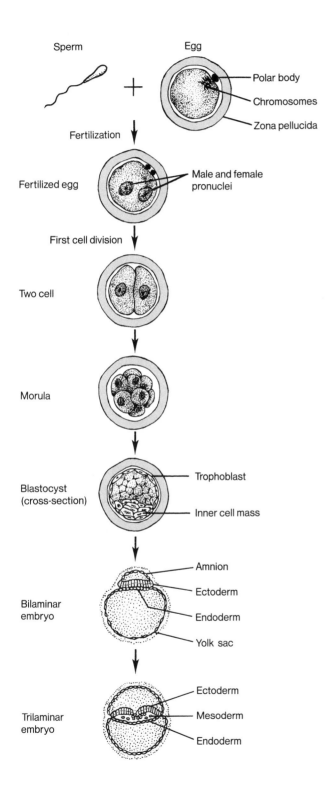

Figure 6.2 *Fertilization and early stages of embryonic development. Development through formation of the germ layers is illustrated.*

Sperm

Egg
- Polar body
- Chromosomes
- Zona pellucida

Fertilization

Fertilized egg
- Male and female pronuclei

First cell division

Two cell

Morula

Blastocyst (cross-section)
- Trophoblast
- Inner cell mass

Bilaminar embryo
- Amnion
- Ectoderm
- Endoderm
- Yolk sac

Trilaminar embryo
- Ectoderm
- Mesoderm
- Endoderm

questions of biology, and the mechanisms that control differentiation are far from understood. Nonetheless, enough is now known about the regulation of cell growth and differentiation that the general principles governing these processes have become apparent.

First, almost all cells (e.g., liver cells and muscle cells) contain the same genetic material, termed the **genome.** The entire genetic complement is replicated at each cell division and passed on to each daughter cell. What distinguishes a liver cell from a muscle cell is not what genes they contain, but what genes are active (i.e., actually expressed) in each cell type.

The human genome can be viewed as a blueprint, providing full specifications for the development of a human being. The complete genome is composed of about 50,000 individual **genes,** each of which specifies one component of cellular function. In combination, these genes direct the entire developmental program of the organism. Importantly, however, not all genes are active in all cells or at all times, and the regulation of gene expression is a critical aspect of cell function. For each gene to act, the corresponding region of DNA must first be copied into another kind of molecule: **ribonucleic acid (RNA)** (Fig. 6.3). The RNA functions as a messenger (**mRNA**), carrying the information from individual genes in the nucleus out into the cytoplasm of the cell. Once in the cyto-

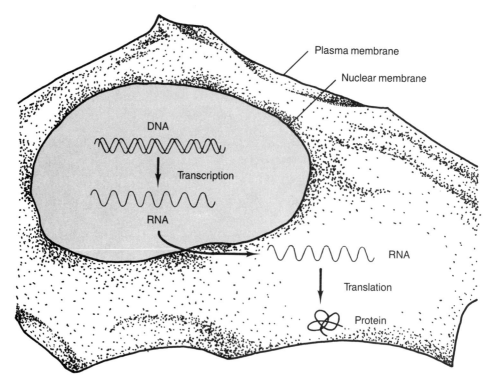

Figure 6.3 Gene expression. DNA in the nucleus is transcribed to RNA, which is exported to the cytoplasm and translated to yield the specified protein.

plasm, the RNA in turn directs the synthesis of a third kind of molecule: a **protein.** Proteins are the active working constituents of the cell, which are directly involved in performing all of the cell's tasks. Each gene encodes, or directs the synthesis of, a unique protein molecule. Gene expression, then, involves **transcription** of DNA to RNA, followed by **translation** of the RNA to proteins, which actively carry out the task specified by the gene in question. In most cases, the expression of a gene is regulated by determining whether or not it is transcribed to RNA.

Some genes are expressed in all cells and function in processes that are common to cells of many different types, such as the synthesis of molecules that are universal cell constituents. However, other genes are expressed only in particular kinds of differentiated cells. For example, some genes are specifically expressed in liver cells, whereas others are specifically expressed in muscle cells. Genes of this type are concerned with the specialized functions of differentiated cells, such as the contractile activity of muscle cells. Distinct patterns of gene expression thus govern each cell's developmental fate and function. In this context, cell differentiation can be viewed as a process of genetic regulation, in which specific programs of gene expression are established for each differentiated cell type.

Hormones and Growth Factors

How, then, is the cell's program of gene expression determined? At least in part, gene expression is regulated by the external factors with which a cell comes in contact. In some cases, such external factors are chemicals (e.g., **hormones**) that are produced and secreted by one cell and then interact with another cell to alter its pattern of gene expression and subsequent behavior. Such secreted factors thus represent signals by which cells can communicate with each other, allowing one cell to regulate the growth or differentiation of another.

A good example of the regulation of differentiation by secreted factors is provided by the formation of blood cells (Fig. 6.4). Both red blood cells (**erythrocytes**) and the different kinds of white cells (**lymphocytes, monocytes, macrophages, granulocytes,** and **platelets**) arise from the same kind of precursor cell in the bone marrow. The formation of each of these distinct kinds of blood cells is regulated by specific secreted factors (**growth factors**), which stimulate the development of particular differentiated cell types. The growth and differentiation of red blood cells, for example, is stimulated by a protein called **erythropoietin,** which is produced in the kidney. The synthesis of erythropoietin is itself regulated by the levels of circulating erythrocytes in blood. Thus, a drop in the number of circulating erythrocytes signals the kidney to produce more erythropoietin, which in turn stimulates the formation of more red blood cells in the bone marrow. The production and differentiation of other kinds of blood cells are similarly regulated by protein factors. For example, exposure of precursor cells in the bone marrow to one factor (**granulocyte colony-stimulating factor or G-CSF**) causes them to differentiate to granulocytes, whereas a different

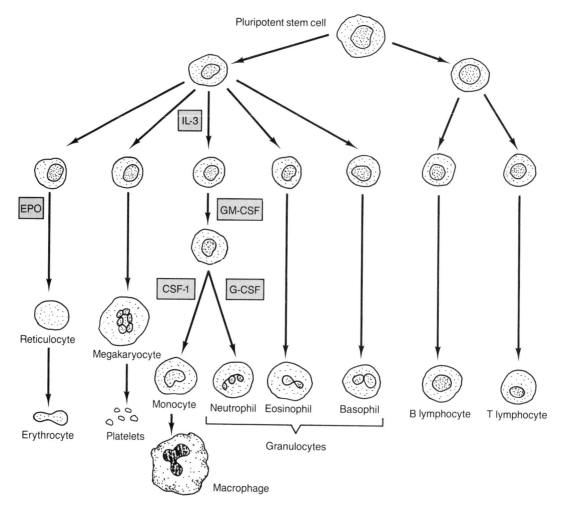

Figure 6.4 *Differentiation of blood cells. A pluripotent stem cell in the bone marrow gives rise to all of the different kinds of blood cells. The development of these different cell types is controlled by a series of specific factors, only some of which are indicated in the figure.* **Interleukin-3 (IL-3)** *stimulates a stem cell giving rise to erythrocytes, megakaryocytes, macrophages, and granulocytes (neutrophils, eosinophils, and basophils).* **Erythropoietin (EPO)** *stimulates the growth and differentiation of cells of the erythrocyte lineage.* **Granulocyte-macrophage colony-stimulating factor (GM-CSF)** *stimulates cells giving rise to both macrophages and granulocytes.* **Colony-stimulating factor-1 (CSF-1)** *and* **granulocyte colony-stimulating factor (G-CSF)** *stimulate cells giving rise specifically to macrophages and granulocytes, respectively.*

factor (**colony-stimulating factor-1 or CSF-1**) induces cells to differentiate to macrophages. Each of these extracellular factors binds to specific molecules (**growth factor receptors**) on the cell surface. Signals are then transmitted from the cell surface to the nucleus, leading to changes in gene expression that direct differentiation along a specified pathway.

Cell Contacts

In addition to being regulated by soluble secreted factors, cells can be similarly affected by direct contact with neighboring cells or with the **extracellular matrix** (the insoluble meshwork of secreted material that fills the space between cells in tissues). Such direct communication between cells and other tissue components is particularly important during the formation of tissues and organs, in which cells must interact closely with one another. For example, adhesive interactions between cells and components of the extracellular matrix are mediated by a family of cell surface receptors called **integrins.** Their function is critical in regulating the migration of cells during embryonic development, leading to the formation of tissues and organs composed of the appropriate differentiated cell types.

Cell Division During Embryogenesis

Cell division, as well as cell differentiation, is carefully controlled during embryogenesis. Cell growth and division are obviously required to generate the mass of cells that make up the adult organism. In addition, controlled cell division is required for formation of the basic body shape. In particular, the body plan of an organism is generated by the coordinated growth and division of groups of cells. For example, limb formation requires not only that cells differentiate to form bone, muscle, nerve, and skin, but also that all of these cell types are produced in the proper quantities and location to generate an arm or a leg, as the case may be. Both cell division and cell differentiation must therefore be coordinately controlled during development. As in the case of differentiation, cell division can be regulated by secreted factors that diffuse between cells, as well as by contacts between neighboring cells or with the extracellular matrix. In addition, the differentiation and reproduction of cells are interrelated processes, so many cells no longer divide once they become fully differentiated.

Programmed Cell Death

Perhaps surprisingly, **programmed cell death** (**apoptosis**) is also critically important during embryonic development. For example, the formation of fingers or toes involves the death of those cells located in the tissue between the digits. In addition, many tissues and organs are formed transiently during embryogenesis, later to be eliminated by programmed cell death as development proceeds. Cell death is thus an intrinsic part of the developmental program of some cell types. It is coordinated along with cell division and differentiation during embryonic development.

CELL DIVISION IN ADULTS

Rapid cell division is characteristic of early development, but in most cases, it ceases in the adult. Rather than continuing to grow, an adult organism needs only to maintain already formed tissues and organs. This does not mean, however, that adult cells are no longer capable of dividing. On the contrary, most adult cells are able to reproduce as required to replace cells that have been lost due to injury or death.

Adult Cells Divide at Different Rates

Cells in the adult can be grouped into three general categories with respect to their reproductive activities (Table 6.1). Some types of differentiated cells, including nerve cells and the cells of cardiac muscle, have permanently lost the capacity for cell division. Death of these cells represents irreversible damage to the organism, since new cells cannot be formed to replace those that have been lost. In addition, since differentiated cells of this type are no longer able to divide, they cannot become cancerous. Cancers of these cells (e.g., neuroblastoma) therefore occur only in children.

Other types of cells, however, divide occasionally in adults, as called upon to replace cells that have been lost. Cells of this type include fibroblasts and other connective tissue cells, as well as epithelial cells of a number of internal organs, such as the lung, liver, uterus, and kidney. Consistent with their capacity for cell division, such cells can give rise to cancers in adult life. Connective tissue cells, particularly fibroblasts, proliferate rapidly to repair injured tissue, such as that resulting from a cut. Some epithelial cells, such as those of the liver and kidney, normally divide only rarely, but they are triggered to divide rapidly to replace damaged tissue in the event of injury to those organs. Other types of epithelial cells (e.g., the cells of the uterine endometrium) divide rapidly in response to hormonal stimulation. The epithelial cells lining other tissues, such as the lung, also divide occasionally, as required to replace cells that have been lost from the surface.

Finally, some cell types continue to divide rapidly throughout life. These include blood-forming cells in the bone marrow, hair follicle cells, male germ cells, and the epithelial cells of the skin and digestive tract. Continuous division of these cells is required to replace mature differentiated cells that have short lifespans. For example, some white blood cells, such as granulocytes, and the epithelial cells that line the intestine live for only a few days. Continual division

Table 6.1 *Cell Division in Adults*

Frequency of Cell Division	Cell Types
Never divide	Nerve cells, cardiac muscle cells
Continuously divide	Blood-forming cells, intestinal epithelial cells, skin cells, hair follicle cells, male germ cells
Occasionally divide	Epithelial cells of most tissues: lung, larynx, oral cavity, pharynx, esophagus, stomach, liver, gall bladder, kidney, urinary bladder, uterus, ovary, prostate, breast, pancreas, thyroid Connective tissue cells: fat cells, fibroblasts, bone cells, cartilage cells, endothelial cells of blood vessels

of precursor cells, called **stem cells,** is therefore required to maintain the populations of these differentiated cells at a constant level. Stem cells are relatively undifferentiated cells that divide to form one daughter that remains a stem cell and a second daughter that differentiates into a mature functional cell (e.g., a granulocyte), which is programmed to die soon after. For cells of this type, cell death is an integral part of the program of differentiation, and the rate of division of stem cells is precisely regulated to match the rate at which differentiated cells die.

Rates of Cell Division Match Rates of Cell Death

The rates at which cells divide in adult organisms are thus carefully regulated to maintain a constant cell population. In tissues with constantly dividing stem cell populations, the rate of cell division is matched with that of programmed cell death. In other tissues, such as the liver, cells proliferate only as required to replace cells that have been lost as a result of injury or other damage. In these cases, cell division continues only until the damaged tissue has been replaced. The reproduction of individual cells is thus rigorously controlled to meet the needs of the organism as a whole.

THE CELL CYCLE

Stages of the Cell Cycle

The reproduction of a cell involves cell growth, duplication of the genetic material, and division of one parental cell into two daughter cells. Since each daughter cell can also reproduce, the entire process can be viewed as the life cycle of a cell, which begins anew following completion of each round of cell division. The **cell cycle** is divided into four stages, during which different kinds of cellular events take place (Fig. 6.5).

   The actual division of one cell into two, **mitosis,** is designated the **M phase** of the cell cycle. Following mitosis, the cell enters the **G1** (gap 1) stage. During the G1 stage, the cell is metabolically active and increases in size, but it is not directly involved in replication. Following G1, the cell enters the **S** (synthesis) phase, during which DNA replication occurs and the genetic material is duplicated. Following S, the cell contains two copies of its genetic complement (e.g., 92 human chromosomes). It then enters the **G2** (gap 2) stage, during which it prepares for division. The cell then completes the cycle, and mitosis again results in the formation of two daughter cells, each containing one copy of the genetic material (46 human chromosomes).

The Cell Cycle and Growth Control

For a typical rapidly proliferating cell, the total cell cycle time might correspond to about 20 hours, with approximately 1 hour for M, 8 hours for G1, 8 hours for

Figure 6.5   *The cell cycle. M, mitosis; G1, gap 1; S, DNA synthesis; G2, gap 2.*

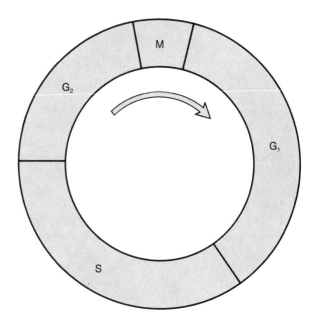

S, and 3 hours for G2. However, as emphasized above, cell division in the body is rigorously controlled, and different kinds of cells vary widely in their rates of replication. Almost all of this variation occurs in the G1 phase of the cell cycle. Thus, the lengths of S, G2, and M are nearly constant for different kinds of cells, whereas G1 can vary from a few hours for rapidly proliferating cells (e.g., intestinal epithelial cells) to many weeks for cells that divide only rarely (e.g., liver cells). Moreover, those differentiated cells that have lost the capacity for further division, such as nerve cells, contain only one copy of the genetic material and thus can be viewed as being permanently arrested in the G1 phase of the cycle.

   The rate at which a cell proliferates is thus determined by the length of time it spends in G1. For slowly dividing cells, progression through G1 is not a continuous process. Rather, these cells become arrested in G1 and do not continue progressing through the cell cycle unless an external signal induces the cell to resume proliferation. Such arrested cells appear to have entered a **quiescent** state, which is generally distinguished from the G1 phase of the cycle of actively proliferating cells and is referred to as **G0**. For example, connective tissue cells, such as skin fibroblasts, remain arrested in G0 unless a signal triggers resumed proliferation, as might be needed to repair damage resulting from a cut or other injury.

   Since most cells, like skin fibroblasts, do not replicate continuously, whether or not they divide is determined by a decision that is made during the G1 phase of each cell cycle. At this point, the cell either becomes arrested in G0 or proceeds through G1 to duplicate its DNA (S phase). With only a few exceptions, cells that progress through G1 are committed to proceed through the rest of the cell cycle and undergo mitosis. For most cell types, therefore, reproduction is regulated in the G1 phase of each cycle by the decision to enter G0 or to proceed through G1 and the rest of the division process.

REGULATION OF CELL DIVISION BY EXTRACELLULAR FACTORS

The problem of how the reproduction of most cells is regulated can thus be considered in terms of what causes a cell in G1 to become quiescent, or, conversely, what causes a cell arrested in G0 to resume active proliferation. The factors that determine this decision depend on the physiological role played by proliferation of each cell type. For example, cells of the uterine endometrium are stimulated to divide by estrogen, which is produced by the ovary during each menstrual cycle. The proliferation of these cells prepares the uterus for implantation of an embryo in the event of pregnancy.

The reproduction of other cell types is similarly determined by the milieu of external factors to which the cell is exposed. Such factors include contact with other cells and the extracellular matrix, as well as exposure to secreted substances such as hormones and growth factors. A good example is provided by skin fibroblasts, which (as noted above) are stimulated to divide as required to repair injury resulting from a wound. One of the substances that stimulates proliferation of these cells is a growth factor called **platelet-derived growth factor** (**PDGF**). PDGF is stored in blood platelets and is released during blood clotting. It then stimulates proliferation of skin cells (fibroblasts) in the neighborhood of the clot, thus leading to regrowth of the damaged tissue and healing of the wound. A number of growth factors, like PDGF, stimulate the proliferation of different kinds of cells, whereas other factors inhibit cell proliferation (Table 6.2). The division of specific cell types can thus be regulated in response to the needs of the whole organism.

TRANSMISSION OF EXTERNAL SIGNALS TO THE NUCLEUS

Growth Factor Receptors

Most growth factors, like PDGF, act by binding to specific receptors (growth factor receptors) on the surface of their target cells. Skin fibroblasts, for example, express a receptor for PDGF on their surface and can therefore be stimulated by extracellular PDGF. On the other hand, the proliferation of liver cells is not affected by PDGF, because they do not express its receptor. The pattern of receptors on the surface of a cell thus determines the growth factors to which it can respond, so that individual growth factors stimulate the proliferation only of specific, physiologically appropriate cell types.

Table 6.2 Representative Growth Factors

Growth Factor	Effect on Cells
Platelet-derived growth factor (PDGF)	Stimulates division of connective tissue cells
Epidermal growth factor (EGF)	Stimulates division of a variety of cell types
Fibroblast growth factor (FGF)	Stimulates division of a variety of cell types
Insulin-like growth factor (IGF)	Stimulates division of a variety of cell types
Keratinocyte growth factor (KGF)	Stimulates division of epithelial cells
Nerve growth factor (NGF)	Development of sympathetic neurons
Transforming growth factor-β (TGF-β)	Inhibits division of epithelial cells
Erythropoietin (EPO)	Stimulates growth and differentiation of erythrocyte precursors
Interleukin-2 (IL-2)	Stimulates growth of T lymphocytes
Interleukin-3 (IL-3)	Stimulates growth of hematopoietic stem cells
Interleukin-4 (IL-4)	Stimulates growth of B lymphocytes
Colony-stimulating factor-1 (CSF-1)	Stimulates growth and differentiation of macrophage precursors
Granulocyte colony-stimulating factor (G-CSF)	Stimulates growth and differentiation of granulocyte precursors
Granulocyte-macrophage colony-stimulating factor (GM-CSF)	Stimulates growth and differentiation of precursors to both granulocytes and macrophages

The response of a cell to a growth factor is thus triggered by the factor's binding to a cell surface receptor. Exposure of a quiescent (G0) skin fibroblast to PDGF, for example, induces the cell to reenter the cell cycle and initiate DNA synthesis (S phase) 10–15 hours later. The stimulus provided by growth factor binding at the cell surface must therefore be transmitted through the cytoplasm to the nucleus, leading to DNA synthesis and cell division.

Intracellular Signal Transduction

The process by which a signal is transmitted from the cell surface to the nucleus is referred to as **intracellular signal transduction** (Fig. 6.6). Growth factor binding activates a cell surface receptor, which then initiates a series of intracellular biochemical reactions. These reactions sequentially modify the function of target molecules, localized first in the plasma membrane, then in the cytoplasm, and finally in the nucleus. Changes in the expression of critical regulatory genes are thereby induced, and the cell reenters the proliferative cycle. The signal initiated at the surface of a cell by an extracellular growth factor is thus transmitted via a regulatory network to the cell nucleus, leading to changes in the cell's program of gene expression, DNA synthesis, and cell proliferation.

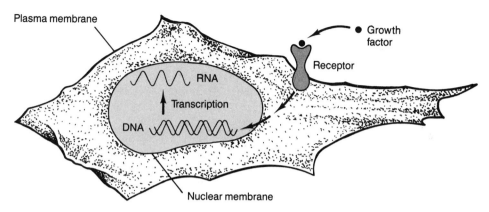

Figure 6.6   *Intracellular signal transduction. Binding of an extracellular growth factor activates its receptor, which initiates a series of reactions that transmit a signal to the nucleus, ultimately resulting in changes in gene expression.*

SUMMARY

The division and differentiation of cells are carefully regulated during embryonic development to generate the tissues and organs that make up the adult organism. In adults, most cells are capable of dividing as required to replace cells that have been lost due to injury or death, and some cells continue to reproduce rapidly in order to maintain populations of differentiated cells that have short lifespans. Both cell differentiation and cell division are regulated by a variety of external signals, including contact with neighboring cells, contact with the extracellular matrix, and secreted growth factors and hormones. Both stimulatory and inhibitory factors serve to regulate the proliferation of specific cell types in response to the needs of the whole organism. Most such factors bind to receptors on the cell surface, generating signals that are then transduced to the nucleus and alter the programs of gene expression that ultimately control cell behavior.

KEY TERMS

epithelial cell
fibroblast
lymphocyte
nucleus
cytoplasm
cytoskeleton
plasma membrane
DNA

chromosomes

genome

gene

cell division

cell differentiation

gene expression

RNA

protein

transcription

translation

hormone

growth factor

growth factor receptor

extracellular matrix

programmed cell death

apoptosis

stem cell

cell cycle

mitosis

M

G1

S

G2

quiescent

G0

platelet-derived growth factor (PDGF)

intracellular signal transduction

REFERENCES AND FURTHER READING

The Organization of Cells

Fawcett, D.W. 1986. *Bloom and Fawcett: a textbook of histology.* 11th ed. W.B. Saunders, Philadelphia.

Prescott, D.M. 1988. *Cells.* Jones and Bartlett, Boston.

Normal Development: From Egg to Organism

Clark, S.C., and Kamen, R. 1987. The human hematopoietic colony-stimulating factors. *Science* 236:1229–1237.

Darnell, J., Lodish, H., and Baltimore, D. 1990. *Molecular cell biology.* 2nd ed. Scientific American Books, New York.

Fuchs, E. 1990. Epidermal differentiation: the bare essentials. *J. Cell Biol.* 111:2807–2814.

Hay, E.D., ed. 1981. *Cell biology of extracellular matrix.* Plenum Press, New York.

Hynes, R.O. 1987. Integrins: a family of cell surface receptors. *Cell* 48:549–554.

Metcalf, D. 1985. The granulocyte-macrophage colony-stimulating factors. *Science* 229: 16–22.

Metcalf, D. 1989. The molecular control of cell division, differentiation commitment and maturation in haemopoietic cells. *Nature* 339:27–30.

Ruoslahti, E., and Pierschbacher, M.C. 1987. New perspectives in cell adhesion: RGD and integrins. *Science* 238:491–497.

Sadler, T.W. 1990. *Langman's medical embryology.* 6th ed. Williams and Wilkins, Baltimore.

Saunders, J.W., Jr. 1966. Death in embryonic systems. *Science* 154:604–612.

Slack, J.M.W. 1983. *From egg to embryo: determinative events in early development.* Cambridge University Press, Cambridge, England.

Sporn, M.B., and Roberts, A.B. 1991. Interactions of retinoids and transforming growth factor-β in regulation of cell differentiation and proliferation. *Mol. Endo.* 5:3–7.

Strickland, S., and Mahdavi, V. 1978. The induction of differentiation in teratocarcinoma stem cells by retinoic acid. *Cell* 15:393–403.

Watson, J.D., Hopkins, N.H., Roberts, J.W., Steitz, J.A., and Weiner, A.M. 1987. *Molecular biology of the gene.* 4th ed. Benjamin/Cummings, Menlo Park.

Cell Division in Adults

Baserga, R. 1985. *The biology of cell reproduction.* Harvard University Press, Cambridge.

Fawcett, D.W. 1986. *Bloom and Fawcett: a textbook of histology.* 11th ed. W.B. Saunders, Philadelphia.

Jandl, J.H. 1991. *Blood: pathophysiology.* Blackwell Scientific Publications, Boston.

Metcalf, D. 1988. *The molecular control of blood cells.* Harvard University Press, Cambridge.

Prescott, D.M. 1976. *Reproduction of eukaryotic cells.* Academic Press, New York.

The Cell Cycle

Baserga, R. 1985. *The biology of cell reproduction.* Harvard University Press, Cambridge.

Pardee, A.B. 1989. G1 events and regulation of cell proliferation. *Science* 246:603–608.

Pardee, A.B., Dubrow, R., Hamlin, J.L., and Kletzien, R.F. 1978. Animal cell cycle. *Ann. Rev. Biochem.* 47:715–750.

Regulation of Cell Division by Extracellular Factors

Carpenter, G., and Cohen, S. 1979. Epidermal growth factor. *Ann. Rev. Biochem.* 48:193–216.

Clark, S.C., and Kamen, R. 1987. The human hematopoietic colony-stimulating factors. *Science* 236:1229–1237.

Cross, M., and Dexter, T.M. 1991. Growth factors in development, transformation, and tumorigenesis. *Cell* 64:271–280.

Finch, P.W., Rubin, J.S., Miki, T., Ron, D., and Aaronson, S.A. 1989. Human KGF is FGF-related with properties of a paracrine effector of epithelial cell growth. *Science* 245:752–755.

Folkman, J., and Klagsbrun, M. 1987. Angiogenic factors. *Science* 235:442–447.

Metcalf, D. 1989. The molecular control of cell division, differentiation commitment and maturation in haemopoietic cells. *Nature* 339:27–30.

Ross, R., Raines, E.W., and Bowen-Pope, D.F. 1986. The biology of platelet-derived growth factor. *Cell* 46:155–169.

Sporn, M.B., and Roberts, A.B. 1988. Peptide growth factors are multifunctional. *Nature* 332:217–219.

Transmission of External Signals to the Nucleus

Cochran, B.H. 1985. The molecular action of platelet-derived growth factor. *Adv. Cancer Res.* 45:183–216.

Rosengurt, E. 1986. Early signals in the mitogenic response. *Science* 234:161–166.

Chapter 7

Differences Between Cancer Cells and Normal Cells

THE BASIC DEFECT in cancer cells is that they proliferate in an uncontrolled fashion, rather than being regulated by the signals that control normal cell reproduction. As discussed in the preceding chapter, control of cell proliferation is a complex process, in which both cell growth and differentiation are coordinated by an interplay of stimulatory and inhibitory signals. Such signals include secreted hormones and growth factors, as well as contacts of cells with both their neighbors and the extracellular matrix. The generalized loss of growth control characteristic of cancer cells is the net result of aberrant responses to a variety of factors that regulate normal cell growth and differentiation. This is reflected in a number of properties that distinguish cancer cells from their normal counterparts. Taken together, these characteristics of cancer cells provide a description of malignancy at the cellular level.

GROWTH OF CELLS IN CULTURE

Culture of Animal Cells

The ultimate definition of a cancer cell is malignancy in an animal. However, the analysis of discrete properties of both cancer cells and normal cells has been greatly facilitated by the ability to grow animal cells in culture, outside of the intact organism. This has allowed scientists to identify systematically the various factors that control cell growth and differentiation and to begin to determine how such factors may act. Many different kinds of cells can be grown in culture by maintaining them at body temperature in media that contain nutrients and appropriate growth factors to stimulate proliferation of the cell type being studied. The use of such cell cultures has allowed researchers not only to analyze the growth requirements and behavior of normal cells but also to compare the properties of normal and cancer cells under carefully defined experimental conditions.

Density-Dependent Inhibition of Cell Proliferation

The basic unregulated growth of cancer cells in an animal is directly reflected by differences in the growth of cancer cells and normal cells in culture (Fig. 7.1). When normal cells are propagated in culture, they divide until they reach a finite

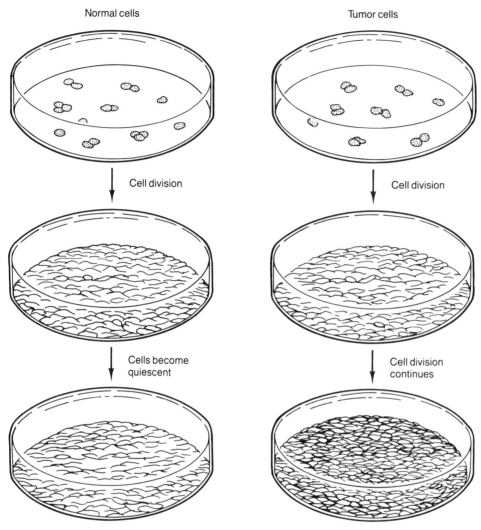

Normal cells

Tumor cells

Cell division

Cell division

Cells become
quiescent

Cell division
continues

Figure 7.1 *Growth of normal cells and tumor cells in culture. Normal cells in culture divide until they cover the available surface area of the culture dish and then become quiescent. In contrast, tumor cells continue dividing in an uncontrolled manner.*

cell density. They then stop dividing and become quiescent, arrested in the G0 stage of the cell cycle (see chapter 6). Normal cells in culture thus display controlled proliferation that is regulated by cell density, seemingly mimicking the tightly regulated growth of normal cells in an organism (such as the controlled proliferation of skin fibroblasts during wound healing).

Most tumor cells, on the other hand, fail to display the **density-dependent inhibition** of proliferation that is characteristic of normal cells in culture. Instead, tumor cells in culture generally continue growing until they die as a result of exhaustion of essential nutrients in the culture media or production of excess toxic waste products. Rather than responding to the signals that cause normal cells to become quiescent and enter G0, neoplastic cells continue their progression through the cell cycle. The proliferation of tumor cells in culture thus appears to be an unregulated process, mimicking their characteristic loss of growth control in animals.

REDUCED GROWTH FACTOR REQUIREMENTS OF CANCER CELLS

The Availability of Growth Factors Limits Cell Growth

As discussed in chapter 6, a variety of growth factors control the proliferation of different kinds of cells. For some cell types, including skin fibroblasts, the availability of growth factors is the major determinant of their proliferative capacity in culture. Thus, the density at which normal cells become quiescent is proportional to the amount of growth factor(s) available. The unregulated proliferation of many tumor cells, conversely, is due at least in part to reduced growth factor requirements.

Some Tumor Cells Produce Autocrine Growth Factors

Cells normally respond to growth factors that are produced by other cell types. Such growth factors provide signals to stimulate the proliferation of target cells as called upon to meet a physiological need. The release of PDGF from platelets to stimulate the proliferation of fibroblasts as required for wound healing is an example of such a normal signaling process. In contrast, some tumor cells produce growth factors that stimulate their own proliferation (Fig. 7.2). In such

Figure 7.2 Growth factor production by cancer cells. Some cancer cells produce growth factors to which they also respond, resulting in continuous autostimulation of cell division.

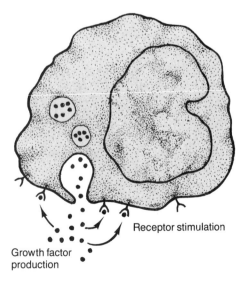

Receptor stimulation

Growth factor production

cases, abnormal growth factor production leads to continual autostimulation of cell division (**autocrine growth stimulation**), and the tumor cells proliferate in the absence of growth factors from other (physiologically normal) sources. Some sarcomas, for example, produce PDGF, thus continuously driving their own uncontrolled proliferation.

Reduced Growth Factor Requirements and Malfunction of Cellular Regulatory Systems

In other cases, tumor cells do not themselves produce growth factors, but reduced growth factor dependence results from the malfunction of other cellular regulatory systems. In these instances, intracellular signals that would normally be initiated by growth factor binding to a cell surface receptor are aberrantly generated within the cell in the absence of growth factor stimulation. For example, as discussed in chapter 6, growth factors act via cell surface receptors that, when activated by growth factor binding, initiate a series of intracellular reactions, ultimately leading to cell division. In some tumor cells, growth factor receptors are altered so that their function is no longer properly regulated (Fig. 7.3). In such cases, the receptors are continuously active, so a proliferative signal is generated even in the absence of stimulation by the appropriate growth factor. Consequently, the tumor cell is continually stimulated to divide, independent of the extracellular growth factors that would be needed to signal normal cell proliferation.

ALTERED ASSOCIATIONS WITH NEIGHBORING CELLS AND THE EXTRACELLULAR MATRIX

Regulation by Cell Contact

In addition to their reduced requirements for soluble growth factors, cancer cells differ from their normal counterparts in their interactions with other components of tissues and organs, including both neighboring cells and the extracellular matrix. As discussed in chapter 6, the **extracellular matrix** is the insoluble secreted material that adheres to the surface of cells and fills the space between cells in tissues. The proliferation of normal cells is regulated by the interaction of specific cell surface receptors with molecules in the extracellular matrix and on the surfaces of neighboring cells. Such regulation of cell growth by direct contact with other cells and tissue components is critical to the development of multicellular organs, the function of which is dependent on the orderly association and coordinated activities of their individual component cells. In general, however, tumor cells are less adhesive than normal cells and consequently are not as stringently regulated by their interactions with either neighboring cells or the extracellular matrix. The growth of cancer cells is therefore unrestrained by these associations, contributing to the ability of malignant cells to invade surrounding normal tissues and metastasize to distant body sites. These

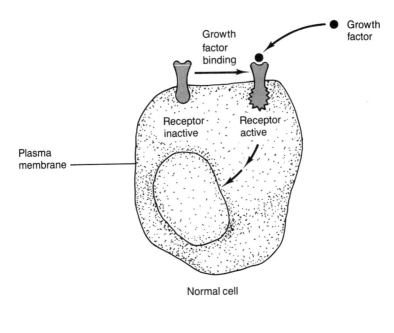

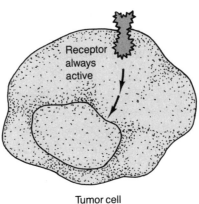

Figure 7.3   Abnormal growth factor receptors stimulate cancer cell proliferation. Normal growth factor receptors are activated by growth factor binding; only then do they signal the cell to divide. In contrast, growth factor receptors of some cancer cells are active in the absence of growth factor binding, so they continuously signal cell proliferation.

differences between normal and malignant cells in animals are also reflected by several aspects of their behavior in culture.

Contact Inhibition of Movement

One striking difference between normal and neoplastic cells in culture is illustrated by the phenomenon of **contact inhibition** (Fig. 7.4). Normal fibroblasts

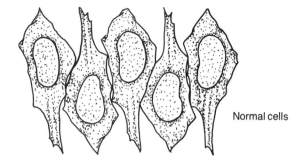

Normal cells

Figure 7.4   *Contact inhibition. The movement of normal fibroblasts in culture is inhibited by contact with neighboring cells, so normal cells form an orderly array on the surface of a culture dish. The movement of tumor cells, in contrast, is not inhibited by cell contact, so tumor cells migrate over one another and grow in a disordered, multilayered pattern.*

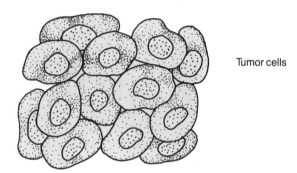

Tumor cells

move across the surface of a culture dish, which is coated with secreted extracellular matrix, but cease such movement when they make contact with a neighboring cell. Once in contact, normal cells adhere to each other, eventually forming an array of cells aligned in an orderly layer on the culture dish surface. Tumor cells, in contrast, do not display such contact inhibition of movement. Instead, they continue moving after contact with their neighbors, migrating over adjacent cells and growing in disordered, multilayered patterns.

Anchorage Dependence

Another distinction between many normal and neoplastic cells is the relationship between growth and attachment to a surface. Normal fibroblasts and epithelial cells require attachment to a solid substrate, such as a culture dish surface, in order to proliferate. This requirement of normal cells for attachment to a surface is referred to as **anchorage dependence.** Tumor cells, in contrast, frequently lack this requirement for surface attachment and can proliferate even when suspended in semisolid media.

The loss of contact inhibition and the anchorage-independent proliferation of tumor cells in culture seem to reflect their capacity for invasion and metastasis in animals. Both of these properties are correlated with the reduced adhesiveness of neoplastic cells. In at least some cases, this arises from a reduction in the level of cell surface receptors (e.g., integrins) for extracellular matrix components. These receptors apparently transmit signals that arise from the interactions of cells with the extracellular matrix, thus serving to regulate cell movement and proliferation in intact tissues as well as in cell cultures. Loss of this regulation is likely to be an important factor contributing to the abnormal proliferation, invasiveness, and metastatic potential of malignant cells.

PROTEASE SECRETION AND ANGIOGENESIS

Proteases, Invasion, and Metastasis

Two additional properties of malignant cells also affect their interactions with other tissue components and therefore play important roles in the growth and metastasis of cancer cells in intact animals. First, malignant cells generally secrete **proteases,** which are enzymes that digest other proteins. Metastasis necessarily involves the passage of cancer cells through the walls of tissues and blood vessels, which are composed of insoluble proteins. The extracellular matrix of connective tissues, for example, is composed largely of **collagen,** a protein secreted by fibroblasts. The secretion of proteases by cancer cells is thought to be a major factor contributing to their ability to dissolve extracellular materials and penetrate through surrounding normal tissues. Indeed, a variety of studies indicate that the metastatic potential of different cancer cells is closely correlated with the level of proteases that they secrete.

Angiogenesis in Tumor Growth and Metastasis

Angiogenesis is the formation of new blood vessels, which are critical to the growth of a tumor beyond the size of about a million cells. Once a tumor reaches this size, its further expansion requires the recruitment of new blood vessels to provide oxygen and nutrients to the proliferating tumor cells. Such blood vessels are formed in response to growth factors secreted by the tumor cells. These growth factors stimulate the proliferation of **endothelial cells** in the walls of capillaries in surrounding tissue, resulting in the outgrowth of new capillaries into the tumor.

Angiogenesis is thus critical to providing a source of nutrients needed to support the growth of a primary tumor beyond a minimal size. In addition, however, angiogenesis may play a direct role in metastasis. In particular, it appears that the actively proliferating capillaries formed in response to angiogenic stimulation are easily penetrated by tumor cells. Thus, these new vessels provide a ready opportunity for tumor cells to enter the circulatory system and thereby to spread to distant body sites. Finally, angiogenesis is required for the growth of metastatic tumors, just as for the growth of primary tumors. The ability of cancer cells to secrete factors that induce formation of new blood vessels is therefore a critical determinant of both tumor growth and metastasis in intact animals.

DEFECTIVE DIFFERENTIATION AND IMMORTALITY

Blocked Differentiation

Another general characteristic of cancer cells is that they fail to differentiate normally. Such defective differentiation is closely related to their abnormal proliferation, since (as discussed in chapter 6) most fully differentiated cells either cease division or divide only slowly. Rather than carrying out their normal differentiation program, cancer cells generally fail to progress to the fully differentiated state. Instead, they are characteristically blocked at an early stage of differentiation, consistent with their active proliferation.

The leukemias (cancers of blood cells) provide good examples of the relationship between defective differentiation and malignancy (Fig. 7.5). As indicated in chapter 6, several different kinds of blood cells are derived from the division of a common stem cell type (called the pluripotential stem cell) in the bone marrow. Descendants of these cells then become committed to specific differentiation pathways. For example, some cells differentiate to form erythrocytes (mature red blood cells), whereas others differentiate to form other kinds of blood cells, such as lymphocytes, granulocytes, and macrophages. All of these cell types undergo several rounds of division as they differentiate, but once they become fully differentiated, cell division ceases. Leukemic cells, in contrast, fail to differentiate normally. Instead, they become arrested at an early stage of maturation (e.g., erythroblasts), at which they retain their capacity for proliferation and continue to reproduce.

Programmed Cell Death

As discussed in chapter 6, **programmed cell death** is an integral part of the differentiation of a number of cell types, including blood cells. Since many kinds of fully differentiated cells are continually being replaced by division of stem cells, the regulation of cell death is as critical to maintaining a constant cell population as is the regulation of cell proliferation. Coincident with their failure to differentiate normally, cancer cells generally fail to undergo programmed cell death, instead continuing to proliferate.

Senescence

The increased lifespan of many cancer cells is also seen in their failure to undergo **senescence.** In addition to undergoing programmed cell death as part of their differentiation program, most normal cells have only limited proliferative capacity in culture, before they become senescent and cease proliferation. For example, normal human fibroblasts can usually be grown for 50 to 100 cell divisions, after which they stop growing and die. In contrast, many tumor cells will continue to grow indefinitely in culture. This capacity for continuous proliferation, **immortality,** is another property that distinguishes many neoplastic cells from their normal counterparts.

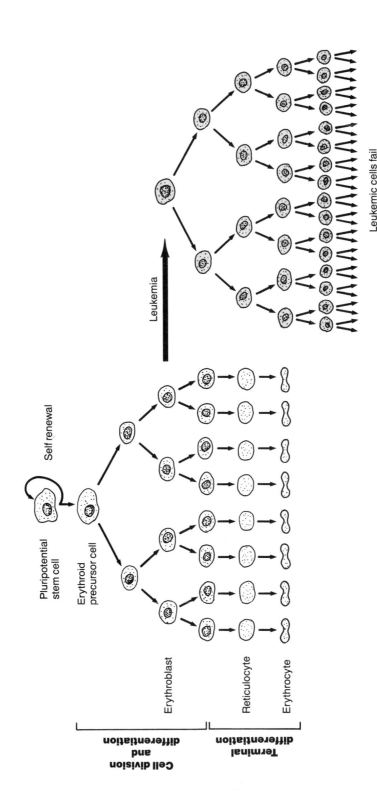

Figure 7.5 Defective cell differentiation in erythroleukemia. Erythrocytes, as well as other blood cell types (see Fig. 6.4), are derived from a common pluripotential stem cell in the bone marrow. Normal erythrocyte precursor cells (erythroblasts) undergo several rounds of cell division as they differentiate, but cell division ceases at the terminal stages of the differentiation process (reticulocytes and erythrocytes). In contrast, the differentiation of leukemic cells is blocked at an early stage, at which they continue to proliferate.

The defective differentiation and prolonged lifespan of cancer cells is thus closely coupled to their capacity for unregulated cell proliferation. In many cases, the progressive growth of malignant cells in an organism is a combined effect of continuous uncontrolled cell division and the failure of cancer cells to undergo normal cell differentiation and programmed cell death.

GENETIC INSTABILITY OF CANCER CELLS

Cancer Cells Are Frequently Aneuploid

An additional noteworthy characteristic of neoplastic cells is that their genetic material is less stable than that of normal cells. Cancer cells usually have abnormal numbers or arrangements of chromosomes (**aneuploidy**), resulting from chromosome duplications, from the loss of part or all of one member of a chromosome pair, or from **translocations** in which a portion of one chromosome is rearranged to a different chromosome. In addition, **gene amplification,** resulting in increased numbers of gene copies, occurs at least 1,000 times more frequently in tumor cells than in normal cells. This genetic instability of cancer cells is likely to play an important role in both the development and subsequent clinical behavior of many neoplasms.

Genetic Instability and Tumor Progression

As discussed in chapter 2, the development of cancer is a multistep process, in which a single initially altered cell becomes malignant through a series of gradual progressive changes. This process of tumor progression represents the stepwise accumulation of multiple defects in cellular regulatory mechanisms, which together result in malignancy. Each of the steps in the process is thought to result from a genetic alteration that leads to increased proliferative capacity or metastatic potential of the cancer cells. The genetic instability of tumor cells may increase the frequency of such alterations, thereby enhancing the process of tumor progression and playing a critical role in the development of malignant neoplasms.

Genetic Instability and Drug Resistance

The genetic instability of tumor cells is important not only in the development of cancer but also in its treatment. As will be discussed in detail in chapter 14, one of the problems commonly encountered in cancer chemotherapy is the development of **drug resistance.** In particular, many tumors respond initially to a given therapeutic agent but then become resistant to the drug during the course of treatment. This results from the outgrowth of drug-resistant variants in the tumor cell population. In contrast to the initially drug-sensitive tumor cells, which are killed by the chemotherapeutic agent being administered, the replication of such drug-resistant variant cells is unaffected. Consequently, the drug-resistant cells continue to proliferate and gradually emerge as the predominant cell type in an altered tumor cell population, which no longer responds to the chemotherapeutic agent in question. Since drug-resistant tumor cells arise by genetic alterations, particularly gene amplification, the genetic instability of tumor cells poses a major problem in the successful treatment of cancer, as well as in its development.

SUMMARY

The defining characteristics of malignancy in animals are uncontrolled cell proliferation, invasion of surrounding normal tissue, and metastasis to distant body sites. The growth of cancer cells in culture is likewise less well controlled than that of normal cells. Rather than responding to the signals that induce normal cells to stop proliferating and become quiescent, cancer cells divide in an unregulated manner. Comparative studies of cell growth and behavior have led to the recognition of a number of specific properties that distinguish many cancer cells from their normal counterparts (Table 7.1). In general, cancer cells have lower requirements for extracellular growth factors than normal cells, and both the proliferation and migration of cancer cells are less stringently regulated by their contacts with neighboring cells or the extracellular matrix. In addition, cancer cells secrete proteases and angiogenic factors, which contribute to their growth and metastasis in intact organisms. Cancer cells are also generally defective in differentiation and fail to undergo programmed cell death. Instead, their maturation is blocked at early stages of differentiation, compatible with continued cell proliferation. Finally, cancer cells are characterized by genetic instability, which is likely to be important in tumor progression, as well as in the development of resistance to chemotherapeutic agents. These properties of cancer cells directly reflect the progressive growth, invasion, and spread of malignant neoplasms. Cancer as a disease entity can thus be understood as the result of alterations in the behavior of individual tumor cells.

Table 7.1   Properties of Normal Cells and Cancer Cells

Characteristic	Normal Cells	Cancer Cells
Density-dependent inhibition of growth	Present	Absent
Growth factor requirements	High	Low
Contact inhibition	Present	Absent
Anchorage dependence	Present	Absent
Adhesiveness	High	Low
Protease secretion	Low	High
Secretion of angiogenic factors	Low	High
Differentiation	Present	Blocked
Proliferative lifespan	Finite	Indefinite
Genetic material	Stable	Unstable

KEY TERMS

cell culture

growth control

density-dependent inhibition

growth factor requirements

autocrine growth stimulation

cell adhesion

contact inhibition

anchorage dependence

protease

angiogenesis

defective differentiation

programmed cell death

senescence

immortality

genetic instability

aneuploidy

chromosome translocation

gene amplification

REFERENCES AND FURTHER READING

Growth of Cells in Culture

Pollack, R., ed. 1981. *Readings in mammalian cell culture.* 2nd ed. Cold Spring Harbor Laboratory, New York.

Reduced Growth Factor Requirements of Transformed Cells

Betsholtz, C., Johnsson, A., Heldin, C.-H., Westermark, B., Lind, P., Urdea, M.S., Eddy, R., Shows, T.B., Philpott, K., Mellor, A.L., Knott, T.J., and Scott, J. 1986. cDNA sequence and chromosomal localization of human platelet-derived growth factor A-chain and its expression in tumour cell lines. *Nature* 320:695–699.

Dulbecco, R. 1970. Topoinhibition and serum requirement of transformed and untransformed cells. *Nature* 227:802–806.

Holley, R.W., and Kiernan, J.A. 1968. "Contact inhibition" of cell division in 3T3 cells. *Proc. Natl. Acad. Sci. USA* 60:300–304.

Sporn, M.B., and Roberts, A.B. 1985. Autocrine growth factors and cancer. *Science* 313:745–747.

Stiles, C.D., Capone, G.T., Scher, C.D., Antoniades, H.N., Van Wyk, J.J., and Pledger, W.J. 1979. Dual control of cell growth by somatomedins and platelet-derived growth factor. *Proc. Natl. Acad. Sci. USA* 76:1279–1283.

Altered Associations with Neighboring Cells and the Extracellular Matrix

Abercrombie, M., and Heaysman, J.E.M. 1954. Observations on the social behaviour of cells in tissue culture. II. "Monolayering" of fibroblasts. *Exptl. Cell Res.* 6:293–306.

Freedman, V.H., and Shin, S.-I. 1974. Cellular tumorigenicity in nude mice: correlation with cell growth in semi-solid medium. *Cell* 3:355–359.

Plantefaber, L.C., and Hynes, R.O. 1989. Changes in integrin receptors on oncogenically transformed cells. *Cell* 56:281–290.

Ruoslahti, E. 1988. Fibronectin and its receptors. *Ann. Rev. Biochem.* 57:375–413.

Protease Secretion and Angiogenesis

Folkman, J. 1990. What is the evidence that tumors are angiogenesis dependent? *J. Natl. Cancer Inst.* 82:4–6.

Folkman, J., and Klagsbrun, M. 1987. Angiogenic factors. *Science* 235:442–447.

Gottesman, M. 1990. The role of proteases in cancer. *Sem. Cancer Biol.* 1:97–160.

Liotta, L.A., Steeg, P.S., and Stetler-Stevenson, W.G. 1991. Cancer metastasis and angiogenesis: an imbalance of positive and negative regulation. *Cell* 64:327–336.

Weidner, N., Semple, J.P., Welch, W.R., and Folkman, J. 1991. Tumor angiogenesis and metastasis—correlation in invasive breast carcinoma. *N. Engl. J. Med.* 324:1–8.

Defective Differentiation and Immortality

Goldstein, S. 1990. Replicative senescence: the human fibroblast comes of age. *Science* 249:1129–1133.

Hayflick, L., and Moorhead, P.S. 1961. The serial cultivation of human diploid cell strains. *Exptl. Cell Res.* 25:585–621.

Sawyers, C.L., Denny, C.T., and Witte, O.N. 1991. Leukemia and the disruption of normal hematopoiesis. *Cell* 64:337–350.

Todaro, G.J., and Green, H. 1963. Quantitative studies of the growth of mouse embryo cells in culture and their development into established lines. *J. Cell Biol.* 17:299–313.

Genetic Instability of Cancer Cells

Sandberg, A.A. 1980. *The chromosomes in human cancer and leukemia.* Elsevier, Amsterdam.

Stark, G.R., and Wahl, G.M. 1984. Gene amplification. *Ann. Rev. Biochem.* 53:447–491.

Tlsty, T.D. 1990. Normal diploid human and rodent cells lack a detectable frequency of gene amplification. *Proc. Natl. Acad. Sci. USA* 87:3132–3136.

Wright, J.A., Smith, H.S., Watt, F.M., Hancock, M.C., Hudson, D.L., and Stark, G.R. 1990. DNA amplification is rare in normal human cells. *Proc. Natl. Acad. Sci. USA* 87:1791–1795.

Chapter 8

Oncogenes

THE PRECEDING CHAPTERS have discussed the abnormalities in cell growth and differentiation that serve to characterize cancer at the cellular level. The environmental agents and hereditary factors that contribute to the development of human cancer were discussed in chapters 3 through 5. The next four chapters will consider the specific molecular nature of the alterations within cells that are responsible for neoplastic growth. As might be expected from their unregulated proliferation, cancer cells arise as a result of defects in the molecular mechanisms that control fundamental aspects of cell growth and differentiation.

One of the basic features of cancer cells, already noted in chapter 1, is that they divide to form more cancer cells, virtually never giving rise to normal progeny. The abnormalities that result in neoplastic growth, therefore, are stably inherited at the cellular level. This suggests that alterations in critical regulatory genes are responsible for neoplastic transformation. The application of molecular biology to cancer research has led to the identification of two distinct classes of such genes, called **oncogenes** and **tumor suppressor genes,** and to an initial understanding of their roles in both normal cell growth and the development of malignancy.

TUMOR VIRUSES AND VIRAL ONCOGENES

Tumor Viruses and Cell Transformation

The first insights into the molecular alterations that induce neoplastic growth were obtained from studies of viruses that cause cancer in experimental animals. As discussed in chapter 4, several kinds of viruses cause human cancers. A number of viruses also cause cancer in other species, including chickens, mice, and rats, which can easily be studied in the laboratory. For several reasons, research on these viruses (called **tumor viruses** or **oncogenic viruses**) has been extremely important to understanding the molecular alterations that lead to tumor development.

First, the induction of cancer by tumor viruses is a reproducible phenomenon, making it amenable to systematic scientific analysis. In particular, some viruses are extremely potent carcinogenic agents; susceptible animals develop tumors within days after inoculation. Such viruses can also be readily studied in cultured cells, eliminating the need for work with intact animals. When normal cells in culture are infected with these tumor viruses, they acquire the altered growth properties characteristic of cancer cells—a process known as cell **transformation.** The development of experimental systems for studying cell transformation in culture was a major advance in cancer research. It allowed scientists to perform experiments in a much simpler setting than in whole animals and therefore to undertake more detailed and probing studies of the transformation process.

In addition to the advantages provided by the ability of tumor viruses to induce transformation of cells in culture, the comparatively small size of viral compared to cellular genomes has allowed molecular analysis of viruses to be readily undertaken. As discussed in chapter 6, the genomes of human and other mammalian cells are extremely complex, being composed of about 50,000 individual genes. Because of this complexity, the task of identifying which of these genes might be functioning aberrantly in cancer cells was initially daunting. On the other hand, many tumor viruses contain 10 genes or fewer, so determining which of these tumor virus genes functioned to induce cell transformation seemed a comparatively simple scientific undertaking. The availability of experimental systems to study virus-induced transformation, combined with the small size of viral genomes, thus provided a major impetus for research directed toward identifying viral genes responsible for causing cancer. Such studies, which began nearly 30 years ago, led to the first identification of specific genes that could induce neoplasia—the **viral oncogenes**.

Acutely Transforming Retroviruses

Retroviruses, a large family of viruses that includes the human T-cell lymphotropic viruses and human immunodeficiency virus (see chapter 4), proved to be particularly informative objects of study for molecular biologists interested in cancer. Importantly, different retroviruses vary substantially in their transforming potency (Fig. 8.1). Some retroviruses, called **acutely transforming viruses,** induce tumors very rapidly in infected animals and efficiently transform cells in culture. A good example of this type of virus is **Rous sarcoma virus** (RSV), first isolated from a chicken tumor in 1911, which causes sarcomas in infected birds and transforms chick embryo cells in culture. Other retroviruses, however, are much less oncogenic. These viruses induce neoplastic transformation only rarely, if at all, even though they infect the same cells as the acutely transforming viruses. For example, although avian leukosis virus (ALV) is closely related to RSV, it replicates in chick embryo cells without inducing transformation.

The *src* Gene of RSV

The disparity in the transforming potential of the acutely transforming viruses (e.g., RSV) and their less oncogenic relatives (e.g., ALV) led scientists to try to understand the basis for this difference in oncogenicity. Because of their small size, it was possible to make direct molecular comparisons of the relevant viral genomes. Such studies revealed that RSV contained a single additional gene that was not present in ALV (Fig. 8.2). This additional gene did not function in virus replication, but it was specifically responsible for the ability of RSV to induce both transformation of cells in culture and sarcomas in infected birds. Because RSV causes *sarcomas*, the gene responsible for its oncogenicity—the first viral oncogene to be identified—was called *src.*

Oncogenes of Other Acutely Transforming Retroviruses

A total of more than 40 different acutely transforming retroviruses have been isolated from several species, including chickens, turkeys, mice, rats, cats, and

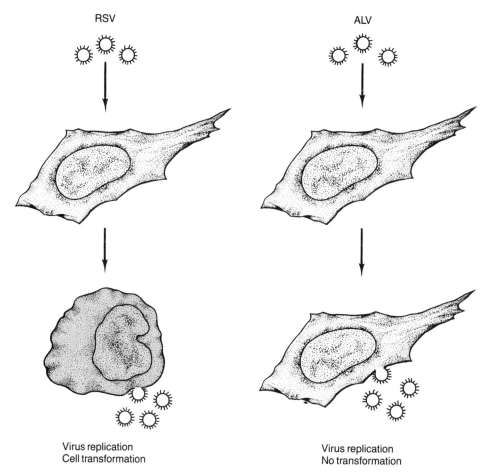

RSV

ALV

Virus replication
Cell transformation

Virus replication
No transformation

Figure 8.1 Cell transformation by Rous sarcoma virus (RSV) and avian leukosis virus (ALV). Both RSV and ALV infect and replicate in chicken embryo fibroblasts, but only RSV induces cell transformation.

monkeys. All of these viruses, like RSV, contain at least one viral oncogene (and in some cases two) that is not involved in virus replication but is responsible for the viruses' tumorigenicity. In several cases, different viruses contain the same oncogene. However, more than two dozen distinct oncogenes have been identified to date among the acutely transforming retroviruses (Table 8.1). These oncogenes are designated by three-letter names that usually refer either to the

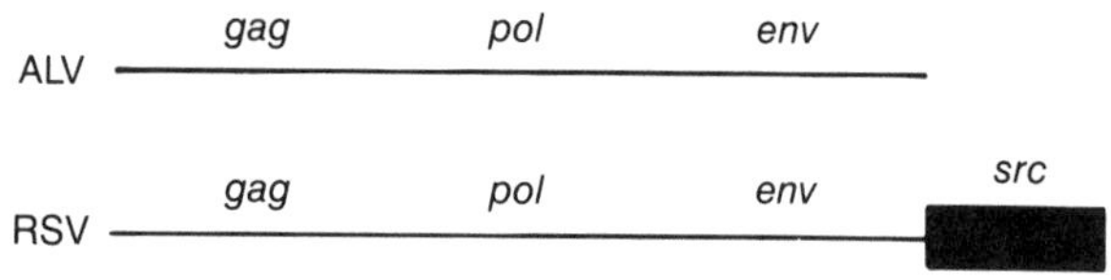

Figure 8.2 The genomes of ALV and RSV. Both ALV and RSV contain three genes (called *gag*, *pol*, and *env*), which are responsible for virus replication. RSV contains an additional gene, *src*, which is responsible for cell transformation in culture and induction of sarcomas in birds.

ALV ————— *gag* ——— *pol* ——— *env* —————

RSV ————— *gag* ——— *pol* ——— *env* ——— **■ *src***

Table 8.1 *Oncogenes of Acutely Transforming Retroviruses*

Oncogene	Virus	Species
abl	Abelson leukemia	Mouse
cbl	Cas NS-1	Mouse
crk	CT10 sarcoma	Chicken
*erb*A	Avian erythroblastosis-ES4	Chicken
*erb*B	Avian erythroblastosis-ES4	Chicken
ets	Avian erythroblastosis-E26	Chicken
fes	Gardner-Arnstein feline sarcoma	Cat
fgr	Gardner-Rasheed feline sarcoma	Cat
fms	McDonough feline sarcoma	Cat
fos	FBJ murine osteogenic sarcoma	Mouse
fps	Fujinami sarcoma	Chicken
jun	Avian sarcoma-17	Chicken
kit	Hardy-Zuckerman feline sarcoma	Cat
maf	Avian sarcoma AS42	Chicken
mos	Moloney sarcoma	Mouse
mpl	Myeloproliferative leukemia	Mouse
myb	Avian myeloblastosis	Chicken
myc	Avian myelocytomatosis	Chicken
raf	3611 murine sarcoma	Mouse
*ras*H	Harvey sarcoma	Rat
*ras*K	Kirsten sarcoma	Rat
rel	Reticuloendotheliosis	Turkey
ros	UR2 sarcoma	Chicken
sea	Avian erythroblastosis-S13	Chicken
sis	Simian sarcoma	Monkey
ski	Avian SK	Chicken
src	Rous sarcoma	Chicken
yes	Y73 sarcoma	Chicken

type of neoplasm induced by the virus, to the species of animal the virus infects, or to the scientist who first isolated the virus. For example, the designation of the oncogene of Abelson leukemia virus, ***abl,*** is derived from the name of the scientist responsible for its isolation (Herbert *Ab*elson) and the type of neoplasm the virus induces (*l*eukemia). The introduction of each of these genes into a cell by virus infection efficiently induces neoplastic transformation, providing clear evidence that cancer can result from the action of specific genes in tumor cells.

Other Viral Oncogenes

Oncogenes have also been identified in several other kinds of viruses (Table 8.2). These oncogenes differ from those of the acutely transforming retroviruses in

Table 8.2 Oncogenes of Other Viruses

Oncogene	Virus	Species in Which Tumors Induced
BNLF-1	Epstein-Barr virus	Human
E1A, E1B	Adenovirus	Rat, hamster
E5, E6	Bovine papillomavirus	Cow
E6, E7	Human papillomavirus	Human
T antigens	Polyomavirus, SV40	Hamster, mouse
tax	Human T-cell lymphotropic virus (HTLV)	Human

that they generally function in virus replication as well as in induction of transformation. Among the viruses that cause human cancer (see chapter 4), oncogenes have been identified in the papillomaviruses (the **E6** and **E7** oncogenes), Epstein-Barr virus (the **BNLF**-1 oncogene), and the human T-cell lymphotropic viruses (the **tax** oncogenes). Still other oncogenes have been identified in viruses that cause cancer in other animal species—for example, the *E1A* and *E1B* oncogenes of adenoviruses and the *T* antigen oncogenes of SV40 and polyomavirus. Hepatitis B virus may also contain an oncogene (the **X** gene), although it is alternatively possible that the induction of liver carcinomas by this virus is a consequence of chronic tissue damage leading to continual cell proliferation.

Viral oncogenes are thus responsible for the ability of many viruses to cause cancer. When these viruses infect cells, their genes become stably incorporated into the cellular chromosomes and are passed on to daughter cells along with the rest of the cell's genetic material (Fig. 8.3). The viral oncogenes thereby become a permanent addition to the genome of the infected cell. Their expression induces abnormal growth of both the initially infected cell and its progeny, leading to the development of a cancer.

PROTO-ONCOGENES IN NORMAL CELLS

Although their discovery provided the first insights into the molecular events leading to the development of cancer, viral oncogenes *per se* are not responsible for most human tumors. As discussed in chapter 4, viruses cause about 20% of human cancers worldwide. Thus, although viruses are obviously important carcinogenic agents, the majority of human cancers apparently arise from other causes, including radiation and the various chemical carcinogens discussed in chapter 3. Importantly, however, the alterations caused by these carcinogens result in the development of cancer by mechanisms that are fundamentally similar, at the cellular and molecular levels, to the action of tumor viruses. Consequently, studies of the viral oncogenes have not only helped to elucidate the nature of virus-induced cancers, but have also paved the way to identifying the molecular events that underlie the development of cancers caused by a wide variety of other carcinogenic agents.

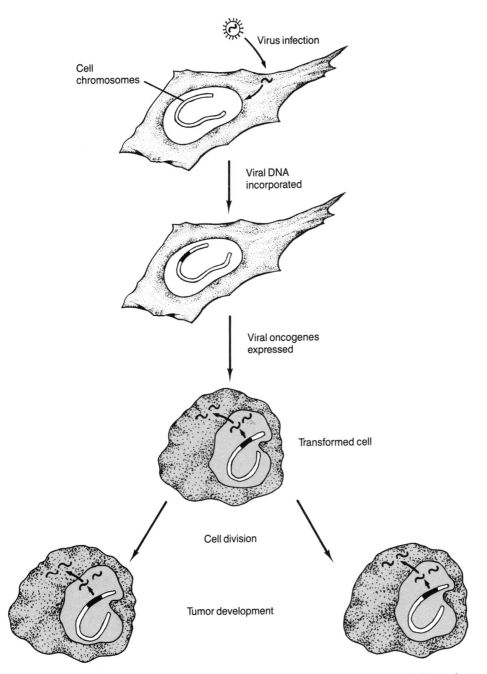

Figure 8.3 Induction of cancer by viruses with oncogenes. Tumor viruses infect cells and become stably integrated into the cell genome. Expression of viral oncogenes then induces cell transformation, and division of the transformed cells leads to tumor development.

The Origin of Acutely Transforming Retroviruses

Our understanding of the fundamental relationship between viral oncogenes and non-virus-induced cancers came from considering the origin of acutely transforming retroviruses, such as RSV. As discussed above, the oncogenes of these viruses are responsible only for tumor induction—they do not play any direct role in virus replication. Since most viruses are streamlined to reproduce as efficiently as possible, the existence of viral oncogenes that were not an integral part of the viruses' life cycle was an unexpected curiosity. Scientists were therefore led to question where the retroviral oncogenes originated and how they became incorporated into viral genomes.

The first clue to the origin of oncogenes came from the way in which the acutely transforming retroviruses were initially isolated (Fig. 8.4). Several viruses of this type have been isolated in laboratory studies, so the events involved in their genesis occurred under controlled conditions. These viruses were obtained from rare tumors of animals that had been inoculated with nonacutely transforming viruses that, like ALV, did not contain oncogenes. A small fraction of animals that were chronically infected with such viruses developed tumors, from which new, more highly oncogenic viruses were then isolated. These new acutely transforming viruses now contained oncogenes that were responsible for their increased carcinogenic potential.

At some point in their generation, therefore, the acutely transforming viruses had acquired oncogenes that were not present in their parental viruses. This suggested the possibility that the retroviral oncogenes were actually derived

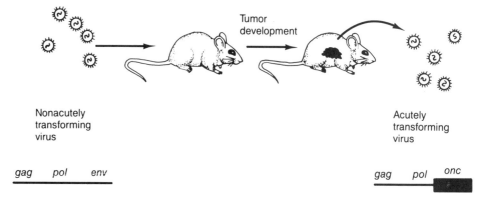

Figure 8.4 Isolation of an acutely transforming retrovirus. The acutely transforming virus was isolated from a rare tumor that developed in a mouse that had been inoculated with a nonacutely transforming virus. The nonacutely transforming virus contains only the *gag, pol,* and *env* genes needed for virus replication. In contrast, the acutely transforming virus has acquired a new viral oncogene, *onc,* which is responsible for its increased carcinogenic potential. In this example, the *onc* gene has been incorporated in place of *env* in the acutely transforming virus genome. The resulting acutely transforming virus is consequently defective in replication but can grow in cells that are simultaneously infected with a nonacutely transforming virus that supplies the *env* function.

from genes of the cells in which the parental viruses replicated. According to this hypothesis, the acutely transforming retroviruses arose as a result of the incorporation of host cell genes into viral genomes, yielding new recombinant viruses containing oncogenes and displaying strikingly increased pathogenicity.

Retroviral Oncogenes Were Derived from Proto-Oncogenes

The critical prediction of this hypothesis is that cells normally contain genes that are closely related to the retroviral oncogenes. This was directly demonstrated in a classical experiment, performed in 1976, in which it was shown that normal chicken cells contained genetic information very closely related to the *src* oncogene of RSV. These and similar experiments have firmly established that the oncogenes of acutely transforming retroviruses are derived from closely related genes of normal cells. This discovery is central to our present understanding of the molecular nature of carcinogenesis, since it clearly implies that cells normally contain genes with the potential for inducing neoplastic transformation. These normal cell genes from which the retroviral oncogenes were derived are called **proto-oncogenes.** Their importance is illustrated by the award of the 1989 Nobel Prize in Physiology or Medicine to their discoverers, J. Michael Bishop and Harold Varmus.

DIFFERENCES BETWEEN ONCOGENES AND PROTO-ONCOGENES

Oncogenes Are Altered Forms of Proto-Oncogenes

It is critical to emphasize that proto-oncogenes and oncogenes are related but not identical. Proto-oncogenes are normal cell genes that are part of the regularly inherited genetic complement of all vertebrates (including humans) and, in many cases, are also present in lower organisms such as insects and yeast. They function as important regulatory genes, playing major roles in a number of aspects of normal cell growth and differentiation. The oncogenes, in contrast, are not present in normal cells, and their function is decidedly abnormal: the induction of cancer. The oncogenes thus represent altered pathological forms of the proto-oncogenes. As a consequence of these alterations, the expression of oncogenes leads to abnormal cell proliferation and tumor formation.

Oncogenes Differ from Proto-Oncogenes in Regulation of Their Expression

Several differences between oncogenes and proto-oncogenes account for the ability of oncogenes to induce neoplastic transformation, rather than functioning normally (Fig. 8.5). In the first place, viral oncogenes differ from the corresponding proto-oncogenes in the regulation of their expression. As discussed in chapter 6, regulated programs of gene expression are critical to the control of normal

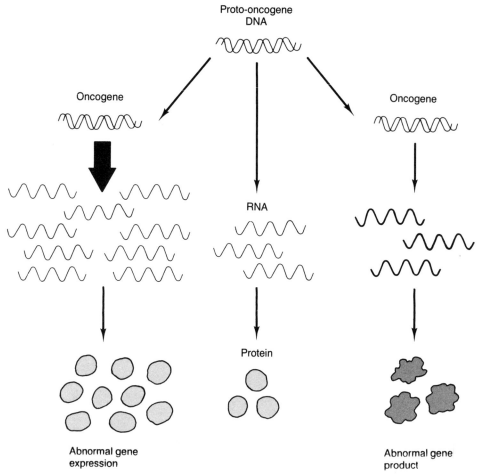

Figure 8.5 Differences between oncogenes and proto-oncogenes. Oncogenes differ from their normal counterparts (proto-oncogenes) both in the regulation of gene expression and in the structure and function of their encoded proteins. In some cases (left side), expression of abnormally high levels of a normal gene product is sufficient to induce cell transformation. In other cases (right side), transformation results from aberrant function of a structurally abnormal oncogene product. Many oncogenes differ from proto-oncogenes in both respects.

cell growth and differentiation. The expression of proto-oncogenes, like all other normal cell genes, is thus controlled to meet the needs of the cell in the context of the whole organism. This normal control of proto-oncogene expression is lost in the retroviral oncogenes. Since the oncogenes are expressed as part of viral genomes, they are no longer subject to the cellular mechanisms that regulate proto-oncogene expression. Consequently, the viral oncogenes are frequently expressed at higher levels, and often in different types of cells, than their normal counterparts. In many cases, such abnormalities of gene expression are sufficient to induce neoplastic transformation.

Proteins Encoded by Oncogenes Frequently Differ from Those Encoded by Proto-Oncogenes

In addition, the protein products encoded by the viral oncogenes often differ significantly from those of the corresponding proto-oncogenes. As will be discussed in detail in chapter 11, many proto-oncogenes encode proteins that act to regulate normal cell proliferation. The function of such proto-oncogenes is controlled not only by regulation of gene expression, but also by regulation of the activity of their protein products within the cell. However, many oncogenes encode altered proteins that no longer respond to normal cellular regulatory signals. For example, several proto-oncogene and oncogene proteins function in intracellular signaling pathways that, when activated, lead to cell proliferation. The activity of the proto-oncogene proteins is controlled by the appropriate physiological signals, such as the binding of a growth factor to its receptor. In contrast, the corresponding oncogene proteins function in an unregulated manner, so they are active and drive cell proliferation even in the absence of growth factor stimulation. Aberrant function of such oncogene proteins thus results in abnormal cell division, thereby contributing to the development of malignancy.

The retroviral oncogenes thus represent altered versions of proto-oncogenes, which differ from their normal cell progenitors in both the regulation of gene expression and the structure and function of their protein products. These alterations are responsible for the ability of the viral oncogenes to induce neoplastic transformation. The proto-oncogenes, conversely, constitute a group of normal cell genes with the potential of being converted into oncogenes. Their discovery set the stage for elucidating the kinds of genetic damage that are responsible for the development of human tumors.

ONCOGENES IN HUMAN CANCER

The identification of proto-oncogenes raised the question as to whether non-virus-induced tumors, which include the majority of human cancers, could be caused by alterations in normal cell genes. Since normal cell proto-oncogenes could be converted to potent oncogenes in viruses, was it possible that oncogenes could similarly arise as a result of genetic alterations within a cell? Could chemical carcinogens, for example, act to convert proto-oncogenes to oncogenes, thus driving the process of malignant transformation? To approach these questions, scientists sought to identify cellular genes that were capable of inducing neoplastic transformation and that had been specifically altered in tumor cells to activate their transforming potential.

Detection of Cellular Oncogenes by Gene Transfer

Such genes, **cellular oncogenes**, were initially detected by **gene transfer** assays, in which the genetic material (DNA) of cancer cells was tested for the ability to induce neoplastic transformation of sensitive recipient cells in culture. The first human tumor oncogene was discovered in a bladder carcinoma in 1981. The DNA of this bladder carcinoma efficiently induced transformation of recipient mouse cells, indicating that it contained a biologically active cellular oncogene (Fig. 8.6). More than a dozen different human oncogenes, activated in cancers of many different types, have subsequently been identified in such experiments.

Figure 8.6 *Demonstration of an oncogene in human cancer. DNA extracted from a human bladder carcinoma induced neoplastic transformation of recipient mouse cells. Transformation of these cells resulted from the incorporation of an oncogene derived from the human tumor DNA.*

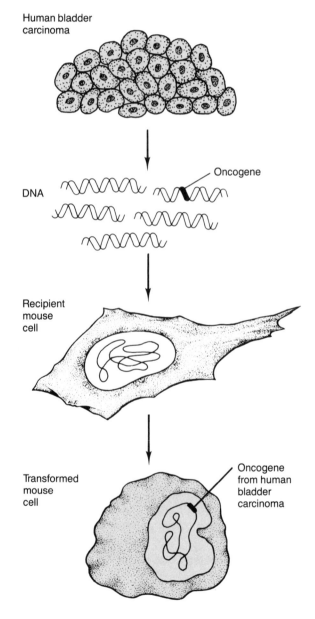

Human Tumor Oncogenes Are Formed from Proto-Oncogenes During Carcinogenesis

Like the oncogenes of retroviruses, oncogenes in human tumors are altered, abnormally functioning versions of their normal cell homologs (proto-oncogenes). Biologically active oncogenes are found in cancer cells, but not in normal cells of the same individual. Thus, oncogenes themselves are not genetically transmitted from parent to child. Instead, they are generated specifically in tumor cells as a consequence of genetic alterations suffered by proto-oncogenes during the process of carcinogenesis (Fig. 8.7). As in the case of retroviral oncogenes, the oncogenes found in human tumors differ from proto-oncogenes either in aberrant regulation of gene expression or in unregulated function of the oncogene products. Once an active oncogene is formed, it acts to drive abnormal cell proliferation, thus contributing to development of a neoplasm.

The *ras* Oncogenes and Point Mutations

The oncogenes most frequently involved in human cancers are members of the ***ras*** gene family, which consists of three related genes designated *ras*H, *ras*K, and *ras*N. These human *ras* genes are similar to oncogenes that were initially identified in acutely transforming retroviruses isolated from rats. The gene family designation, *ras*, thus derives from *rat sarcoma* virus. The individual gene designations refer to the names of the discoverers for *ras*H (*Har*vey) and *ras*K (*Kirsten*), or to the type of tumor in which the *ras*N oncogene was first detected (a *neuroblastoma*). In human tumors, mutations that alter only a single nucleotide of the *ras* gene sequences (**point mutations**) suffice to convert normally functioning *ras* proto-oncogenes to potent oncogenes that efficiently induce neoplastic transformation (Fig. 8.8). These mutations alter single amino acids at critical regulatory positions of the *ras* proteins, with the result that the *ras* oncogene proteins function in an uncontrolled manner. As will be further discussed in chapter 11, the *ras* proto-oncogene proteins are involved in intracellular transmission of proliferative signals. Their activity within the cell is tightly controlled; they normally function as part of the response initiated by growth factor binding to cell surface receptors. The proteins encoded by mutated *ras* oncogenes, however, are no longer properly controlled by cellular regulatory signals. Instead, the *ras* oncogene proteins function in a continuous, unregulated fashion, leading to abnormal stimulation of cell proliferation even in the absence of the appropriate growth factors.

Distinct members of the *ras* family are encountered as active oncogenes in different types of human cancers. The *ras*K oncogene is most frequently involved in carcinomas, being found in about 25% of lung carcinomas, 50% of colon carcinomas, and 90% of pancreatic carcinomas. The *ras*N oncogene, in contrast, is more frequently involved in leukemias and lymphomas. For example, about 25% of acute myeloid and lymphoid leukemias contain *ras*N oncogenes. All

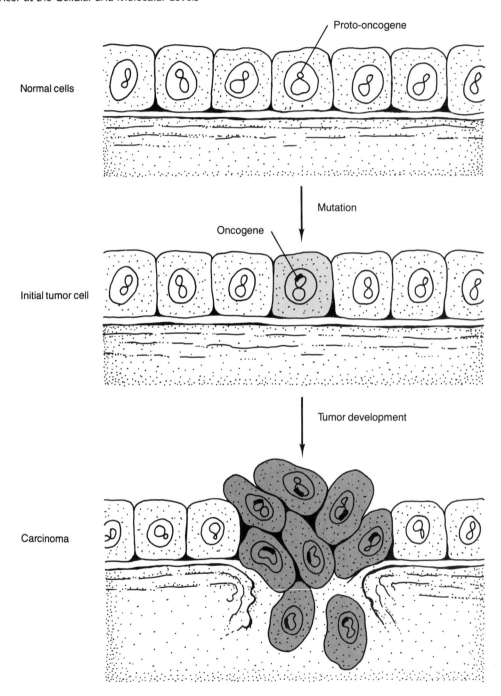

Figure 8.7 Oncogenes are formed during carcinogenesis. Normal cells contain proto-oncogenes, not oncogenes. Oncogenes are formed as a result of mutations suffered by proto-oncogenes during carcinogenesis. Once an oncogene is formed, it drives abnormal cell proliferation, resulting in tumor development.

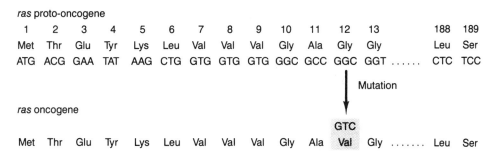

ras proto-oncogene

1	2	3	4	5	6	7	8	9	10	11	12	13		188	189
Met	Thr	Glu	Tyr	Lys	Leu	Val	Val	Val	Gly	Ala	Gly	Gly		Leu	Ser
ATG	ACG	GAA	TAT	AAG	CTG	GTG	GTG	GTG	GGC	GCC	GGC	GGT	......	CTC	TCC

Mutation

ras oncogene

GTC

Met Thr Glu Tyr Lys Leu Val Val Val Gly Ala Val Gly .... Leu Ser

Figure 8.8 Formation of *ras* oncogenes. Single nucleotide changes, resulting in alterations of critical amino acids, are sufficient to convert *ras* proto-oncogenes to oncogenes. In the example shown here, the change of a G to a T in a *ras* gene alters amino acid 12 from glycine (gly) to valine (val) in the *ras* protein. This results in a *ras* oncogene protein, which functions in an unregulated manner to drive abnormal cell proliferation.

three members of the *ras* oncogene family are found frequently in thyroid carcinomas. Given the relative frequencies of these neoplasms (see chapter 1), *ras* oncogenes appear to play a role in the development of about 15–20% of all cancers in the United States. In addition, other more distantly related genes (such as *gsp* and *gip*) can also be converted to oncogenes by mutations similar to those first identified in the *ras* oncogenes. The *gsp* and *gip* oncogenes seem to be involved particularly in tumors of hormone-responsive cells, such as ovarian carcinomas and cancers of the pituitary and adrenal glands.

Activation of Oncogenes by Chromosome Translocation and DNA Rearrangement

The single nucleotide changes responsible for the formation of *ras* and related oncogenes are not the only way in which proto-oncogenes can be converted to oncogenes in human tumors. As noted in chapter 7, many cancer cells display abnormal arrangements of their genetic material, which can often be detected as abnormalities in the structure of their chromosomes. Genetic rearrangements of this sort frequently result in the formation of oncogenes in human tumors. For example, in virtually all Burkitt's lymphomas, a region of chromosome 8 is rearranged to chromosome 2, 14, or 22 (Fig. 8.9). These rearrangements (**chromosome translocations**) activate the c-*myc* proto-oncogene (which was first discovered in a chicken retrovirus) by removing it from its normal environment on chromosome 8 and inserting it instead into loci on chromosomes 2, 14, or 22. Importantly, these abnormal loci are the regions in which the antibody (**immunoglobin**) genes, which are actively expressed in Burkitt's lymphoma cells, are located. The result of these rearrangements is unregulated expression of c-*myc*, which is sufficient to drive abnormal cell proliferation and contribute to lymphoma development.

Additional chromosome translocations occur regularly in other leukemias and lymphomas, activating the oncogenes *abl*, *bcl*-**2**, *E2A*, and *RAR*. Follicular

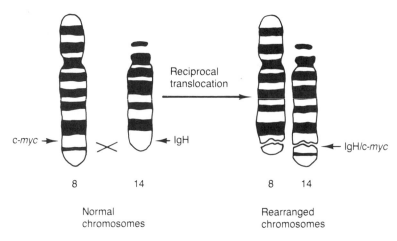

Reciprocal
translocation

c-myc →

← IgH

← IgH/c-myc

8   14

8   14

Normal
chromosomes

Rearranged
chromosomes

Figure 8.9 Activation of the c-myc proto-oncogene by chromosome translocation. The c-myc proto-oncogene is translocated from chromosome 8 to the immunoglobulin heavy-chain (IgH) locus on chromosome 14 in Burkitt's lymphomas. This translocation results in abnormal c-myc expression.

B-cell lymphomas regularly involve activation of bcl-2, which, like myc, occurs as a result of unregulated gene expression. Conversion of the abl proto-oncogene to an oncogene, however, involves formation of an altered gene product (Fig. 8.10). Chronic myelogenous leukemias are characterized by a chromosomal abnormality, known as the **Philadelphia chromosome,** which results from an exchange of genetic material between chromosomes 9 and 22. This leads to a gene rearrangement in which the abl proto-oncogene from chromosome 9 becomes fused with another gene, called **bcr,** on chromosome 22. The result is production of an abnormal protein, encoded by the fused bcr/abl oncogene, which functions in an uncontrolled manner, leading to the development of leukemia. Similar gene rearrangements, leading to the formation of aberrantly functioning oncogene proteins, are also involved in the generation of several other oncogenes, including the E2A and RAR oncogenes in acute lymphocytic and promyelocytic leukemias, respectively, as well as the **ret** and **trk** oncogenes in human thyroid carcinomas.

Gene Amplification

Another mechanism by which oncogenes are activated in human cancers is **gene amplification** (Fig. 8.11). As noted in chapter 7, gene amplification occurs frequently in cancer cells, resulting in increased numbers of gene copies. This, in turn, leads to elevated levels of expression of the amplified gene. Amplification of a proto-oncogene, then, results in increased expression of a protein that acts to drive cell proliferation. Members of the myc oncogene family (c-myc, N-myc, and L-myc) are frequently amplified in a variety of tumors, especially breast and lung carcinomas and neuroblastomas. A particularly interesting example is pro-

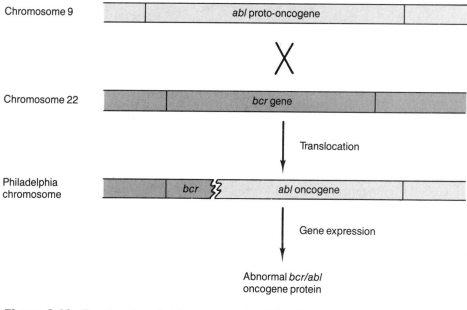

Figure 8.10   Translocation of *abl* in chronic myelogenous leukemia. The *abl* proto-oncogene is translocated from chromosome 9 to chromosome 22, forming the Philadelphia chromosome. This translocation results in fusion of *abl* with another gene, *bcr*, leading to expression of an abnormal *bcr/abl* oncogene protein.

vided by amplification of the N-*myc* gene in neuroblastomas, which occurs specifically in rapidly progressing, highly malignant neoplasms. In these tumors, amplification of N-*myc* appears to be a frequent event in tumor progression, closely associated with the development of increasing malignancy. Amplification of another oncogene, **erbB-2,** seems to be similarly associated with the progression of breast and ovarian carcinomas. A number of other oncogenes have also been detected as amplified sequences in human tumors, although their amplifi-

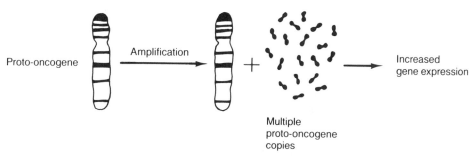

Figure 8.11   Proto-oncogene amplification. Gene amplification results in the formation of multiple copies of a proto-oncogene, leading to increased gene expression.

cation occurs less regularly than that of the *myc* genes and *erb*B-2. However, since gene amplification is a relatively common event in cancer cells, it may be that amplification and increased expression of proto-oncogenes frequently contribute to tumor progression.

SUMMARY

Oncogenes were first discovered as specific viral genes that were responsible for neoplastic transformation of virus-infected cells. Subsequent studies showed that viral oncogenes originated from genes of normal cells, the proto-oncogenes, which had become incorporated into viral genomes. Proto-oncogenes are normal cell genes that play important roles in normal cell growth and differentiation. The oncogenes, in contrast, are altered versions of proto-oncogenes, which are abnormal in both the regulation of gene expression and the structure and function of their protein products. Oncogenes are not only present in viruses, but can also be formed in human cancer cells as a result of damage to proto-oncogenes during the process of carcinogenesis. A number of different oncogenes are thus activated in human tumors as a consequence of point mutations, rearrangements, or amplification of proto-oncogenes (Table 8.3). The identification of oncogenes in a significant fraction of human tumors clearly indicates their importance in the development of human cancers.

Table 8.3   *Oncogenes in Human Tumors*

Oncogene	Types of Cancer	Activation Mechanism
abl	Chronic myelogenous leukemia, acute lymphocytic leukemia	Translocation
bcl-2	Follicular B-cell lymphoma	Translocation
E2A	Acute lymphocytic leukemia	Translocation
*erb*B-2	Breast and ovarian carcinoma	Amplification
gip	Adrenal cortical and ovarian carcinomas	Point mutation
gsp	Pituitary tumors	Point mutation
c-*myc*	Burkitt's and other B-cell lymphomas Breast and lung carcinomas	Translocation Amplification
L-*myc*	Lung carcinoma	Amplification
N-*myc*	Neuroblastoma, lung carcinoma	Amplification
RAR	Acute promyelocytic leukemia	Translocation
*ras*H	Thyroid carcinoma	Point mutation
*ras*K	Colon, lung, pancreatic, and thyroid carcinomas	Point mutation
*ras*N	Acute myelocytic and lymphocytic leukemias, thyroid carcinoma	Point mutation
ret	Thyroid carcinoma	DNA rearrangement
trk	Thyroid carcinoma	DNA rearrangement

KEY TERMS

oncogene
tumor virus
oncogenic virus
cell transformation
viral oncogene
retrovirus
acutely transforming virus
Rous sarcoma virus (RSV)
src
abl
E6
E7
BNLF-1
tax
X gene
proto-oncogene
cellular oncogene
ras
point mutation
gsp
gip
myc
chromosome translocation
bcl-2
E2A
RAR
Philadelphia chromosome
ret
trk
gene amplification
*erb*B-2

REFERENCES AND FURTHER READING

General References

Cooper, G.M. 1990. *Oncogenes.* Jones and Bartlett, Boston.

Weiss, R., Teich, N., Varmus, H., and Coffin, J., eds. 1985. *Molecular biology of tumor viruses: RNA tumor viruses.* 2nd ed. Cold Spring Harbor Laboratory, New York.

Tumor Viruses and Viral Oncogenes

Botchan, M., Grodzicker, T., and Sharp, P.A., eds. 1986. *Cancer cells*, Vol. 4, *DNA tumor viruses*. Cold Spring Harbor Laboratory, New York.

Chisari, F.V., Klopchin, K., Moriyama, T., Pasquinelli, C., Dunsford, H.A., Sell, S., Pinkert, C.A., Brinster, R.L., and Palmiter, R.D. 1989. Molecular pathogenesis of hepatocellular carcinoma in hepatitis B virus transgenic mice. *Cell* 59:1145–1156.

Fahreus, R., Rymo, L., Rhim, J.S., and Klein, G. 1990. Morphological transformation of human keratinocytes expressing the *LMP* gene of Epstein-Barr virus. *Nature* 345: 447–449.

Munger, K., Phelps, W.C., Bubb, V., Howley, P.M., and Schlegel, R. 1989. The *E6* and *E7* genes of the human papillomavirus type 16 together are necessary and sufficient for transformation of primary human keratinocytes. *J. Virol.* 63:4417–4421.

Nerenberg, M., Hinrichs, S.H., Reynolds, R.K., Khoury, G., and Jay, G. 1987. The *tat* gene of human T-lymphotropic virus type 1 induces mesenchymal tumors in transgenic mice. *Science* 237:1324–1329.

Nishizawa, M., Kataoka, K., Goto, N., Fujiwara, K.T., and Kawai, S. 1989. v-*maf*, a viral oncogene that encodes a "leucine zipper" motif. *Proc. Natl. Acad. Sci. USA* 86:7711–7715.

Souyri, M., Vigon, I., Penciolelli, J.-F., Heard, J.-M., Tambourin, P., and Wendling, F. 1990. A putative truncated cytokine receptor gene transduced by the myeloproliferative leukemia virus immortalizes hematopoietic progenitors. *Cell* 63:1137–1147.

Takada, S., and Koike, K. 1990. Trans-activation function of a 3′ truncated *X* gene-cell fusion product from integrated hepatitis B virus DNA in chronic hepatitis tissues. *Proc. Natl. Acad. Sci. USA* 87:5628–5632.

Tanaka, A., Takahashi, C., Yamaoka, S., Nosaka, T., Maki, M., and Hatanaka, M. 1990. Oncogenic transformation by the *tax* gene of human T-cell leukemia virus type I *in vitro*. *Proc. Natl. Acad. Sci. USA* 87:1071–1075.

Wang, D., Liebowitz, D., and Kieff, E. 1985. An EBV membrane protein expressed in immortalized lymphocytes transforms established rodent cells. *Cell* 43:831–840.

Wilson, J.B., Weinberg, W., Johnson, R., Yuspa, S., and Levine, A.J. 1990. Expression of the *BNLF*-1 oncogene of Epstein-Barr virus in the skin of transgenic mice induces hyperplasia and aberrant expression of keratin 6. *Cell* 61:1315–1327.

Proto-Oncogenes in Normal Cells

Stehelin, D., Varmus, H.E., Bishop, J.M., and Vogt, P.K. 1976. DNA related to the transforming gene(s) of avian sarcoma viruses is present in normal avian DNA. *Nature* 260:170–173.

Differences Between Oncogenes and Proto-Oncogenes

Blair, D.G., Oskarsson, M., Wood, T.G., McClements, W.L., Fischinger, P.J., and Vande Woude, G.F. 1981. Activation of the transforming potential of a normal cell sequence: a molecular model for oncogenesis. *Science* 212:941–943.

Kmiecik, T.E., and Shalloway, D. 1987. Activation and suppression of pp60$^{c\text{-}src}$ transforming ability by mutation of its primary sites of tyrosine phosphorylation. *Cell* 49:65–73.

Takeya, T., and Hanafusa, H. 1983. Structure and sequence of the cellular gene homologous to the RSV *src* gene and the mechanism for generating the transforming virus. *Cell* 32:881–890.

Oncogenes in Human Cancer

Alcalay, M., Zangrilli, D., Pandolfi, P.P., Longo, L., Mencarelli, A., Giacomucci, A., Rocchi, M., Biondi, A., Rambaldi, A., Lo Loco, R., Diverio, D., Donti, E., Grignani, F., and Pelicci, D.G. 1991. Translocation breakpoint of acute promyelocytic leukemia lies within the retinoic acid receptor-α locus. *Proc. Natl. Acad. Sci. USA* 88:1977–1981.

Alitalo, K., and Schwab, M. 1986. Oncogene amplification in tumor cells. *Adv. Cancer Res.* 47:235–281.

Barbacid, M. 1987. *ras* genes. *Ann. Rev. Biochem.* 56:779–827.

Bongarzone, I., Pierotti, M.A., Monzine, N., Mondellini, P., Manenti, G., Donghi, R., Pilotti, S., Grieco, M., Santoro, M., Fusco, A., Vecchio, G., and Della Porta, G. 1989. High frequency of activation of tyrosine kinase oncogenes in human papillary thyroid carcinoma. *Oncogene* 4:1457–1462.

Borrow, J., Goddard, A.D., Sheer, D., and Solomon, E. 1990. Molecular analysis of acute promyelocytic leukemia breakpoint cluster region on chromosome 17. *Science* 249: 1577–1580.

Bos, J.L. 1989. *ras* oncogenes in human cancer: a review. *Cancer Res.* 49:4682–4689.

Brodeur, G.M., Seeger, R.C., Schwab, M., Varmus, H.E., and Bishop, J.M. 1984. Amplification of N-*myc* in untreated human neuroblastomas correlates with advanced disease stage. *Science* 224:1121–1124.

Der, C.J., Krontiris, T.G., and Cooper, G.M. 1982. Transforming genes of human bladder and lung carcinoma cell lines are homologous to the *ras* of Harvey and Kirsten sarcoma viruses. *Proc. Natl. Acad. Sci. USA* 79:3637–3640.

de Thé, H., Chomienne, C., Lanotte, M., Degos, L., and Dejean, A. 1990. The t(15;17) translocation of acute promyelocytic leukaemia fuses the retinoic acid receptor α gene to a novel transcribed locus. *Nature* 347:558–561.

Grieco, M., Santoro, M., Berlingieri, M.T., Melillo, R.M., Donghi, R., Bongarzone, I., Pierotti, M.A., Della Porta, G., Fusco, A., and Vecchio, G. 1990. PTC is a novel rearranged form of the *ret* proto-oncogene and is frequently detected *in vivo* in human thyroid papillary carcinomas. *Cell* 60:557–563.

Krontiris, T.G., and Cooper, G.M. 1981. Transforming activity of human tumor DNAs. *Proc. Natl. Acad. Sci. USA* 78:1181–1184.

Leder, P, Battey, J., Lenoir, G., Moulding, C., Murphy, W., Potter, H., Stewart, T., and Taub, R. 1983. Translocations among antibody genes in human cancer. *Science* 222:765–771.

Lyons, J., Landis, C.A., Harsh, G., Vallar, L., Grunewald, K., Feichtinger, H., Duh, Q.-Y., Clark, O.H., Kawasaki, E., Bourne, H.R., and McCormick, F. 1990. Two G protein oncogenes in human endocrine tumors. *Science* 249:655–659.

Mellentin, J.D., Murre, C., Donlon, T.A., McCaw, P.S., Smith, S.D., Carroll, A.J., McDonald, M.E., Baltimore, D., and Cleary, M.L. 1989. The gene for enhancer binding proteins E12/E47 lies at the t(1;19) breakpoint in acute leukemias. *Science* 246:379–382.

Nourse, J., Mellentin, J.D., Galili, N., Wilkinson, J., Stanbridge, E., Smith, S.D., and Cleary, M.L. 1990. Chromosomal translocation t(1;19) results in synthesis of a homeobox fusion mRNA that codes for a potential chimeric transcription factor. *Cell* 60:535–545.

Shih, C., Padhy, L.C., Murray, M., and Weinberg, R.A. 1981. Transforming genes of carcinomas and neuroblastomas introduced into mouse fibroblasts. *Nature* 290:261–264.

Shtivelman, E., Lifshitz, B., Gale, R.P., and Canaani, E. 1985. Fused transcript of *abl* and *bcr* genes in chronic myelogenous leukemia. *Nature* 315:550–554.

Slamon, D.J., Godolphin, W., Jones, L.A., Holt, J.A., Wong, S.G., Keith, D.E., Levin, W.J., Stuart, S.G., Udove, J., Ullrich, A., and Press, M.F. 1989. Studies of the *HER-2/neu* proto-oncogene in human breast and ovarian cancer. *Science* 244:707–712.

Tabin, C.J., Bradley, S.M., Bargmann, C.I., Weinberg, R.A., Papageorge, A.G., Scolnick, E.M., Dhar, R., Lowy, D.R., and Chang, E.H. 1982. Mechanism of activation of a human oncogene. *Nature* 300:143–149.

Tsujimoto, Y., Cossmann, J., Jaffee, E., and Croce, C.M. 1985. Involvement of the *bcl*-2 gene in human follicular lymphoma. *Science* 228:1440–1443.

Chapter 9
Tumor Suppressor Genes

THE ONCOGENES are one of two distinct classes of genes that contribute to the development of human cancer. They arise from proto-oncogenes as a result of genetic alterations that either increase gene expression or lead to uncontrolled function of the oncogene-encoded proteins. As will be discussed in chapter 11, the products of most proto-oncogenes act to stimulate normal cell proliferation, and the unregulated action of the corresponding oncogene proteins leads to the abnormal proliferation characteristic of cancer cells.

The other class of genes that are important in carcinogenesis, the tumor suppressor genes, represent the opposite side of the coin of cellular growth control. Whereas oncogenes stimulate cell proliferation and tumor development, the tumor suppressor genes act to inhibit these processes. If oncogenes are viewed as gas pedals, acting to accelerate the growth of cancer cells, tumor suppressor genes can be viewed as the brakes, acting to slow down cell growth. In many neoplasms, genetic alterations lead to the loss or inactivation of tumor suppressor genes, thereby eliminating inhibitors of cell proliferation. Activation of oncogenes and inactivation of tumor suppressor genes thus represent complementary events in the development of cancer, both contributing to increased cell proliferation and loss of normal growth control.

HYBRIDS BETWEEN NORMAL CELLS AND CANCER CELLS

Normal Cell Genes Suppress Tumor Development

The function of tumor suppressor genes was first illustrated in straightforward experiments in which normal cells and cancer cells were fused with each other (Fig. 9.1). The result is a hybrid cell that contains genes from both its parents. In most cases, such hybrid cells no longer behave as full-fledged cancer cells. Although they retain some characteristics of tumor cells and therefore are not fully normal, hybrids between cancer cells and normal cells are usually unable to form tumors following inoculation into experimental animals. The tumorigenicity of the parental cancer cell is thus inhibited by fusion with a normal cell, suggesting that genes derived from the normal cell parent act to suppress tumor cell growth. Hence the term **tumor suppressor gene.**

Further analysis of hybrids between cancer cells and normal cells has revealed that normal cells contain multiple tumor suppressor genes, and that distinct tumor suppressor genes are involved in different cancers. For example, the ability of some cancer cells to form tumors is inhibited by a tumor suppressor gene present on chromosome 11 of normal cells. Other cancer cells, however, are unaffected by the tumor suppressor gene from chromosome 11. Instead, their ability to form tumors is blocked by other tumor suppressor genes located on different chromosomes of normal cells.

Tumor Suppressor Genes Versus Oncogenes

The action of tumor suppressor genes is thus opposite to that of oncogenes. Tumor suppressor genes are inactivated in cancer cells, and the introduction of

Normal cell   Tumor cell

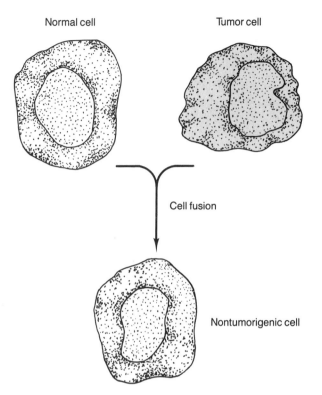

Cell fusion

Nontumorigenic cell

Figure 9.1 *Suppression of tumorigenicity by cell fusion. Hybrids between tumor cells and normal cells are usually nontumorigenic.*

functional tumor suppressor genes from normal cells into cancer cells results in loss of tumorigenicity. In contrast, oncogenes are activated in cancer cells, and their introduction into normal cells leads to tumor formation. Not only do oncogenes and tumor suppressor genes both contribute to the development of human cancers; as discussed in chapter 11, the products of some tumor suppressor genes and oncogenes interact directly to regulate each other's expression and function within the cell.

TUMOR SUPPRESSOR GENES AND INHERITED CANCERS: THE RETINOBLASTOMA PROTOTYPE

As discussed in chapter 8, active oncogenes are not genetically transmitted from parent to child. Rather, they are generated from proto-oncogenes as a result of genetic damage that occurs somatically during the process of tumor development. In contrast, some rare hereditary cancers (see chapter 5) are due to the genetic transmission of defective tumor suppressor genes, which have been inactivated as a consequence of mutation. Children who inherit such inactivated tumor suppressor genes suffer a high likelihood that cancer will develop.

RB and Inheritance of Retinoblastoma

The disease that has provided the most persuasive model for the involvement of tumor suppressor genes in human cancer is retinoblastoma, which was discussed in chapter 5 as the prototype of inherited childhood cancers. Extensive genetic and molecular studies have established that retinoblastoma results from the inactivation of a tumor suppressor gene called *RB*, which is located on human chromosome 13.

The involvement of the *RB* gene in the development of both inherited and noninherited forms of retinoblastoma is illustrated in Figure 9.2. Normal individuals inherit two functional copies of *RB*, one from each parent. In cases of hereditary retinoblastoma, however, a defective copy of the *RB* gene is inherited from the afflicted parent. The child therefore has only one functional *RB* copy, inherited from the normal parent. Importantly, this single normal copy of the *RB* gene is sufficient to prevent immediate tumor formation. It directs the production of enough functional *RB* protein to regulate cell growth, and cells that contain only one rather than two normal copies of *RB* behave normally. Retinoblastoma almost inevitably develops, however, as a result of a second mutation leading to loss or inactivation of this single normal *RB* copy. Frequently, this occurs by loss (**deletion**) of part or all of the copy of chromosome 13 containing the normal *RB* gene (Fig. 9.3). The result is formation of a cell that no longer contains even a single normal *RB* gene. Functional *RB* protein can therefore no longer be produced, and in the absence of *RB* tumor suppressor activity, retinoblastoma develops. Other events may also contribute to the genesis of these neoplasms, but it appears that loss of both copies of *RB* is the critical rate-limiting step in the malignant transformation of the embryonal retinal cells that give rise to this tumor.

The frequency of second mutations leading to loss of a single functional *RB* copy is sufficiently high that almost all children who have inherited a defective *RB* gene copy develop retinoblastoma, usually with multiple tumors in both eyes. Since a defective *RB* copy is transmitted to half of the offspring of a parent with hereditary retinoblastoma, the likelihood that a child of such an affected parent will develop the disease is nearly 50%. The development of noninherited or sporadic retinoblastoma, however, requires the occurrence of two somatic mutations to inactivate both normal *RB* copies in the same cell (Fig. 9.2). This is comparatively unlikely, so retinoblastoma afflicts only about 1 in 40,000 children of normal parents. Inheritance of a defective copy of the *RB* tumor suppressor gene thus results in approximately a 40,000-fold increased risk of retinoblastoma development.

A Functional *RB* Gene Suppresses Tumorigenicity

The critical distinction between *RB* and the oncogenes, consistent with the designation of *RB* as a tumor suppressor gene, is that retinoblastoma develops as a consequence of inactivation or loss of both copies of the normal *RB* gene. Thus, a lack of functional *RB* protein leads to tumor development. Consistent with the predicted activity of a tumor suppressor, it has been demonstrated in experimental culture systems that introduction of a normal, functional *RB* gene reverses the tumorigenicity of retinoblastoma cells (Fig. 9.4). As expected, therefore, normal *RB* acts to inhibit tumor development.

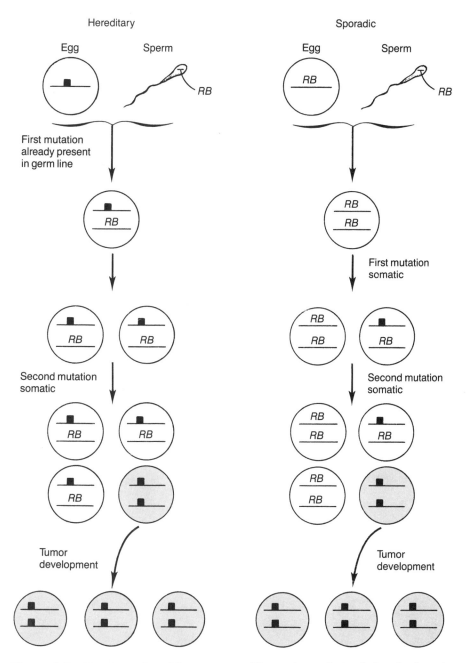

Figure 9.2 *Mutations of the RB gene during retinoblastoma development. In hereditary retinoblastoma, a defective copy of the RB gene (designated ■) is inherited from the affected parent. A second somatic mutation, which inactivates the single normal* RB copy in a retinal cell, then leads to the development of retinoblastoma. In sporadic retinoblastoma, two normal RB genes are inherited. Retinoblastoma develops only if two somatic mutations inactivate both copies of RB in the same cell.

Figure 9.3 Chromosome deletions in retinoblastoma. The *RB* gene is frequently lost in retinoblastoma by the deletion of the relevant region of chromosome 13.

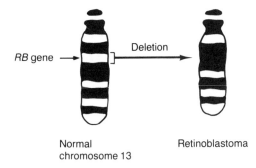

Normal
chromosome 13

Retinoblastoma

INVOLVEMENT OF THE *RB* GENE IN OTHER TUMORS

Although the *RB* tumor suppressor gene was discovered and characterized through studies of inherited retinoblastoma, it also appears to be involved in the development of several other human cancers, including some of the common malignancies of adults. In particular, inactivation of *RB* has been found to occur in osteosarcomas, rhabdomyosarcomas, small cell lung carcinomas, bladder carcinomas, breast carcinomas, and prostate carcinomas. Loss of the *RB* tumor suppressor is particularly frequent in small cell lung carcinomas, occurring in nearly all of these tumors. Approximately 25% of lung cancers are of the small cell type, which therefore accounts for about 5% of all cancer deaths in the United States. Thus the *RB* tumor suppressor gene, originally discovered in a rare childhood tumor, appears to be involved in a substantial fraction of lethal adult malignancies.

Figure 9.4 Suppression of tumorigenicity by *RB*. Introduction of a normal, functional *RB* gene reverses the tumorigenicity of retinoblastoma cells.

Retinoblastoma
cell

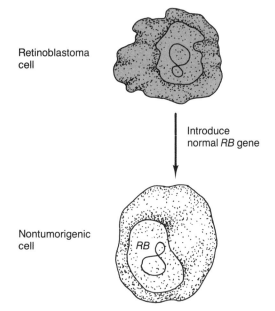

Introduce
normal *RB* gene

Nontumorigenic
cell

OTHER TUMOR SUPPRESSOR GENES

Identification of *WT*1 and *NF*1 Genes in Hereditary Cancers

RB was the first tumor suppressor gene to be isolated and studied at the molecular level, but the identification and characterization of additional tumor suppressor genes has rapidly followed these initial studies (Table 9.1). Several of these genes, like *RB*, have been identified by studies of rare inherited cancers of childhood. Wilms' tumor, for example, is an inherited childhood kidney tumor (also discussed in chapter 5). Like retinoblastoma, the development of Wilms' tumor results from loss of a tumor suppressor gene—in this case, a gene designated **WT1,** which is located on chromosome 11. Another tumor suppressor gene, the **NF**1 gene on chromosome 17, has been isolated by studies of type 1 (von Recklinghausen) neurofibromatosis. This disease is one of the most common human genetic disorders, occurring with an incidence of about 1 in 3,000 people. Afflicted individuals develop multiple benign tumors (**neurofibromas**), which frequently progress to malignant neurofibrosarcomas. Three distinct tumor suppressor genes—*RB, WT1,* and *NF1*—have thus been isolated and characterized by virtue of their involvement in hereditary neoplasms.

Other Genes Responsible for Inherited Cancers

Genes responsible for the inheritance of a number of other cancers have been found to reside on other chromosomes, and it seems likely that at least some of these genes will also be tumor suppressors. These include the genes responsible for several of the inherited cancer syndromes discussed in chapter 5, including the *familial adenomatous polyposis* form of inherited colon cancer (the *FAP* gene on chromosome 5), *neurofibromatosis type 2* (the *NF2* gene on chromosome 22), *Von Hippel-Lindau syndrome* (the *VHL* gene on chromosome 3), *multiple endocrine neoplasia type 1* (the *MEN1* gene on chromosome 11), and *multiple endocrine neoplasia type 2A* (the *MEN2A* gene on chromosome 10). In addition, at least one other gene on chromosome 11, but distinct from *WT1*, appears to be responsible for the inheritance of some Wilms' tumors. The pattern of genetic transmission of each of these cancers is similar to that of retinoblastoma. In

Table 9.1   Tumor Suppressor Genes

Gene	Kinds of Tumors
DCC	Colon/rectum carcinomas
MCC	Colon/rectum carcinomas
*NF*1	Neurofibrosarcomas
p53	Brain tumors, breast, colon/rectum, esophageal, hepatocellular, and lung carcinomas, osteosarcomas, rhabdomyosarcomas, leukemias, and lymphomas
RB	Retinoblastoma, osteosarcoma, rhabdomyosarcoma, bladder, breast, lung, and prostate carcinomas
*WT*1	Wilms' tumor

addition, genetic information from the normal copy of the relevant chromosome (e.g., chromosome 5 in familial adenomatous polyposis) is frequently lost in tumor cells, suggesting that the genes responsible for inheritance of these cancers function analogously to *RB* as tumor suppressor genes. Further understanding of these genes, however, awaits their isolation and characterization as distinct molecular entities—an active area of current cancer research.

Common Involvement of the *p53* Gene in a Variety of Cancers

In addition to the genes involved in rare hereditary cancers, still other tumor suppressor genes have been identified as genes that are frequently lost in noninherited cancers, including some of the common malignancies of adults. Inactivation of the *p53* tumor suppressor gene on chromosome 17 appears to be particularly frequent in human tumors. Loss or inactivation of *p53* has been consistently observed in a variety of different kinds of cancers, including lung, breast, colon/rectum, and hepatocellular carcinomas, osteosarcomas, rhabdomyosarcomas, neurofibrosarcomas, glioblastomas, leukemias, and lymphomas. Interestingly, it has also been recently found that inherited mutations of *p53* are responsible for the Li-Fraumeni cancer family syndrome, which is associated with the development of several different types of cancer, including sarcomas, breast carcinomas, brain tumors, and leukemias. It is particularly noteworthy that inactivation of *p53* apparently occurs in about 50% of all types of lung cancer and in the majority of breast and colon/rectum carcinomas. Since lung, breast, and colon/rectum carcinomas account for approximately one-half of all cancer deaths in the United States (see chapter 1), the *p53* tumor suppressor gene is clearly of critical importance.

The *MCC* and *DCC* Genes in Colon Cancer

Two other tumor suppressor genes, the *MCC* gene on chromosome 5 and the *DCC* gene on chromosome 18, are also frequently inactivated in colon and rectum carcinomas, indicating that they play important roles in the development of this common human malignancy. The *MCC* gene may be identical to the *FAP* gene, which is responsible for inherited familial adenomatous polyposis, although this remains to be established. Loss of genetic material from additional human chromosomes is also common in other types of tumors. For example, deletions affecting chromosome 1 are frequent in neuroblastomas, and deletions affecting chromosome 3 are frequent in all types of lung cancer. It is likely that these chromosome abnormalities indicate the sites of still more tumor suppressor genes awaiting isolation and characterization.

SUMMARY

Tumor suppressor genes, in contrast to oncogenes, normally act to inhibit cell growth and tumor development. In many cancers, tumor suppressor genes are lost or inactivated, so that functional gene products are no longer produced. Such a lack of normal tumor suppressor gene function leads to unregulated cell proliferation and tumor development. Tumor suppressor genes were first identified in rare inherited human cancers, such as retinoblastoma and Wilms' tumor, but these genes are also involved in common adult malignancies. Six distinct tumor suppressor genes (*RB*, *WT*1, *NF*1, *p53*, *MCC*, and *DCC*) have been characterized to date at the molecular level. However, the existence of a number of additional such genes is inferred from studies of both hereditary cancers and gene loss during tumor development. Some of the tumor suppressor genes (including *RB*, *p53*, *MCC*, and *DCC*) are frequently inactivated in the most common types of human cancers, such as breast, lung, and colon/rectum carcinomas, indicating that damage to these genes contributes to a substantial fraction of cancer mortality.

KEY TERMS

tumor suppressor gene
cell fusion
loss of tumorigenicity
inherited cancers
RB
gene loss or inactivation
deletion
***WT*1**
***NF*1**
p53
MCC
DCC

REFERENCES AND FURTHER READING

General References

Cooper, G.M. 1990. *Oncogenes*. Jones and Bartlett, Boston.
Marshall, C.J. 1991. Tumor suppressor genes. *Cell* 64:313–326.
Stanbridge, E.J. 1990. Human tumor suppressor genes. *Ann. Rev. Genet.* 24:615–657.

Hybrids Between Normal Cells and Cancer Cells

Harris, H., Miller, O.J., Klein, G., Worst, P., and Tachibana, T. 1969. Suppression of malignancy by cell fusion. *Nature* 223:363–368.
Stanbridge, E.J. 1976. Suppression of malignancy in human cells. *Nature* 260:17–20.

Tumor Suppressor Genes and Inherited Cancers: The Retinoblastoma Prototype

Friend, S.H., Bernards, R., Rogelj, S., Weinberg, R.A., Rapaport, J.M., Albert, D.M., and Dryja, T.P. 1986. A human DNA segment with properties of the gene that predisposes to retinoblastoma and osteosarcoma. *Nature* 323:643–646.

Fung, Y.-K., Murphree, A.L., T'Ang, A., Qian, J., Hinrichs, S.H., and Benedict, W.F. 1987. Structural evidence for the authenticity of the retinoblastoma gene. *Science* 236:1657–1661.

Huang, H.-J. S., Yee, J.-K., Shew, J.-Y., Chen, P.-L., Bookstein, R., Friedmann, T., Lee, E. Y.-H. P., and Lee, W.-H. 1988. Suppression of the neoplastic phenotype by replacement of the *RB* gene in human cancer cells. *Science* 242:1563–1566.

Knudson, A.G., Jr. 1971. Mutation and cancer: statistical study of retinoblastoma. *Proc. Natl. Acad. Sci. USA* 68:820–823.

Lee, W.-H., Bookstein, R., Hong, F., Young, L.-J., Shew, J.-Y., and Lee, E.Y.-H.P. 1987. Human retinoblastoma susceptibility gene: cloning, identification, and sequence. *Science* 235:1394–1399.

Involvement of the *RB* Gene in Other Tumors

Bookstein, R., Rio, P., Madreperla, S.A., Hong, F., Allred, C., Grizzle, W.E., and Lee, W.H. 1990. Promoter deletion and loss of retinoblastoma gene expression in human prostate carcinoma. *Proc. Natl. Acad. Sci. USA* 87:7762–7766.

Friend, S.H., Horowitz, J.M., Gerber, M.R., Wang, X.-F., Bogenmann, E., Li, F.P., and Weinberg, R.A. 1987. Deletions of a DNA sequence in retinoblastomas and mesenchymal tumors: organization of the sequence and its encoded protein. *Proc. Natl. Acad. Sci. USA* 84:9059–9063.

Harbour, J.W., Lai, S.-L., Whang-Peng, J., Gazdar, A.F., Minna, J.D., and Kaye, F.J. 1988. Abnormalities in structure and expression of the human retinoblastoma gene in SCLC. *Science* 241:353–357.

Horowitz, J.M., Park, S.-H., Bogenmann, E., Cheng, J.-C., Yandell, D.W., Kaye, F.J., Minna, J.D., Dryja, T.P., and Weinberg, R.A. 1990. Frequent inactivation of the retinoblastoma anti-oncogene is restricted to a subset of human tumor cells. *Proc. Natl. Acad. Sci. USA* 87:2775–2779.

Lee, E. Y.-H.P., To, H., Shew, J.-Y., Bookstein, R., Scully, P., and Lee, W.-H. 1988. Inactivation of the retinoblastoma susceptibility gene in human breast cancers. *Science* 241:218–221.

T'Ang, A., Varley, J.M., Chakraborty, S., Murphree, A.L., and Fung, Y.-K. 1988. Structural rearrangement of the retinoblastoma gene in human breast carcinoma. *Science* 242:263–266.

Other Tumor Suppressor Genes

Ahuja, H., Bar-Eli, M., Advani, S.H., Benchimol, S., and Cline, M.J. 1989. Alterations in the *p53* gene and the clonal evolution of the blast crisis of chronic myelocytic leukemia. *Proc. Natl. Acad. Sci. USA* 86:6783–6787.

Baker, S.J., Fearon, E.R., Nigro, J.M., Hamilton, S.R., Preisinger, A.C., Jessup, J.M., van Tuinen, P., Ledbetter, D.H., Barker, D.F., Nakamura, Y., White, R., and Vogelstein, B. 1989. Chromosome 17 deletions and *p53* gene mutations in colorectal carcinomas. *Science* 244:217–221.

Bressac, B., Kew, M., Wands, J., and Ozturk, M. 1991. Selective G to T mutations of *p53* gene in hepatocellular carcinoma from southern Africa. *Nature* 350:429–431.

Call, K.M., Glaser, T., Ito, C.Y., Buckler, A.J., Pelletier, J., Haber, D.A., Rose, E.A., Kral, A., Yeger, H., Lewis, W.H., Jones, C., and Housman, D.E. 1990. Isolation and characterization of a zinc finger polypeptide gene at the human chromosome 11 Wilms' tumor locus. *Cell* 60:509–520.

Cawthon, R.M., Weiss, R., Xu, G., Viskochil, D., Culver, M., Stevens, J., Robertson, M., Dunn, D., Gesteland, R., O'Connell, P., and White, R. 1990. A major segment of the neurofibromatosis type 1 gene: cDNA sequence, genomic structure, and point mutations. *Cell* 62:193–201.

Cheng, J., and Haas, M. 1990. Frequent mutations in the *p53* tumor suppressor gene in human leukemia T-cell lines. *Mol. Cell. Biol.* 10:5502–5509.

Fearon, E.R., Cho, K.R., Nigro, J.M., Kern, S.E., Simons, J.W., Ruppert, J.M., Hamilton, S.R., Preisinger, A.C., Thomas, G., Kinzler, K.W., and Vogelstein, B. 1990. Identification of a chromosome 18q gene that is altered in colorectal cancers. *Science* 247:49–56.

Haber, D.A., Buckler, A.J., Glaser, T., Call, K.M., Pelletier, J., Sohn, R.L., Douglass, E.D., and Housman, D.E. 1990. An internal deletion within an 11p13 zinc finger gene contributes to the development of Wilms' tumor. *Cell* 61:1257–1269.

Hollstein, M.C., Metcalf, R.A., Welsh, J.A., Montesano, R., and Harris, C.C. 1990. Frequent mutation of the *p53* gene in human esophageal cancer. *Proc. Natl. Acad. Sci. USA* 87:9958–9961.

Hsu, I.C., Metcalf, R.A., Sun, T., Welsh, J.A., Wang, N.J., and Harris, C.C. 1991. Mutational hotspot in the *p53* gene in human hepatocellular carcinomas. *Nature* 350:427–428.

Kinzler, K.W., Nilbert, M.C., Vogelstein, B., Bryan, T.M., Levy, D.B., Smith, K.J., Preisinger, A.C., Hamilton, S.R., Hedge, P., Markham, A., Carlson, M., Joslyn, G., Groden, J., White, R., Miki, Y., Mitoshi, Y., Nishisho, I., and Nakamura, Y. 1991. Identification of a gene located at chromosome 5q21 that is mutated in colorectal cancers. *Science* 251:1366–1370.

Malkin, D., Li, F.P., Strong, L.C., Fraumeni, J.F., Jr., Nelson, C.E., Kim, D.H., Kassel, J., Gryka, M.A., Bischoff, F.Z., Tainsky, M.A., and Friend, S.H. 1990. Germ line *p53* mutations in a familial syndrome of breast cancer, sarcomas, and other neoplasms. *Science* 250:1233–1238.

Mulligan, L.M., Matlashewski, G.J., Scrable, H.J., and Cavenee, W.K. 1990. Mechanisms of *p53* loss in human sarcomas. *Proc. Natl. Acad. Sci. USA* 87:5863–5867.

Nigro, J.M., Baker, S.J., Preisinger, A.C., Jessup, J.M., Hostetter, R., Cleary, K., Bigner, S.H., Davidson, N., Baylin, S., Devilee, P., Glover, T., Collins, F.S., Weston, A., Modali, R., Harris, C.C., and Vogelstein, B. 1989. Mutations in the *p53* gene occur in diverse human tumour types. *Nature* 342:705–708.

Rose, E.A., Glaser, T., Jones, C., Smith, C.L., Lewis, W.H., Call, K.M., Minden, M., Champagne, E., Bonetta, L., Yeger, H., and Housman, D.E. 1990. Complete physical map of the WAGR region of 11p13 localizes a candidate Wilms' tumor gene. *Cell* 60:495–508.

Slingerland, J.M., Minden, M.D., and Benchimol, S. 1991. Mutation of the *p53* gene in human acute myelogenous leukemia. *Blood* 77:1500–1507.

Sugimoto, K., Toyoshima, H., Sakai, R., Miyagawa, K., Hagiwara, K., Hirai, H., Ishikawa, F., and Takaku, F. 1991. Mutations of the *p53* gene in lymphoid leukemia. *Blood* 77:1153–1156.

Takahashi, T., Nau, M.M., Chiba, I., Birrer, M.J., Rosenberg, R.K., Vinocour, M., Levitt, M., Pass, H., Gazdar, A.F., and Minna, J.D. 1989. *p53*: a frequent target for genetic abnormalities in lung cancer. *Science* 246:491–494.

Viskochil, D., Buchberg, A.M., Xu, G., Cawthon, R.M., Stevens, J., Wolff, R.K., Culver, M., Carey, J.C., Copeland, N.G., Jenkins, N.A., White, R., and and O'Connell, P. 1990. Deletions and a translocation interrupt a cloned gene at the neurofibromatosis type 1 locus. *Cell* 62:187–192.

Wallace, M.R., Marchuk, D.A., Andersen, L.B., Letcher, R., Odeh, H.M., Saulino, A.M., Fountain, J.W., Brereton, A., Nicholson, J., Mitchell, A.L., Brownstein, B.H., and Collins, F.S. 1990. Type 1 neurofibromatosis gene: identification of a large transcript disrupted in three NF1 patients. *Science* 249:181–186.

Chapter 10

Genetic Alterations in Tumor Development

THE DEVELOPMENT OF CANCER, as discussed in chapter 2, is a complex multistep
process in which normal cells become malignant through a series of progressive
alterations. Cancers arise from single cells that have begun to proliferate abnormally. Progression to malignancy then occurs as the tumor cells undergo a series
of stepwise transitions, gradually leading to the acquisition of increased proliferative capacity, invasiveness, and metastatic potential.

Each of the steps in tumor progression is thought to be the result of a genetic
alteration leading to increased cell growth. Although researchers have not yet
mapped out the complete series of events involved in the pathogenesis of any
human cancer, it is generally estimated that four to six discrete steps are required
for the development of most malignant tumors. Both the activation of oncogenes
and the inactivation of tumor suppressor genes have been shown to be important steps in the initiation and progression of many human tumors; accumulated
damage to several such genes eventually results in the loss of growth control
characteristic of fully malignant cancer cells. The roles of oncogenes and tumor
suppressor genes in the development of human cancers are the subject of this
chapter, and the functions of these genes in control of cell proliferation are
discussed in chapter 11.

MUTATION OF *ras* AND *p53* GENES BY RADIATION AND CHEMICAL CARCINOGENS

As discussed in chapter 3, risk factors for many human cancers include radiation
and chemical carcinogens. How do these agents lead to the development of
cancer, and how are their effects on cells related to the activities of oncogenes
and tumor suppressor genes?

Many Carcinogens Act by Damaging DNA

Radiation and many chemical carcinogens are known to damage DNA, resulting
in a high frequency of mutations, DNA rearrangements, and chromosome abnormalities. It is therefore reasonable to suspect that these agents cause cancer by

inducing genetic damage, leading to abnormalities in critical cell regulatory genes. The relationship between genetic damage and the development of cancer is further illustrated by the fact that individuals with inherited deficiencies in maintenance of their genetic material suffer a high cancer incidence (see chapter 5). Patients with xeroderma pigmentosum, for example, are unable to repair the genetic damage caused by ultraviolet irradiation from sunlight. Consequently, individuals with this disease suffer an extremely high incidence of skin cancer. The relationship of genetic damage to the action of radiation and chemical carcinogens clearly suggests the possibility that these agents might cause cancer by inducing mutations that result in either the formation of oncogenes or the inactivation of tumor suppressor genes. Direct evidence that radiation and chemical carcinogens can act in just this way has come from studies of carcinogenesis in experimental animals, particularly from analysis of *ras* oncogenes in radiation-induced and chemically induced tumors of mice and rats, as well as from analysis of mutations of the *p53* tumor suppressor gene in human liver cancers.

Tumor Initiation and Promotion

One of the classic experimental models used to study chemical carcinogenesis is the induction of skin tumors in mice (Fig. 10.1). Analysis of the development of these tumors first defined two distinct stages of carcinogenesis, termed **initia-**

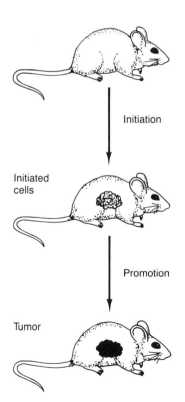

Initiation

Initiated cells

Promotion

Tumor

Figure 10.1   *Two-stage carcinogenesis in mouse skin. Initiation is an irreversible change, presumably resulting from carcinogen-induced mutations. Tumor development, however, requires subsequent treatment with a tumor promoter.*

tion and **promotion**, which were introduced in chapter 2. The first step, initiation, is an irreversible change brought about by a single carcinogen treatment. However, tumors do not develop unless the mice are subsequently treated with a **tumor promoter.** Treatment with tumor promoters can be delayed more than a year after treatment with the initiating carcinogen, indicating the irreversible nature of the initiation step. In contrast, tumor promotion is a reversible process that requires multiple continuing applications of the promoting agent.

As discussed in chapter 3, many carcinogens act as both initiators and promoters, but some carcinogens display only one of these activities. Those carcinogens that act as initiators are also potent **mutagens,** suggesting that the initiation of tumor development results from mutations in critical genes of a carcinogen-treated cell. Tumor promoters, in contrast, are not mutagenic but instead act to stimulate cell proliferation. It is thus thought that the initiation stage of tumor development corresponds to an irreversible genetic alteration (e.g., a mutation) that confers increased proliferative capacity. Development of a tumor, however, requires further stimulation of cell proliferation by application of a promoting agent. This presumably serves to enhance the process of tumorigenesis, with additional mutations occurring more readily in proliferating cells.

Chemical Carcinogens Induce *ras* Gene Mutations in Experimental Animals

Members of the *ras* oncogene family are frequently detected in mouse skin tumors, as well as in other kinds of radiation-induced and chemically induced cancers in experimental animals. In these tumors, the formation of *ras* oncogenes is an early event in the carcinogenic process. Indeed, the mutations that convert the normal *ras* proto-oncogenes to oncogenes appear to be caused directly by the carcinogens that initiate tumor development. For example, breast tumors in rats can be induced by the chemical carcinogen *N*-methyl-*N*-nitrosourea (MNU). The *ras* oncogenes found in these tumors have undergone a mutation changing a guanine residue (G) in the DNA encoding the *ras* proto-oncogene to an adenine (A) in the *ras* oncogene. This kind of mutation, a G to A transition, is precisely the kind of mutation characteristically induced by MNU. In contrast, different types of mutations are found in the *ras* oncogenes of tumors induced by other carcinogens. In each case, these mutations are characteristic of the kinds of DNA damage caused by the carcinogenic agent. In addition, it has been shown that mutations in *ras* oncogenes can be detected as early as two weeks after carcinogen treatment. It thus appears that *ras* oncogenes are direct targets for chemical carcinogens during the initiation phase of tumor development in experimental animals.

Mutations of *ras* Genes in Human Cancers

As discussed in chapter 8, *ras* oncogenes are frequently found in a number of different kinds of human tumors, including colon, rectum, lung, thyroid, and pancreatic carcinomas, as well as leukemias and lymphomas. In colon, rectum, and thyroid carcinomas, as well as in some types of leukemia, *ras* oncogenes are detected in **premalignant** lesions (e.g., colon adenomas), which represent early stages of the disease process. In at least some human cancers, therefore, the formation of *ras* oncogenes also appears to be an early event in tumorigenesis.

It has also been found that *ras* oncogenes are frequently characterized by specific mutations in different types of human cancers. For example, *ras*K oncogenes in lung carcinomas usually have a mutation of a G to a T. In contrast, *ras*K oncogenes in colon cancers usually have a mutation of a G to an A. Such observations are reminiscent of the fact that specific *ras* oncogene mutations are found in experimental animal tumors induced by different chemical carcinogens, leading to the suggestion that different carcinogens induce *ras* mutations in human lung and colon carcinomas.

The possibility that *ras* oncogenes are direct targets for carcinogens in human cancers is also supported by studies of melanomas. About 20% of human melanomas have been found to contain *ras*N oncogenes. Interestingly, those melanomas with *ras* oncogenes generally develop at body sites (such as the hands and face) that are continuously exposed to sunlight. In addition, the mutations found in the *ras* oncogenes of these tumors are characteristic of those induced by ultraviolet radiation. These findings therefore suggest that *ras* oncogenes in human melanomas are generated by mutations induced directly by the carcinogenic agent—in this case, solar ultraviolet irradiation.

Aflatoxin-Induced Mutations Inactivate the *p53* Gene in Hepatocellular Carcinomas

The clearest example of the interaction of a carcinogen with a specific gene in human cancer is provided by studies of mutations of the *p53* tumor suppressor gene in hepatocellular carcinomas. Aflatoxin, in addition to hepatitis B virus infection, is a major risk factor for hepatocellular carcinoma in southern Africa and China. As discussed in chapter 3, aflatoxin is a liver carcinogen in animals, as well as a potent mutagen that induces mutations of G to T. Inactivation of the *p53* tumor suppressor gene occurs in about 50% of hepatocellular carcinomas from southern African and Chinese patients. In these tumors, *p53* inactivation specifically results from a G to T mutation. Since *p53* genes in other types of tumors (e.g., lung, colon, or breast carcinomas) are inactivated by different mutations, it appears that *p53* inactivation in hepatocellular carcinomas is a direct result of aflatoxin mutagenesis.

BURKITT'S LYMPHOMA: THE COMBINED ROLES OF EPSTEIN-BARR VIRUS AND THE *c-myc* ONCOGENE

Two-Step Development of Burkitt's Lymphomas

The development of Burkitt's lymphomas provides a good example of the sequential roles of two different oncogenes in tumorigenesis (Fig. 10.2). As dis-

Figure 10.2 Two-step development of Burkitt's lymphomas. Infection with Epstein-Barr virus (EBV) leads to abnormal proliferation of lymphocytes. However, the development of malignancy also requires activation of c-*myc* by chromosome translocation.

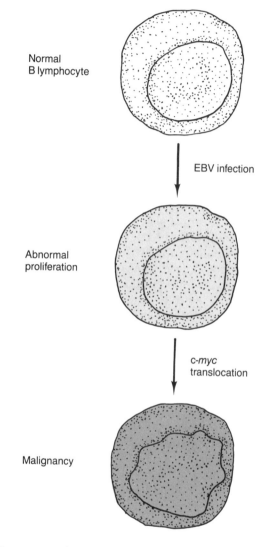

Normal B lymphocyte

EBV infection

Abnormal proliferation

c-*myc* translocation

Malignancy

cussed in chapter 4, Burkitt's lymphoma in Africa is caused by infection with Epstein-Barr virus (EBV). EBV carries at least one oncogene, *BNLF*-1, which can induce abnormal proliferation of cultured lymphocytes. However, infection with EBV is not by itself sufficient to produce a malignancy. Indeed, infection with EBV frequently either has no pathologic consequence or causes mononucleosis.

It thus appears that additional events, beyond EBV infection and expression of the viral *BNLF*-1 oncogene, are necessary for lymphomagenesis. One such event in the pathogenesis of Burkitt's lymphoma appears to be activation of the cellular oncogene, c-*myc*. As discussed in chapter 8, translocations of c-*myc* from its normal locus on chromosome 8 to loci on chromosomes 2, 14, or 22 occur in virtually all Burkitt's lymphomas. These translocations result in the insertion of c-*myc* in proximity to one of the immunoglobulin genes, leading to inappropriate, deregulated c-*myc* expression. Both the viral oncogene *BNLF*-1 and the cellular oncogene c-*myc* thus appear to play important roles in lymphomagenesis.

Effects of Epstein-Barr Virus and c-*myc* on Human Lymphocytes

Further insight into the possible roles of EBV and c-*myc* in the pathogenesis of Burkitt's lymphomas has come from experiments with cultured human lymphocytes. Infection of normal lymphocytes with EBV induces continuous abnormal cell proliferation, but such EBV-infected lymphocytes do not form tumors when inoculated into mice. Introduction of an active c-*myc* oncogene into these cells, however, converts them to the fully malignant state. It thus appears that EBV infection and c-*myc* activation represent distinct molecular events in a two-step process of lymphomagenesis. Tumor development is initiated by EBV infection, which leads to increased lymphocyte proliferation. Translocation of c-*myc* within such rapidly growing cells then appears to result in their subsequent progression to malignancy.

ONCOGENE AMPLIFICATION AND TUMOR PROGRESSION

Amplification of oncogenes frequently occurs in tumors, perhaps because the characteristic genomic instability of cancer cells (see chapter 7) makes gene amplification a frequent event. In at least some tumors, it appears that such oncogene amplification is closely associated with progression to increasing malignancy.

Neuroblastomas and N-*myc*

Neuroblastomas have provided the prototypical example of the relationship between oncogene amplification and tumor progression. Neuroblastomas are commonly divided into four clinical stages, based on tumor size and the extent to which the tumor has spread from its site of origin (Table 10.1). Amplification of the N-*myc* oncogene occurs in about half of stage III and IV tumors, but only rarely in stage I and II tumors. In addition, the behavior of individual neuroblastomas is closely correlated with N-*myc* amplification. For example, those stage II tumors with amplified N-*myc* genes have a much greater likelihood of further progression and metastasis than stage II tumors in which N-*myc* ampli-

Table 10.1   Amplification of N-*myc* in Neuroblastomas

Stage	Description	N-*myc* Amplification (Percent of Tumors)
I	Tumor confined to organ of origin	<10
II	Tumor extending beyond organ of origin but not crossing the midline	12
III	Tumor extending beyond midline	65
IV	Widespread disease	48

Data from R.C. Seeger et al., *N. Engl. J. Med.* 313:1111–1116, 1985.

fication has not occurred. Similarly, stage III and IV tumors containing amplified N-*myc* genes progress more rapidly than those that do not. The amplification of N-*myc* thus appears to be a critical event in the pathogenesis of neuroblastomas, closely correlated with the development of increasing malignancy.

Amplification of *erb*B-2 in Breast and Ovarian Carcinomas

Amplification of another oncogene, *erb*B-2, appears to be similarly correlated with the malignancy of breast and ovarian carcinomas. Amplified copies of *erb*B-2 are found in 25–30% of both tumor types. In both cases, those tumors with *erb*B-2 amplification appear to be more aggressive than those without. For example, in one detailed study, the median survival time for patients with ovarian carcinomas lacking *erb*B-2 amplification was approximately 5 years. This was reduced to approximately 2.5 years for patients whose tumors contained 2–5 amplified *erb*B-2 copies and to less than 1 year for patients whose tumors contained more than 5 amplified copies of *erb*B-2. Similarly, in breast carcinomas, *erb*B-2 amplification is more frequent in advanced-stage tumors and correlates with reduced patient survival.

Members of the *myc* Gene Family Are Frequently Amplified in Small Cell Lung Carcinomas

A number of other oncogenes are also amplified in several types of tumors, including amplification of three different members of the *myc* gene family (c-*myc*, N-*myc*, and L-*myc*) in small cell lung carcinomas. Although the prognostic significance of amplification of other oncogenes is not yet as clear as that discussed for N-*myc* and *erb*B-2 above, it seems likely that at least some additional cases of oncogene amplification will similarly be associated with tumor progression.

COLON/RECTUM CARCINOMA: A MODEL FOR ACCUMULATED GENETIC DAMAGE IN TUMOR DEVELOPMENT

Alterations of *ras*, *MCC*, *DCC*, and *p53* During the Development of Colon/Rectum Carcinomas

The development of colon and rectum carcinomas was discussed in chapter 2 to illustrate the multiple stages involved in the pathogenesis of a common human malignancy. These tumors also serve as the best characterized example of the role played by multiple genetic alterations in carcinogenesis. Formation of *ras*K oncogenes and inactivation or loss of three different tumor suppressor genes—*MCC*, *p53*, and *DCC*—are frequent events in the development of these neoplasms, occurring in a high fraction of all colon/rectum carcinomas examined. Moreover, because lesions representing several stages of the development of these neoplasms are regularly obtained as surgical specimens, it has been possible to correlate these genetic alterations with a series of steps in tumor progression (Fig. 10.3).

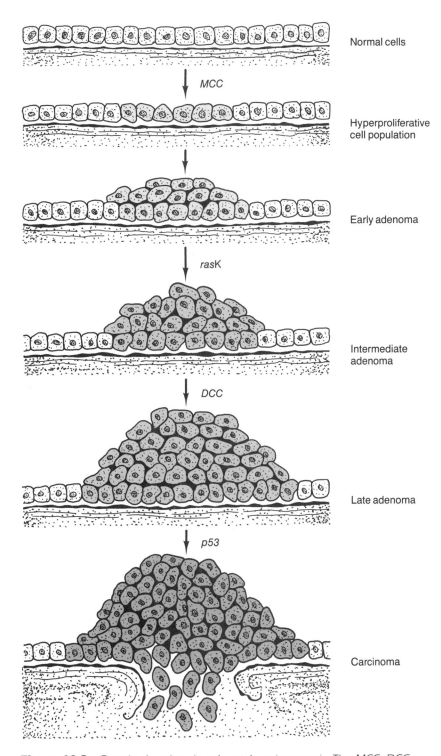

Normal cells

MCC

Hyperproliferative
cell population

Early adenoma

rasK

Intermediate
adenoma

DCC

Late adenoma

p53

Carcinoma

Figure 10.3 *Genetic alterations in colorectal carcinogenesis. The MCC, DCC, and p53 tumor suppressor genes and the rasK oncogene appear to be involved most frequently in the indicated stages of development of colorectal carcinomas.*

Inactivation of *MCC* tumor suppressor genes and formation of *ras*K oncogenes are both early events in colon/rectum carcinogenesis. Genetic transmission of defective copies of the *FAP* gene on chromosome 5, which may be the same as the *MCC* tumor suppressor gene, is responsible for inherited familial adenomatous polyposis. The inactivation of *FAP* is thought to result in abnormally high levels of colon cell proliferation, leading to the development of hundreds of premalignant adenomas in the colons of patients suffering from this disease. Inactivation of the same gene probably also occurs in patients with noninherited colon/rectum carcinomas, apparently at an early stage of the disease process. Similarly, *ras*K appears to be involved in early stages of carcinogenesis, since *ras*K oncogenes are frequently detected in premalignant adenomas (Fig. 10.3). The *MCC* and *ras*K genes thus appear to be involved in early stages of the initiation and progression of colon/rectum cancer.

In contrast to *MCC* and *ras*K, inactivation of the *p53* and *DCC* tumor suppressor genes apparently occurs at later stages of neoplasm development. Thus, inactivation of *p53* and *DCC* is only rarely detected in early- and intermediate-stage adenomas. Rather, inactivation of these genes is characteristic only of advanced adenomas and is found even more frequently in fully malignant carcinomas. This timing suggests that the loss of these tumor suppressor genes is important in the later stages of progression to malignancy, rather than in earlier stages of tumor formation.

The Importance of Accumulated Genetic Alterations

The order of genetic alterations discussed above and illustrated in Figure 10.3 is that most frequently observed, but it is not invariant. For example, in a few cases, inactivation of *p53* is observed in early adenomas, and it can occur prior to loss of *MCC*. Perhaps of greatest significance is the fact that accumulated damage to multiple genes is clearly associated with the progression of these tumors from early premalignant lesions to metastatic carcinomas. This is illustrated in Figure 10.4, which depicts the average number of combined alterations to the *MCC*, *ras*K, *p53*, and *DCC* genes which are detected at different stages of colon/rectum carcinogenesis. Increasing numbers of genetic alterations are apparent at each stage of tumor development, with approximately a sixfold higher frequency of accumulated abnormalities in carcinomas than in early adenomas. The multiple steps in colon/rectum carcinogenesis can thus be clearly correlated with the accumulation of abnormalities in at least these four distinct growth regulatory genes.

INVOLVEMENT OF MULTIPLE ONCOGENES AND TUMOR SUPPRESSOR GENES IN OTHER HUMAN NEOPLASMS

Colon/rectum carcinomas are currently the best characterized example of the combined roles of multiple oncogenes and tumor suppressor genes in the development of human cancer. However, multiple genetic abnormalities are also frequently present in other kinds of human cancer (Table 10.2). It would be surpris-

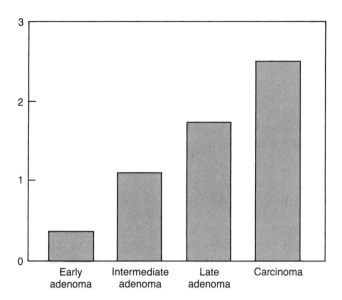

Figure 10.4 *Accumulated genetic alterations during pathogenesis of colorectal carcinomas. The average number of combined alterations to the DCC, MCC, p53, and rasK genes in a series of tumors representing the indicated stages of colorectal carcinogenesis is indicated. (Data are plotted from E.R. Fearon and B. Vogelstein, Cell 61:759–767, 1990).*

ing if the accumulation of such multiple gene defects did not similarly correlate with stages of tumor development.

Progression of Neuroblastomas Involves Loss of a Tumor Suppressor Gene in Addition to Amplification of N-*myc*

The example of Burkitt's lymphoma, involving EBV (containing the viral onco-gene *BNLF*-1) and the cellular oncogene c-*myc*, has been discussed above, as has the role of amplification of N-*myc* in the development of neuroblastomas. Neuroblastomas are also characterized by loss of a region of chromosome 1, which is thought to contain a tumor suppressor gene whose inactivation is important in development of these neoplasms. Loss of the putative chromosome 1 tumor suppressor, like amplification of N-*myc*, is generally seen in more ad-vanced and aggressively growing stage III and IV tumors. Both activation of the N-*myc* oncogene and inactivation of the chromosome 1 tumor suppressor gene thus seem to be correlated with progression of neuroblastomas to advanced stages of increasing malignancy.

The Roles of *NF*1 and *p53* in the Development of Neurofibrosarcomas

As discussed in chapter 9, type 1 (von Recklinghausen) neurofibromatosis is a genetic disease resulting from inheritance of a defective copy of the *NF*1 tumor

Table 10.2 Tumors Involving Multiple Oncogenes and Tumor Suppressor Genes

Tumor	Oncogenes	Tumor Suppressor Genes
Breast carcinoma	c-*myc*, *erb*B-2	*RB*, *p53*, chromosomes 1, 3, 11, 18
Burkitt's lymphoma	*BNLF*-1, c-*myc*	
Colon/rectum carcinoma	*ras*K	*DCC*, *MCC*, *p53*
Lung carcinoma	*ras*K, c-*myc*, L-*myc*, N-*myc*	*RB*, *p53*, chromosome 3
Neuroblastoma	N-*myc*	chromosome 1
Neurofibrosarcoma		*NF1*, *p53*

suppressor gene. Individuals with this disorder develop multiple neurofibro-mas—benign tumors that can, however, progress to malignant neurofibro-sarcomas. Interestingly, the progression of these neurofibromas to malignancy is strongly associated with the inactivation of a second tumor suppressor gene, *p53*. This suggests a clear parallel between the development of neurofibro-sarcomas in patients with type 1 neurofibromatosis and colon carcinoma in patients with familial adenomatous polyposis. In the latter disease, inheritance of a defective tumor suppressor gene results in the formation of multiple benign colon polyps (adenomas). The progression of these lesions to malignancy is then associated with additional genetic events, including inactivation of *p53*. Thus, in both familial adenomatous polyposis and type 1 neurofibromatosis, inherited abnormalities in a tumor suppressor gene lead to the growth of multiple benign tumors. Progression to malignancy then involves subsequent genetic alterations in both cases, including *p53* inactivation.

Accumulation of Multiple Genetic Alterations in Breast and Lung Carcinomas

Multiple genetic lesions are also clearly accumulated during the development of breast and lung carcinomas. In breast cancers, these include inactivation of the *RB* and *p53* tumor suppressor genes, as well as amplification of the c-*myc* and *erb*B-2 oncogenes. In addition, breast carcinomas frequently display losses of regions of chromosomes 1, 3, 11, and 18, very likely signaling inactivation of additional tumor suppressor genes at these loci. These genes remain to be explic-itly identified, although the *DCC* gene is a possible candidate for the relevant tumor suppressor gene on chromosome 18.

As discussed in chapter 2, lung cancers can be classified as small cell carcino-mas and non-small cell carcinomas (which include adenocarcinomas, squamous cell carcinomas, and large cell carcinomas). All types of lung cancer are associ-ated with loss or inactivation of *p53* and a tumor suppressor gene on chromo-some 3. In addition, other genetic alterations occur more specifically in different types of lung cancer. Inactivation of the *RB* tumor suppressor gene occurs in nearly all small cell carcinomas, which also display frequent amplification of one of the members of the *myc* gene family—c-*myc*, L-*myc*, and N-*myc*. On the other hand, *ras*K oncogenes are most frequently involved in adenocarcinomas, where their activity appears to be correlated with more aggressive tumor growth and a poorer prognosis for the patient.

SUMMARY

The development of human cancer is a multistep process in which cells gradually progress to malignancy. Both oncogene activation and tumor suppressor gene loss or inactivation represent steps in the initiation and progression of human tumors. In several human cancers, as in experimental animal systems, the formation of *ras* oncogenes appears to occur early in tumor development, perhaps resulting directly from mutations induced by radiation or chemical carcinogens. The *p53* tumor suppressor gene likewise appears to be a direct target for carcinogen-induced mutations in human liver cancers. Other oncogenes, such as N-*myc* in neuroblastomas and *erb*B-2 in breast and ovarian carcinomas, may be involved in the subsequent progression of tumors to increasing malignancy. Colorectal cancers have provided a particularly clear model for the role of multiple oncogenes and tumor suppressor genes in the development of a common human malignancy. In these tumors, inactivation of the *MCC* tumor suppressor gene and formation of the *ras*K oncogene occur at early stages of carcinogenesis, leading to formation of premalignant adenomas. Inactivation of two additional tumor suppressor genes, *p53* and *DCC*, then leads to the subsequent progression of adenomas to malignant carcinomas. Multiple oncogenes and tumor suppressor genes are likewise involved in other human cancers, including those of the breast and lung, further indicating the importance of accumulated damage to critical cell regulatory genes in the pathogenesis of human malignancies.

KEY TERMS

carcinogen
initiation
promotion
tumor promoter
mutagen
ras
p53
premalignant
adenoma
Burkitt's lymphoma
Epstein-Barr virus
BNLF-1
c-*myc*

amplification

tumor progression

neuroblastoma

N-*myc*

*erb*B-2

L-*myc*

colon/rectum carcinogenesis

MCC

DCC

FAP

familial adenomatous polyposis

neurofibromatosis

*NF*1

neurofibroma

neurofibrosarcoma

RB

REFERENCES AND FURTHER READING

General References

Cooper, G.M. 1990. *Oncogenes*. Jones and Bartlett, Boston.

Hunter, T. 1991. Cooperation between oncogenes. *Cell* 64:149–270.

Mutation of *ras* and *p53* Genes by Radiation and Chemical Carcinogens

Balmain, A., Ramsden, M., Bowden, G.T., and Smith, J. 1984. Activation of the mouse cellular Harvey-*ras* gene in chemically induced benign skin papillomas. *Nature* 307:658–660.

Barbacid, M. 1987. *ras* genes. *Ann. Rev. Biochem.* 56:779–827.

Bos, J.L. 1989. *Ras* oncogenes in human cancer: a review. *Cancer Res.* 49:4682–4689.

Bos, J.L., Toksoz, D., Marshall, C.J., Verlaan-de Vries, M., van Boom, J.H., van der Eb, A.J., and Vogelstein, B. 1987. Prevalence of *ras* gene mutations in human colorectal cancers. *Nature* 327:293–297.

Bressac, B., Kew, M., Wands, J., and Ozturk, M. 1991. Selective G to T mutations of *p53* gene in hepatocellular carcinoma from southern Africa. *Nature* 350:429–431.

Forrester, K., Almoguera, C., Han, K., Grizzle, W.E., and Perucho, M. 1987. Detection of high incidence of K-*ras* oncogenes during human colon tumorigenesis. *Nature* 327:298–303.

Hsu, I.C., Metcalf, R.A., Sun, T., Welsh, J.A., Wang, N.J., and Harris, C.C. 1991. Mutational hotspot in the *p53* gene in human hepatocellular carcinomas. *Nature* 350:427–428.

Kumar, R., Sukumar, S., and Barbacid, M. 1990. Activation of *ras* oncogenes preceding the onset of neoplasia. *Science* 248:1101–1104.

Slebos, R.J.C., Kibbelaar, R.E., Dalesio, O., Kooistra, A., Stam, J., Meijer, C.J.L.M., Wagenaar, S.S., Vanderschueren, R.G.J.R.A., van Zandwijk, N., Mooi, W.J., Bos, J.L., and Rodenhuis, S. 1990. K-*ras* oncogene activation as a prognostic marker in adenocarcinoma of the lung. *N. Engl. J. Med.* 323:561–565.

Van't Veer, L.J., Burgering, B.M.T., Versteeg, R., Boot, A.J.M., Ruiter, D.J., Osanto, S., Schrier, P.I., and Bos, J.L. 1989. N-*ras* mutations in human cutaneous melanoma from sun-exposed body sites. *Mol. Cell. Biol.* 9:3114–3116.

Zarbl, H., Sukumar, S., Arthur, A.V., Martin-Zanca, D., and Barbacid, M. 1985. Direct mutagenesis of Ha-*ras*-1 oncogenes by *N*-nitroso-*N*-methylurea during initiation of mammary carcinogenesis in rats. *Nature* 315:382–385.

Burkitt's Lymphoma: The Combined Roles of Epstein-Barr Virus and the c-*myc* Oncogene

Lombardi, L., Newcomb, E.W., and Dalla-Favera, R. 1987. Pathogenesis of Burkitt lymphoma: expression of an activated c-*myc* oncogene causes the tumorigenic conversion of EBV-infected human B lymphoblasts. *Cell* 49:161–170.

Oncogene Amplification and Tumor Progression

Johnson, B.E., Ihde, D.C., Makuch, R.W., Gazdar, A.F., Carney, D.N., Oie, H., Russell, E., Nau, M.M., and Minna, J.D. 1987. *myc* family oncogene amplification in tumor cell lines established from small cell lung cancer patients and its relationship to clinical status and course. *J. Clin. Invest.* 79:1629–1634.

Seeger, R.C., Brodeur, G.M., Sather, H., Dalton, A., Siegel, S.E., Wong, K.Y., and Hammond, D. 1985. Association of multiple copies of the N-*myc* oncogene with rapid progression of neuroblastomas. *N. Engl. J. Med.* 313:1111–1116.

Slamon, D.J., Godolphin, W., Jones, L.A., Holt, J.A., Wong, S.G., Keith, D.E., Levin, W.J., Stuart, S.G., Udove, J., Ullrich, A., and Press, M.F. 1989. Studies of the *HER*-2/*neu* proto-oncogene in human breast and ovarian cancer. *Science* 244:707–712.

Colon/Rectum Carcinoma: A Model for Accumulated Genetic Damage in Tumor Development

Fearon, E.R., and Vogelstein, B. 1990. A genetic model for colorectal tumorigenesis. *Cell* 61:759–767.

Kinzler, K.W., Nilbert, M.C., Vogelstein, B., Bryan, T.M., Levy, D.B., Smith, K.J., Preisinger, A.C., Hamilton, S.R., Hedge, P., Markham, A., Carlson, M., Joslyn, G., Groden, J., White, R., Miki, Y., Mitoshi, Y., Nishisho, I., and Nakamura, Y. 1991. Identification of a gene located at chromosome 5q21 that is mutated in colorectal cancers. *Science* 251:1366–1370.

Involvement of Multiple Oncogenes and Tumor Suppressor Genes in Other Human Neoplasms

Cropp, C.S., Lidereau, R., Campbell, G., Champene, M.H., and Callahan, R. 1990. Loss of heterozygosity on chromosomes 17 and 18 in breast carcinoma: two additional regions identified. *Proc. Natl. Acad. Sci. USA* 87:7737–7741.

Escot, C., Theillet, C., Lidereau, R., Spyratos, F., Champeme, M.H., Gest, J., and Callahan, R. 1986. Genetic alteration of the c-*myc* proto-oncogene (*MYC*) in human primary breast carcinomas. *Proc. Natl. Acad. Sci. USA* 85:4834–4838.

Fong, C.-T., Dracopoli, N.C., White, P.S., Merrill, P.T., Griffith, R.C., Housman, D.E., and Brodeur, G.M. 1989. Loss of heterozygosity for the short arm of chromosome 1 in human neuroblastomas: correlation with N-*myc* amplification. *Proc. Natl. Acad. Sci. USA* 86:3753–3757.

Harbour, J.W., Lai, S.-L., Whang-Peng, J., Gazdar, A.F., Minna, J.D., and Kaye, F.J. 1988. Abnormalities in structure and expression of the human retinoblastoma gene in SCLC. *Science* 241:353–357.

Johnson, B.E., Ihde, D.C., Makuch, R.W., Gazdar, A.F., Carney, D.N., Oie, H., Russell, E., Nau, M.M., and Minna, J.D. 1987. *myc* family oncogene amplification in tumor cell lines established from small cell lung cancer patients and its relationship to clinical status and course. *J. Clin. Invest.* 79:1629–1634.

Kok, K., Osinga, J., Carritt, B., Davis, M.B., van der Hout, A.H., van der Veen, A.Y., Landsvater, R.M., de Leij, L.F.M.H., Berendsen, H.H., Postmus, P.E., Poppema, S., and Buys, C.H.C.M. 1987. Deletion of a DNA sequence at the chromosomal region 3p21 in all major types of lung cancer. *Nature* 330:578–581.

Lee, E.Y.-H.P., To, H., Shew, J.-Y., Bookstein, R., Scully, P., and Lee, W.-H. 1988. Inactivation of the retinoblastoma susceptibility gene in human breast cancers. *Science* 241:218–221.

Menon, A.G., Anderson, K.M., Riccardi, V.M., Chung, R.Y., Whaley, J.M., Yandell, D.W., Farmer, G.E., Freiman, R.N., Lee, J.K., Li, F.P., Barker, D.F., Ledbetter, D.H., Kleider, A., Martuza, R.L., Gusella, J.F., and Seizinger, B.R. 1990. Chromosome 17p deletions and *p53* gene mutations associated with the formation of malignant neurofibrosarcomas in von Recklinghausen neurofibromatosis. *Proc. Natl. Acad. Sci. USA* 87:5435–5439.

Nigro, J.M., Baker, S.J., Preisinger, A.C., Jessup, J.M., Hostetter, R., Cleary, K., Bigner, S.H., Davidson, N., Baylin, S., Devilee, P., Glover, T., Collins, F.S., Weston, A., Modali, R., Harris, C.C., and Vogelstein, B. 1989. Mutations in the *p53* gene occur in diverse human tumour types. *Nature* 342:705–708.

Slebos, R.J.C., Kibbelaar, R.E., Dalesio, O., Kooistra, A., Stam, J., Meijer, C.J.L.M., Wagenaar, S.S., Vanderschueren, R.G.J.R.A., van Zandwijk, N., Mooi, W.J., Bos, J.L., and Rodenhuis, S. 1990. K-*ras* oncogene activation as a prognostic marker in adenocarcinoma of the lung. *N. Engl. J. Med.* 323:561–565.

Takahashi, T., Nau, M.M., Chiba, I., Birrer, M.J., Rosenberg, R.K., Vinocour, M., Levitt, M., Pass, H., Gazdar, A.F., and Minna, J.D. 1989. *p53*: a frequent target for genetic abnormalities in lung cancer. *Science* 246:491–494.

T'Ang, A., Varley, J.M., Chakraborty, S., Murphree, A.L., and Fung, Y.-K.T. 1988. Structural rearrangement of the retinoblastoma gene in human breast carcinoma. *Science* 242:263–266.

Weston, A., Willey, J.C., Modali, R., Sugimura, H., McDowell, E.M., Resau, J., Light, B., Haugen, A., Mann, D.L., Trump, B.F., and Harris, C.C. 1989. Differential DNA sequence deletions from chromosomes 3, 11, 13, and 17 in squamous-cell carcinoma, large-cell carcinoma, and adenocarcinoma of the human lung. *Proc. Natl. Acad. Sci. USA* 86:5099–5103.

Chapter 11

Functions of Oncogene and Tumor Suppressor Gene Products

THE FUNDAMENTAL DIFFERENCE between normal cells and cancer cells, as empha-sized in previous chapters, is control of cell proliferation. Normal cell growth and differentiation are stringently regulated to meet the needs of the whole organism, whereas cancer cells are characterized by their autonomous, uncon-trolled self-reproduction. Since cancer cells arise as a result of alterations in oncogenes and tumor suppressor genes, it is apparent that the protein products of these genes can serve as critical regulators of cell behavior.

As discussed in chapter 6, the growth and differentiation of normal cells are regulated by a variety of external signals, including secreted hormones, growth factors, and contact with other cells and the extracellular matrix. In most cases, these factors act by binding to specific receptor molecules on the target cell surface. The interaction of an extracellular molecule with its receptor then initi-ates a series of reactions inside the target cell, which serve to transmit a signal from the cell surface to the nucleus. This ultimately results in alterations in gene expression, which bring about the programmed change in cell behavior. As will be discussed in this chapter, the proteins encoded by both oncogenes and tumor suppressor genes function as key regulatory elements at multiple levels of such signaling pathways. Abnormalities in their activities result in a breakdown of the normal regulation of cell proliferation, thereby leading to malignancy.

ONCOGENES AND GROWTH FACTORS

Platelet-Derived Growth Factor and the *sis* Oncogene

A number of secreted proteins act as **growth factors,** serving to induce the proliferation of appropriate target cells. One example, previously discussed in chapter 6, is **platelet-derived growth factor** (**PDGF**), which is stored in blood platelets and stimulates the growth of skin fibroblasts during wound healing. In another guise, PDGF is an oncogene, clearly illustrating how a defect in the mechanisms that control normal cell growth can lead to malignancy.

The relationship of PDGF to cancer was discovered in the course of studying the oncogene, called *sis*, of a virus that causes sarcomas in monkeys. The story is a good illustration of the importance of the unanticipated in scientific discovery. One group of scientists was studying the *sis* oncogene, hoping to learn how it caused cancer. At the same time, other scientists were studying PDGF, hoping to learn how it stimulated normal cells to divide. Surprisingly, when the results of these independent research groups were compared, it became apparent that *sis* and PDGF were one and the same. This serendipitous discovery showed, for the first time, that cancer could result from the aberrant activity of a normal cell growth factor.

Autocrine Growth Stimulation

How does PDGF, a normal cell protein, cause malignancy? In the case of PDGF (and other growth factors that act as oncogenes), cancer results from expression of the gene in the wrong type of cell (Fig. 11.1). Growth factors are normally made by one kind of cell, secreted into extracellular fluids, and then act on a different kind of cell. PDGF, for example, is normally released by blood platelets to act on fibroblasts. The important principle is that the type of cell that makes a growth factor normally does not respond to that factor. Rather, growth factors are normally produced by one cell in order to signal proliferation of a different cell type. When a growth factor acts as an oncogene, this regulation goes awry, and a cell begins to produce a growth factor to which it also responds. For example, when a virus carrying the *sis* oncogene infects a skin fibroblast, that infected cell begins to produce PDGF. Since the fibroblast also expresses the PDGF receptor and responds to PDGF, abnormal production of PDGF results in continual stimulation of cell proliferation, eventually leading to cancer.

This situation, in which a cell responds to a growth factor that it also produces, is called **autocrine growth stimulation**. It pertains not only to PDGF, but also to a wide variety of other growth factors that can similarly act as oncogenes. They include the oncogenes *hst* and *int*-2, members of the **fibroblast growth factor** family, which are sometimes amplified in breast cancers and other human tumors, as well as the growth factor **interleukin-3,** which is translocated in some human leukemias.

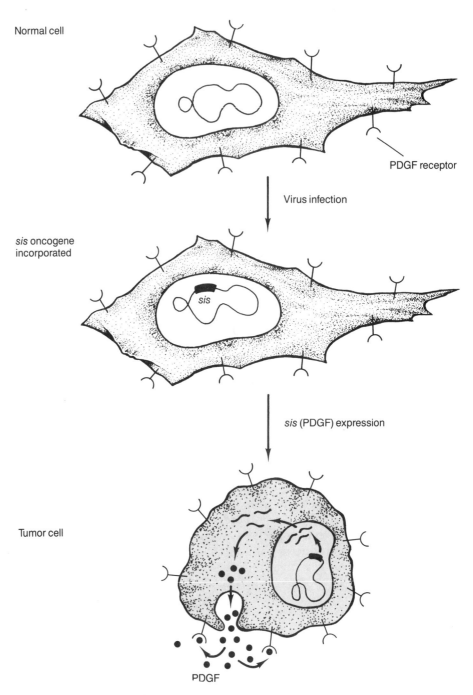

Normal cell

PDGF receptor

Virus infection

sis oncogene
incorporated

sis (PDGF) expression

Tumor cell

PDGF

Figure 11.1 *Transformation by the* sis *oncogene. A virus carrying the* sis *oncogene infects a fibroblast, which expresses the receptor for PDGF. The infected cell then produces PDGF (the* sis *gene product), resulting in autostimulation of cell proliferation and tumor development.*

GROWTH FACTOR RECEPTORS

Extracellular growth factors act by binding to specific receptor molecules on the surface of their target cells. Such growth factor binding stimulates activity of the receptor molecule, so that a signal is generated inside the cell, ultimately leading to cell division. The receptors thus serve as external sensors, acting to transmit information from outside the cell to its interior. Since growth factors can act as oncogenes, it may be expected that abnormal activity of **growth factor receptors** can likewise induce neoplastic cell proliferation.

Structure of Growth Factor Receptors

The receptors for many growth factors have similar structural organizations (Fig. 11.2). Part of the receptor, called the **extracellular domain**, is exposed to the external environment on the outside of the cell surface. It is this part of the receptor that binds extracellular growth factors. A second part of the receptor, the **transmembrane domain**, passes through the plasma membrane, serving to connect the extracellular domain with the third part of the receptor, the **intracellular domain**, which is inside the cell. The binding of a growth factor to the extracellular domain causes the receptor molecule to change its shape, resulting in activation of the intracellular domain. Thus, by traversing the plasma membrane, the receptor conveys information from the outside to the inside of the cell.

Conversion of Growth Factor Receptors to Oncogenes

Growth factor receptors can be converted into oncogenes by either of two kinds of alterations in their normal function. First, abnormally high expression of a recep-

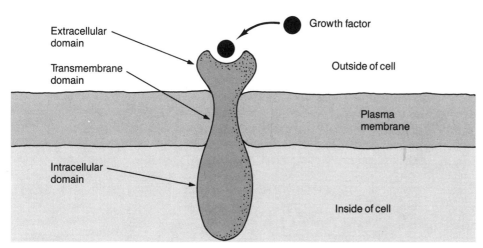

Figure 11.2 Organization of growth factor receptors. The receptors consist of extracellular domains, which bind growth factors, transmembrane domains, which span the plasma membrane, and intracellular domains, which interact with target molecules inside the cell.

tor can result in an exaggerated cellular response to growth factor binding, leading to excess cell proliferation. Amplification of the *erb*B-2 oncogene in breast and ovarian carcinomas is a good example of the role of increased expression of a growth factor receptor in tumor progression (see chapter 10). Alternatively, a number of growth factor receptors become oncogenes as a consequence of structural changes in the receptor molecules. In these cases, the receptor is altered such that the intracellular domain is active even in the absence of growth factor binding. Frequently, this occurs as a consequence of DNA rearrangements that result in loss of the extracellular domain and lead to unregulated activity of the intracellular portion of the receptor molecule (Fig. 11.3). The *ret* and *trk* oncogenes are frequently generated by this mechanism in human thyroid carcinomas (chapter 8).

ONCOGENES AND INTRACELLULAR SIGNAL TRANSDUCTION

Growth factor binding to a receptor serves to generate a signal at the surface of the target cell. The next step in transmission of this signal to the cell nucleus (**intracellular signal transduction**) is the interaction of the intracellular domain of a receptor with signal-transducing molecules inside the cell. These molecules are activated by the receptor and, in turn, function to activate other intracellular signal transducers. A series of such interactions serves to carry a signal through the cytoplasm to the cell nucleus, ultimately resulting in changes in gene expression and cell division.

Many Growth Factor Receptors Are Protein-Tyrosine Kinases

Many growth factor receptors, including the PDGF receptor and the receptors corresponding to the *erb*B-2, *ret*, and *trk* oncogenes (see chapter 8), are members

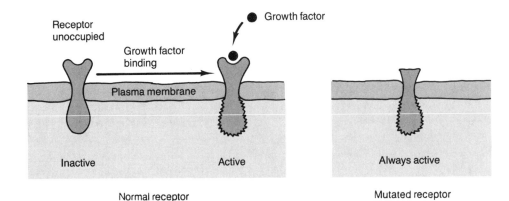

Figure 11.3 Conversion of a growth factor receptor to an oncogene. The normal growth factor receptor (a proto-oncogene) is activated by growth factor binding. In contrast, the mutated receptor (an oncogene) has lost its extracellular domain and is active even in the absence of growth factor stimulation.

of a family of proteins that function as **protein-tyrosine kinases**. This means that the intracellular domains of these receptors are enzymes that activate their target molecules by inducing a specific chemical reaction—namely, the attachment of phosphate to the amino acid tyrosine (Fig. 11.4). The name protein-tyrosine kinase reflects this activity. A **kinase** is any enzyme that transfers phosphate groups (HPO_4^{-2}) from **ATP** to another molecule. If the recipient molecule is itself a protein, as is the case here, then the enzyme that catalyzes the phosphate transfer is called a **protein kinase.** If the recipient protein (or target) is phosphorylated on the amino acid tyrosine, the enzyme catalyzing the reaction is a protein-tyrosine kinase. Other protein kinases, which will be discussed below, catalyze the phosphorylation of the amino acids serine and threonine in their target proteins. These enzymes are therefore called **protein-serine/threonine kinases** (Fig. 11.4).

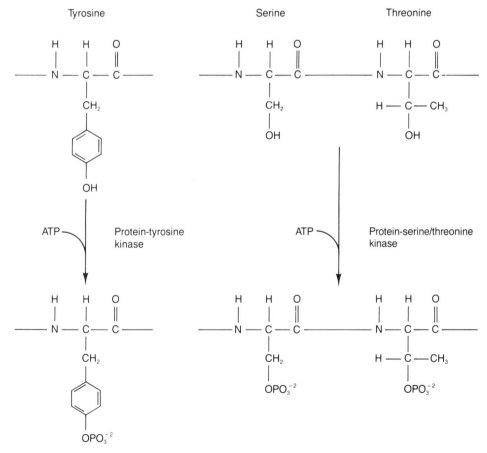

Figure 11.4 The activity of protein kinases. Protein-tyrosine kinases and protein-serine/threonine kinases catalyze the transfer of phosphate groups from ATP to the indicated amino acids of their target proteins.

The Role of Protein Phosphorylation in Signal Transduction

Protein phosphorylation is a very common mechanism for transmitting information within cells. The phosphorylation of a protein generally alters its activity, so protein kinases often serve to regulate the function of their target molecules. In addition, a series of protein kinases can act together to elicit a physiological response. For example, one protein kinase might activate two other protein kinases, each of which in turn might activate still additional kinases or other cell regulatory molecules. The process of signal transduction from growth factor receptors exemplifies such interactions.

Growth Factor Receptors and Inositol Phospholipids

The PDGF receptor is an example of a growth factor receptor with protein-tyrosine kinase activity. PDGF binding initiates the phosphorylation of at least four distinct intracellular targets (Fig. 11.5). Two of these targets are enzymes that regulate the metabolism of a group of molecules called **inositol phospholipids**. These molecules, in turn, regulate the activity of several different protein-serine/threonine kinases within the cell. For example, one of the proteins phosphorylated by the PDGF receptor is the enzyme **phospholipase C,** which catalyzes the formation of two compounds that play critical roles in regulating cell division—**diacylglycerol** (DAG) and **inositol triphosphate** (IP₃) (Fig. 11.6).

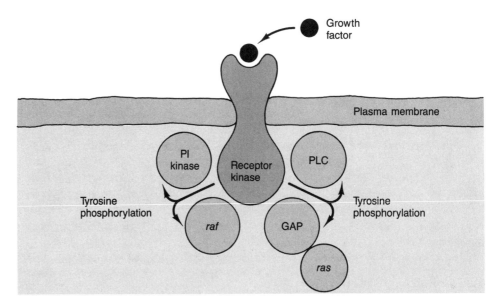

Figure 11.5 Targets of growth factor receptors. The protein-tyrosine kinase growth factor receptors phosphorylate at least four intracellular target molecules. Phosphatidylinositol kinase (PI kinase) and phospholipase C (PLC) regulate the metabolism of inositol phospholipids. The *raf* proto-oncogene product is a protein-serine/threonine kinase. GAP functions to regulate the activity of *ras* proto-oncogenes.

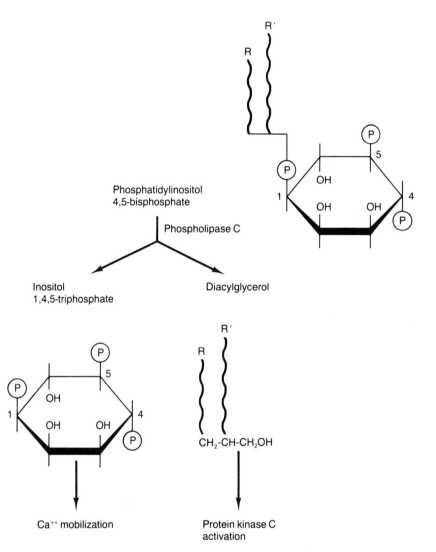

Figure 11.6 *Formation of inositol triphosphate and diacylglycerol. Phospholipase C catalyzes the hydrolysis of phosphatidylinositol 4,5-bisphosphate, generating inositol triphosphate and diacylglycerol. Inositol triphosphate stimulates the release of calcium from intracellular stores, causing an increase in the concentration of calcium in the cytoplasm and activating several calcium-stimulated protein-serine/threonine kinases. Diacylglycerol activates protein kinase C.*

DAG activates the enzyme **protein kinase C,** a protein-serine/threonine kinase. IP_3 causes the concentration of calcium inside the cell to increase, which leads to the activation of several other calcium-stimulated protein-serine/threonine kinases. The importance of this pathway is illustrated by the fact that protein kinase C can itself stimulate cells to divide. Moreover, protein kinase C can act as an oncogene, indicating that unregulated activity of the inositol phospholipid signaling pathway can contribute to tumor formation.

A second enzyme affecting metabolism of inositol phospholipids, **phosphatidylinositol kinase,** is also phosphorylated and activated by the PDGF receptor. This enzyme is thought to participate in the formation of additional kinds of inositol phospholipid molecules, which also serve to signal cell proliferation.

Growth Factor Receptors and the *raf* Proto-Oncogene

The third target phosphorylated by the protein-tyrosine kinase receptors (Fig. 11.5) is a protein-serine/threonine kinase called *raf*. The activity of *raf* is increased as a result of this phosphorylation, and *raf* then acts to transmit the growth factor-initiated signal further by phosphorylating other intracellular target molecules. In addition, *raf* is itself an oncogene, as discussed in chapter 8. It is activated in several viruses by mutations that make the *raf* protein kinase function in an unregulated manner. Thus, physiological activation of *raf* is a normal response to growth factor stimulation. If, on the other hand, *raf* functions in an uncontrolled manner, the result is abnormal cell proliferation and the development of a tumor.

Growth Factor Receptors, GAP, and *ras*

The fourth important molecule phosphorylated in response to growth factor binding is a protein called **GAP** (**GTPase activating protein**), which regulates the activity of *ras* proteins. As discussed in chapters 8 and 10, the three members of the *ras* family are the oncogenes most frequently found in human tumors. In normal cells, the *ras* proteins also appear to be involved in transducing signals from growth factor receptors to intracellular targets that promote cell growth. Although the precise targets of the *ras* proteins have not yet been identified, it is possible that *ras* regulates the metabolism of phospholipids, such as the inositol phospholipids discussed above. In any event, the GAP gene product acts as a negative regulator of *ras* protein activity. The *ras* proteins are controlled by guanine nucleotide binding, alternating between active (GTP bound) and inactive (GDP bound) states (Fig. 11.7). The GAP protein functions to stimulate the hydrolysis of bound GTP to GDP (*ras* GTPase), thereby turning off the signaling activity of normal *ras* proteins. Growth factor binding leads to phosphorylation of the GAP protein by receptor protein-tyrosine kinases (Fig. 11.5). It has been suggested that this inhibits GAP, thereby leading to increased *ras* activity consonant with cell proliferation.

The formation of *ras* oncogenes, which occurs by point mutations in a variety of human tumors, leads to unregulated activity of the *ras* proteins. The oncogene proteins no longer respond to GAP and therefore remain active regardless of growth factor stimulation. Thus, *ras* proteins provide an additional example of a situation in which uncontrolled activity of a normal cell regulatory molecule can lead to tumorigenesis. Similar mutations are responsible for formation of the *gsp* and *gip* oncogenes (see chapter 8), which function analogously to *ras*.

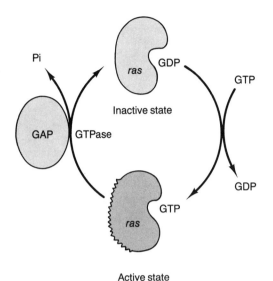

Figure 11.7   Regulation of *ras* proteins by guanine nucleotide binding. In the inactive state, *ras* protein is bound to GDP. Exchange of bound GDP for GTP converts *ras* to the active state. The *ras* activity is terminated by hydrolysis of bound GTP to GDP (GTPase), which is stimulated by GAP.

A number of oncogenes thus serve as intracellular elements in signal transduction pathways. Their activities are normally regulated, via receptors, by extracellular growth factors. Abnormal expression or function of these oncogene proteins can, however, lead to cell growth in the absence of the normal physiological stimulus, resulting in inappropriate cell proliferation and the eventual development of cancer.

REGULATION OF GENE EXPRESSION IN RESPONSE TO GROWTH FACTORS

Transcription Factors Control Gene Expression

The intracellular signals generated by activated growth factor receptors are transmitted to the nucleus, where they induce alterations in gene expression. As noted in chapter 6, gene expression is regulated primarily at the level of transcription of DNA to messenger RNA. This process is controlled by specific proteins, called **transcription factors,** which act either to turn on or turn off the expression of their target genes. The activity of transcription factors is in turn controlled by extracellular molecules, acting through intracellular signal transduction pathways. Transcription factors thus represent the terminal elements in the intracellular signaling process.

The AP-1 Transcription Factor Is Composed of the *fos* and *jun* Proto-Oncogene Proteins

A good example of the regulation of gene expression in the transmission of signals leading to cell division is provided by the transcription factor called **AP-1**

Figure 11.8 *The AP-1 transcription factor in signal transduction. Growth factor binding activates phospholipase C (PLC) via protein-tyrosine phosphorylation, leading to hydrolysis of phosphatidylinositol 4,5-bisphosphate (PIP$_2$) and formation of diacylglycerol (DAG), which activates protein kinase C. Protein-serine/threonine phosphorylation pathways initiated by protein kinase C lead to activation of the AP-1 transcription factor, which is composed of the fos and jun proto-oncogene products. AP-1 then stimulates transcription of critical target genes, ultimately resulting in cell proliferation.*

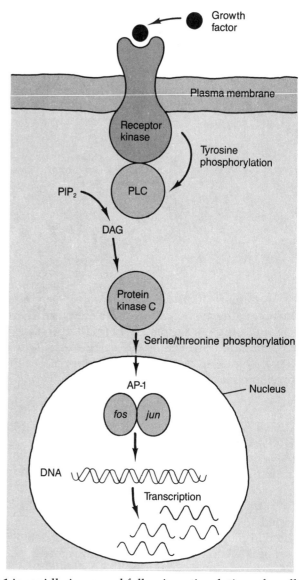

(Fig. 11.8). The activity of AP-1 is rapidly increased following stimulation of a cell with growth factors, such as PDGF, which transmit an intracellular signal through protein kinase C. Following its activation, the AP-1 transcription factor turns on the expression of a series of critical target genes, ultimately leading to cell division.

Clearly, given its central role in regulation of cell proliferation, the activity of AP-1 must be stringently controlled. Perhaps not surprisingly, then, abnormal activity of AP-1 can contribute to tumorigenesis. The complete AP-1 transcription factor is actually composed of two different proteins, both of which are the products of proto-oncogenes (*fos* and *jun*), which were introduced in chapter 8.

Increased expression or unregulated function of either *fos* or *jun* is sufficient to lead to uncontrolled expression of critical genes that are normally regulated by AP-1, resulting in abnormal cell proliferation and tumor development.

The *myc* Proteins Also Appear to Regulate Transcription

A number of other oncogenes function similarly to stimulate expression of genes involved in cell proliferation. The *myc* genes are particularly important members of this group of oncogenes, especially from the standpoint of human cancer (see chapters 8 and 10). Expression of c-*myc*, for example, is normally activated in response to growth factor stimulation, such as PDGF stimulation of fibroblasts. The c-*myc* protein is then thought to turn on the expression of other critical target genes required for cell division. Unregulated expression of members of the *myc* family, on the other hand, can lead to uncontrolled cell proliferation. Prominent examples, discussed in chapters 8 and 10, are the activation of c-*myc* by chromosome translocation in Burkitt's lymphomas and the amplification of N-*myc* in neuroblastomas.

REGULATION OF GENE EXPRESSION BY STEROIDS AND RELATED HORMONES

The preceding sections of this chapter have discussed oncogenes as elements in signal transduction pathways, which lead to cell proliferation in response to extracellular growth factors. These growth factors are secreted proteins that act by binding to receptors on the cell surface. The steroids and related hormones, which are also important regulators of cell growth and differentiation, act by a quite different mechanism. However, abnormalities in their functions are also associated with the development of cancer.

Steroid Hormones Pass Through the Plasma Membrane and Bind to Intracellular Receptors

The **steroids** (including **estrogen, progesterone, testosterone,** and the **glucocorticoids**) and related hormones (including **thyroid hormone** and **retinoic acid**) are small lipid-soluble molecules, in contrast to protein growth factors like PDGF. Consequently, the steroids and related hormones are able to pass freely through the plasma membrane to enter the inside of the cell, rather than binding to receptors on the cell surface (Fig. 11.9). Once inside a responsive cell, they then bind to specific receptor molecules, which act directly to regulate gene expression. The steroid hormones thus bypass the complicated machinery employed for transmission of signals from growth factor receptors to the nucleus. Since the steroid hormone receptors themselves act to regulate gene expression in the nucleus, there are no intermediaries between hormone receptor and alterations in gene expression.

Figure 11.9 Action of steroid hormones. Steroids and related hormones pass freely through the plasma membrane and bind to intracellular receptors, which act directly to regulate gene expression.

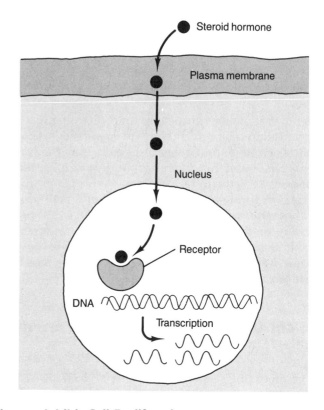

Steroids Can Either Stimulate or Inhibit Cell Proliferation

In some cases, steroid hormones act to stimulate cell growth. For example, estrogen is produced by the ovary and acts to stimulate proliferation of endometrial cells in the uterus. As discussed in chapter 3, excess estrogen (either from medicinal sources or produced by fat cells in obese individuals) is associated with increased risk of endometrial cancer, consistent with its role as a growth factor for these cells. In other instances, however, these hormones, particularly retinoic acid (vitamin A) and thyroid hormone, act to induce cell differentiation, concomitant with a decrease in cell proliferation. Malfunctions of these hormone receptors are also associated with the development of cancer.

The Retinoic Acid Receptor in Acute Promyelocytic Leukemia

In human acute promyelocytic leukemias, the retinoic acid receptor (*RAR*) is translocated from chromosome 17 to chromosome 15 (see chapter 8). This translocation results in a structural alteration in the receptor molecule, although the precise effect of this alteration on the function of the receptor is not yet known. It is noteworthy, however, that administration of retinoic acid is an effective therapy for this disease; it appears to work by inducing terminal differentiation of the leukemic cells. It might be speculated that abnormal function of the translocated receptor (the *RAR* oncogene) interferes with the normal differentiation process, and that this interference can be overcome by additional hormonal administration.

The Thyroid Hormone Receptor in Chicken Erythroleukemia

A presently better understood, but perhaps analogous, situation is the involvement of the thyroid hormone receptor in erythroid (red blood cell) leukemia in chickens. This leukemia is induced by a virus that carries an oncogene, termed *erb*A (see chapter 8), which is an altered version of the thyroid hormone receptor. Expression of this aberrant thyroid hormone receptor interferes with the action of the normal receptor, which is to turn on genes involved in red blood cell differentiation. Consequently, by preventing the normal thyroid hormone receptor from acting, the *erb*A oncogene acts to block the differentiation of erythroid cells. These cells are thus maintained in an undifferentiated state, in which they actively proliferate, resulting in the development of leukemia.

TUMOR SUPPRESSOR GENES ALSO ENCODE TRANSCRIPTIONAL REGULATORS

Proteins Encoded by *RB*, *WT*1, and *p53* Appear to Regulate Gene Expression

The tumor suppressor genes, described in chapter 9, have been discovered more recently than the oncogenes, so scientists currently know less about how they work. However, it appears likely that at least three of these genes—*RB*, *WT*1, and *p53*—encode proteins that also function as regulators of gene expression. Since these genes act to inhibit cell proliferation, they might act, like the thyroid hormone receptor, to turn on expression of genes involved in cell differentiation. Alternatively, or in addition, they might turn off expression of genes that induce cell proliferation.

TGFβ, *RB*, and c-*myc*

In one case, that of *RB*, the function of a tumor suppressor gene product has been linked to the action of an extracellular factor, **TGFβ,** that induces cell differentiation and inhibits cell division (Fig. 11.10). TGFβ is a potent inhibitor of the proliferation of skin epithelial cells (**keratinocytes**). Like the growth factors discussed in earlier sections of this chapter, TGFβ is a protein that acts by binding to a specific cell surface receptor. Activation of this receptor, however, generates a signal that inhibits, rather than stimulates, cell growth. Part of the cellular response to TGFβ appears to be activation of the *RB* tumor suppressor gene product, which then serves to mediate TGFβ-induced inhibition of cell proliferation. One action of *RB*, which may be critical to inhibition of keratinocyte growth, is to turn off expression of c-*myc*. This chain of events clearly

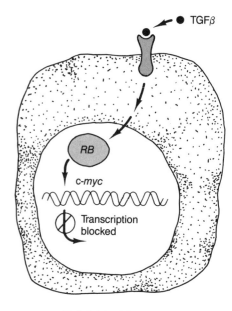

Figure 11.10 Regulation of the c-*myc* proto-oncogene by TGFβ. The proliferation of skin epithelial cells is inhibited by TGFβ, which acts by binding to a cell surface receptor. Part of the response to TGFβ is activation of the *RB* tumor suppressor gene product, which then acts to block transcription of c-*myc*.

Cell division inhibited

illustrates the interplay between oncogenes and tumor suppressor genes in regulation of normal cell proliferation: A growth inhibitory factor activates a tumor suppressor gene product, which, in turn, down-regulates expression of a proto-oncogene.

Interactions Between the Papillomavirus Oncogenes and the *RB* and *p53* Tumor Suppressor Gene Products

A different kind of interaction between oncogene and tumor suppressor gene products is involved in the induction of cancer by several tumor viruses, including the human papillomaviruses (Fig. 11.11). The *E7* oncogene protein of the human papillomaviruses, for example, binds tightly to the *RB* tumor suppressor gene product in infected cells. This binding of *E7* appears to inactivate the function of *RB*, so expression of the *E7* oncogene protein serves to block the normal activity of the *RB* protein in regulating cell growth. The other oncogene of human papillomaviruses, *E6*, similarly appears to inactivate the *p53* tumor suppressor protein. In these cases, then, the viral oncogene proteins function by blocking the action of tumor suppressor gene products.

ras, GAP, AND THE *NF1* TUMOR SUPPRESSOR GENE

Studies of another tumor suppressor gene, *NF1*, point to a distinct tumor suppressor-oncogene interaction at a different level of signal transduction. As discussed above, the activity of *ras* proto-oncogene proteins is normally regu-

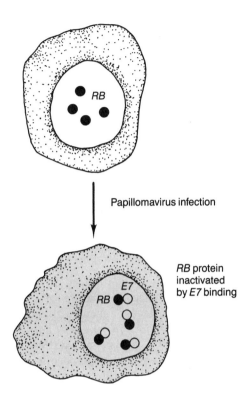

Figure 11.11   Interaction of the *RB* tumor suppressor gene product with the *E7* oncogene protein of human papillomaviruses. The *E7* protein binds to and inactivates the *RB* protein in infected cells. Transformation by the papillomaviruses thus appears to involve inactivation of the *RB* gene product.

Papillomavirus infection

RB protein inactivated by *E7* binding

Cell transformation

lated by a protein called GAP, which acts to turn off *ras* function. Indeed, it appears that inactivation of GAP by stimulated growth factor receptors is responsible for physiological activation of *ras* proto-oncogene proteins—a critical part of the signaling system leading to normal cell proliferation. The *ras* oncogenes, on the other hand, are generated by mutations that render them unresponsive to GAP; their unregulated activity then leads to neoplasia.

Viewed in this way, it might be expected that GAP would be a tumor suppressor gene, since its inactivation would result in unregulated *ras* activity and uncontrolled cell proliferation. The *NF1* tumor suppressor gene, identified in type 1 neurofibromatosis (see chapter 9), may be an example of precisely this relationship. Structural studies of *NF1* have shown that it is closely related to GAP, and further experiments have demonstrated that the *NF1* product can indeed act like GAP to turn off *ras* activity. One interpretation of these results is that both *NF1* and GAP normally act to control the *ras* proto-oncogene products. Inactivation of the *NF1* tumor suppressor gene in type 1 neurofibromatosis would then lead to excess *ras* activity, resulting in the outgrowth of neurofibromas. It is not yet clear, however, whether the situation is really this simple. The GAP and *NF1* proteins are both complicated molecules, which may do more than just regulate *ras*. If this is the case, it is possible that other activities of the *NF1* protein may also be involved in its function as a tumor suppressor gene product.

THE *DCC* TUMOR SUPPRESSOR GENE AND CELL ADHESION

Importance of Cell Interactions in Invasion and Metastasis

This chapter has so far discussed the actions of oncogene and tumor suppressor gene products in controlling cell proliferation and differentiation. A distinct, although not unrelated, aspect of cell behavior is the interaction of cells in tissues with both their neighbors and the surrounding extracellular matrix. As discussed in chapter 7, cancer cells are generally less adhesive than normal cells, and they behave comparatively independently of these interactions. This reduced sensitivity of cancer cells to association with other tissue components appears to be related to the ability of malignant cells to invade surrounding tissues and metastasize to distant body sites.

The *DCC* Gene Product Is Similar to Cell Adhesion Molecules

The *DCC* tumor suppressor gene, discussed in chapters 9 and 10, may function directly in controlling these aspects of cell behavior. Structural analysis of *DCC* has shown that it is a member of a family of cell surface molecules that are involved in cell-cell and cell-matrix interactions. In addition, independent studies have shown that another member of this family of cell surface adhesion molecules (integrins) can act as a tumor suppressor gene in an experimental model system. It thus appears that inactivation of *DCC* may result in reduced cell adhesion, thereby contributing to the development of malignancy. In this regard, it is noteworthy that inactivation of *DCC* generally occurs as a late event in colorectal carcinogenesis (see chapter 10), usually coinciding with acquisition of the capacity to invade surrounding normal tissue.

THE *bcl-2* ONCOGENE AND PROGRAMMED CELL DEATH

As discussed in chapter 7, the failure of cancer cells to undergo programmed cell death, as well as their uncontrolled proliferation, contributes to the development of malignancy. Interestingly, one oncogene, *bcl-*2 (see chapter 8), appears to contribute to the development of lymphomas by blocking programmed cell death, rather than by stimulating cell proliferation. When lymphocytes are deprived of required growth factors, they normally undergo an active process leading to degradation of their DNA and cell death. The *bcl*-2 protein apparently blocks this process, so lymphocytes expressing *bcl*-2 remain alive even in the absence of growth factors required for the survival of their normal counterparts. The *bcl*-2 gene product is localized in the mitochondria, suggesting that the energy-generating mechanisms of the cell may be involved in regulating the process of programmed cell death. However, neither the cellular mechanisms involved in programmed cell death nor the mechanism of action of the *bcl*-2 gene product are yet understood. Nonetheless, the role of *bcl*-2 in lymphomagenesis clearly illustrates the significance of programmed cell death to cancer development.

SUMMARY

Both oncogenes and tumor suppressor genes function as regulators of cell growth and differentiation (Table 11.1). Cell proliferation is frequently stimulated by the binding of extracellular growth factors to cell surface receptors, which then generate intracellular signals leading to changes in gene expression. Oncogenes can function at several levels of this signal transduction process: as growth factors, growth factor receptors, intracellular signaling molecules, and transcriptional regulatory factors. In each case, their abnormal expression or activity leads to excess uncontrolled cell division, eventually resulting in cancer. Receptors for retinoic acid and thyroid hormones directly regulate gene expression, frequently acting to promote cell differentiation rather than cell division. Abnormal versions of these receptors can also act as oncogenes, apparently by blocking normal receptor action and maintaining cells in an undifferentiated, actively proliferating state.

The products of several tumor suppressor genes also act to regulate gene expression, presumably turning on genes involved in cell differentiation or turning off genes involved in cell division. The products of some viral oncogenes directly block the action of these tumor suppressor gene products, illustrating a direct functional interaction between the two classes of growth regulatory molecules. Another interaction between oncogenes and tumor suppressor genes is illustrated by the *NF1* tumor suppressor gene, which may function to regulate the activity of normal *ras* proteins. The product of yet another tumor suppressor gene, *DCC*, is a cell surface molecule that appears to be involved in adhesion of cells to their neighbors or the extracellular matrix, indicating the importance of such interactions in carcinogenesis. Finally, one oncogene, *bcl-2*, acts by blocking programmed cell death rather than by stimulating cell proliferation. Oncogenes and tumor suppressor genes thus regulate several aspects of cell behavior, malfunctions of which lead to the development of cancer.

Table 11.1   Functions of Oncogene and Tumor Suppressor Gene Products

Cellular Function	Representative Gene Products
Growth factors	*sis*, FGF, *hst*, *int-2*, EGF, EPO, IL-2, IL-3, CSF-1, GM-CSF
Growth factor receptors	*erb*B, *erb*B-2, *fms*, *kit*, *met*, *ret*, *ros*, *trk*
Intracellular signal transduction	protein kinase C, *raf*, *mos*, *ras*, NF1, *gsp*, *gip*
Transcriptional regulation	*fos*, *jun*, *myc*, *myb*, *rel*, E2A, RB, *p53*, WT1
Steroid receptors	*erb*A, *RAR*
Cell adhesion	*DCC*
Cell survival	*bcl-2*

KEY TERMS

growth factor
platelet-derived growth factor (PDGF)
sis
autocrine growth stimulation
growth factor receptor
*erb*B-2
ret
trk
intracellular signal transduction
protein-tyrosine kinase
protein-serine/threonine kinase
inositol phospholipids
phospholipase C
diacylglycerol
inositol triphosphate
protein kinase C
phosphatidylinositol kinase
raf
GAP
ras
transcription factor
AP-1
fos
jun
myc
steroid hormones
estrogen
progesterone
testosterone
glucocorticoids
thyroid hormone
retinoic acid
RAR
*erb*A
RB
WT1
p53

TGFβ

E7

E6

*NF*1

DCC

cell adhesion

*bcl-*2

programmed cell death

REFERENCES AND FURTHER READING

General References

Cantley, L.C., Auger, K.R., Carpenter, C., Duckworth, B., Graziani, A., Kapeller, R., and Soltoff, S. 1991. Oncogenes and signal transduction. *Cell* 64:281–302.

Cooper, G.M. 1990. *Oncogenes.* Jones and Bartlett, Boston.

Oncogenes and Growth Factors

Cross, M., and Dexter, T.M. 1991. Growth factors in development, transformation, and tumorigenesis. *Cell* 64:271–280.

Doolittle, R.F., Hunkapiller, M.W., Hood, L.E., Devare, S.G., Robbins, K.C., Aaronson, S.A., and Antoniades, H.N. 1983. Simian sarcoma virus *onc* gene, v-*sis*, is derived from the gene (or genes) encoding a platelet-derived growth factor. *Science* 221:275–277.

Sawyers, C.L., Denny, C.T., and Witte, O.N. 1991. Leukemia and the disruption of normal hematopoiesis. *Cell* 64:337–350.

Sporn, M.B., and Roberts, A.B. 1985. Autocrine growth factors and cancer. *Nature* 313:745–747.

Waterfield, M.D., Scrace, G.T., Whittle, N., Stroobant, P., Johnsson, A., Wasteson, A., Westermark, B., Heldin, C.-H., Huang, J.S., and Deuel, T.F. 1983. Platelet-derived growth factor is structurally related to the putative transforming protein p28^{sis} of simian sarcoma virus. *Nature* 304:35–39.

Growth Factor Receptors

Di Fiore, P.P., Pierce, J.H., Kraus, M.H., Segatto, O., King, C.R., and Aaronson, S.A. 1987. *erb*B-2 is a potent oncogene when overexpressed in NIH/3T3 cells. *Science* 237:178–182.

Downward, J., Yarden, Y., Mayes, E., Scrace, G., Totty, N., Stockwell, P., Ullrich, A., Schlessinger, J., and Waterfield, M.D. 1984. Close similarity of epidermal growth factor receptor and v-*erb*-B oncogene protein sequences. *Nature* 307:521–527.

Martin-Zanca, D., Hughes, S.H., and Barbacid, M. 1986. A human oncogene formed by the fusion of truncated tropomyosin and protein kinase sequences. *Nature* 319:743–748.

Sherr, C.J., Rettenmeier, C.W., Sacca, R., Roussel, M.F., Look, A.T., and Stanley, E.R. 1985. The c-*fms* proto-oncogene product is related to the receptor for the mononuclear phagocyte growth factor, CSF-1. *Cell* 41:665–676.

Takahashi, M., and Cooper, G.M. 1987. *ret* transforming gene encodes a fusion protein homologous to tyrosine kinases. *Mol. Cell. Biol.* 7:1378–1385.

Yarden, Y., and Ullrich, A. 1988. Growth factor receptor tyrosine kinases. *Ann. Rev. Biochem.* 57:443–478.

Oncogenes and Intracellular Signal Transduction

Auger, K.R., Serunian, L.A., Soltoff, S.P., Libby, P., and Cantley, L.C. 1989. PDGF-dependent tyrosine phosphorylation stimulates production of novel polyphosphoinositides in intact cells. *Cell* 57:167–175.

Barbacid, M. 1987. *ras* genes. *Ann. Rev. Biochem.* 56:779–827.

Bourne, H.R., Sanders, D.A., and McCormick, F. 1991. The GTPase superfamily: conserved structure and molecular mechanism. *Nature* 349:117–127.

Hanks, S.K., Quinn, A.M., and Hunter, T. 1988. The protein kinase family: conserved features and deduced phylogeny of the catalytic domains. *Science* 241:42–52.

Hunter, T. 1987. A thousand and one protein kinases. *Cell* 50:823–829.

Kaplan, D.R., Morrison, D.K., Wong, G., McCormick, F., and Williams, L.T. 1990. PDGF β-receptor stimulates tyrosine phosphorylation of GAP and association of GAP with a signaling complex. *Cell* 61:125–133.

Krengel, U., Schlichting, I., Scherer, A., Schumann, R., Frech, M., John, J., Kabsch, W., Pai, E.F., and Wittinghofer, A. 1990. Three-dimensional structures of H-*ras* p21 mutants: molecular basis for their inability to function as signal switch molecules. *Cell* 62:539–548.

Megidish, T., and Mazurek, N. 1989. A mutant protein kinase C that can transform fibroblasts. *Nature* 342:807–811.

Milburn, M.V., Tong, L., deVos, A.M., Brünger, A., Yamaizumi, Z., Nishimura, S., and Kim, S.-H. 1990. Molecular switch for signal transduction: structural differences between active and inactive forms of protooncogenic *ras* proteins. *Science* 247:939–945.

Morrison, D.K., Kaplan, D.R., Escobedo, J.A., Rapp, U.R., Roberts, T.M., and Williams, L.T. 1989. Direct activation of the serine/threonine kinase activity of *raf*-1 through tyrosine phosphorylation by the PDGF β-receptor. *Cell* 58:649–657.

Nishibe, S., Wahl, M.I., Hernández-Sotomayor, S.M.T., Tonks, N.K., Rhee, S.G., and Carpenter, G. 1990. Increase of the catalytic activity of phospholipase C-γ1 by tyrosine phosphorylation. *Science* 250:1253–1256.

Ullrich, A., and Schlessinger, J. 1990. Signal transduction by receptors with tyrosine kinase activity. *Cell* 61:203–212.

Zhang, K., DeClue, J.E., Vass, W.C., Papageorge, A.G., McCormick, F., and Lowy, D.R. 1990. Suppression of c-*ras* transformation by GTPase-activating protein. *Nature* 346:754–756.

Regulation of Gene Expression in Response to Growth Factors

Blackwood, E.M., and Eisenman, R.N. 1991. Max: a helix-loop-helix zipper protein that forms a sequence-specific DNA-binding complex with *myc*. *Science* 251:1211–1217.

Bohmann, D., Bos, T.J., Admon, A., Nishimura, T., Vogt, P.K., and Tjian, R. 1987. Human proto-oncogene c-*jun* encodes a DNA binding protein with structural and functional properties of transcription factor AP-1. *Science* 238:1386–1392.

Boyle, W.J., Smeal, T., Defize, L.H.K., Angel, P., Woodgett, J.R., Karin, M., and Hunter, T. 1991. Activation of protein kinase C decreases phosphorylation of c-*jun* at sites that negatively regulate its DNA-binding activity. *Cell* 64:573–584.

Curran, T., and Franza, B.R., Jr. 1988. *Fos* and *jun*: the AP-1 connection. *Cell* 55:395–397.

Greenberg, M.E., and Ziff, E.B. 1984. Stimulation of 3T3 cells induces transcription of the c-*fos* proto-oncogene. *Nature* 311:433–438.

Kelly, K., Cochran, B.H., Stiles, C.D., and Leder, P. 1983. Cell-specific regulation of the c-*myc* gene by lymphocyte mitogens and platelet-derived growth factor. *Cell* 35:603–610.

Lewin, B. 1991. Oncogenic conversion by regulatory changes in transcription factors. *Cell* 64:303–312.

Prendergast, G.C., Lawe, D., and Ziff, E.B. 1991. Association of myn, the murine homolog of max, with c-*myc* stimulates methylation-sensitive DNA binding and *ras* cotransformation. *Cell* 65:395–407.

Regulation of Gene Expression by Steroids and Related Hormones

Beato, M. 1989. Gene regulation by steroid hormones. *Cell* 56:335–344.

Borrow, J., Goddard, A.D., Sheer, D., and Solomon, E. 1990. Molecular analysis of acute promyelocytic leukemia breakpoint cluster region on chromosome 17. *Science* 249:1577–1580.

Castaigne, S., Chomienne, C., Daniel, M.T., Ballerini, P., Berger, R., Fenaux, P., and Degos, L. 1990. All-*trans* retinoic acid as a differentiation therapy for acute promyelocytic leukemia. I. Clinical results. *Blood* 76:1704–1709.

Chomienne, C., Ballerini, P., Balitrand, N., Daniel, M.T., Fenaux, P., Castaigne, S., and Degos, L. 1990. All-*trans* retinoic acid in acute promyelocytic leukemias. II. In vitro studies: structure-function relationship. *Blood* 76:1710–1717.

Damm, K., Thompson, C.C., and Evans, R.M. 1989. Protein encoded by v-*erb*A functions as a thyroid hormone receptor antagonist. *Nature* 339:593–597.

de Thé, H., Chomienne, C., Lanotte, M., Degos, L., and Dejean, A. 1990. The t(15;17) translocation of acute promyelocytic leukaemia fuses the retinoic acid receptor α gene to a novel transcribed locus. *Nature* 347:558–561.

Evans, R.M. 1988. The steroid and thyroid hormone receptor superfamily. *Science* 240:889–895.

Huang, M., Ye, Y., Chen, S., Chai, J., Lu, J.-X., Zhoa, L., Gu, L., and Wang, Z. 1988. Use of all-*trans* retinoic acid in the treatment of acute promyelocytic leukemia. *Blood* 72:567–572.

Sap, J., Muñoz, A., Schmitt, J., Stunnenberg, H., and Vennström, B. 1989. Repression of transcription mediated at a thyroid hormone response element by the v-*erb*A oncogene product. *Nature* 340:242–244.

Sporn, M.B., and Roberts, A.B. 1991. Interactions of retinoids and transforming growth factor-β in regulation of cell differentiation and proliferation. *Mol. Endo.* 5:3–7.

Zenke, M., Muñoz, A., Sap, J., Vennström, B., and Beug, H. 1990. v-*erb*A oncogene activation entails the loss of hormone-dependent regulator activity of c-*erb*A. *Cell* 61:1035–1049.

Tumor Suppressor Genes Also Encode Transcriptional Regulators

Call, K.M., Glaser, T., Ito, C.Y., Buckler, A.J., Pelletier, J., Haber, D.A., Rose, E.A., Kral, A., Yeger, H., Lewis, W.H., Jones, C., and Housman, D.E. 1990. Isolation and characterization of a zinc finger polypeptide gene at the human chromosome 11 Wilms' tumor locus. *Cell* 60:509–520.

Dyson, N., Howley, P.M., Münger, K., and Harlow, E. 1989. The human papillomavirus-16 *E7* oncoprotein is able to bind to the retinoblastoma gene product. *Science* 243:934–937.

Fields, S., and Jang, S.K. 1990. Presence of a potent transcription activating sequence in the *p53* protein. *Science* 249:1046–1049.

Laiho, M., DeCaprio, J.A., Ludlow, J.W., Livingston, D.M., and Massagué, J. 1990. Growth inhibition by TGF-β linked to suppression of retinoblastoma protein phosphorylation. *Cell* 62:175–185.

Lee, W.-H., Shew, J.-Y., Hong, F.D., Sery, T.W., Donoso, L.A., Young, L.-J., Bookstein, R., and Lee, E.Y.-H.P. 1987. The retinoblastoma susceptibility gene encodes a nuclear phosphoprotein associated with DNA binding activity. *Nature* 319:642–645.

Marshall, C.J. 1991. Tumor suppressor genes. *Cell* 64:313–326.

Moses, H.L., Yang, E.Y., and Pietenpol, J.A. 1990. TGF-β stimulation and inhibition of cell proliferation: new mechanistic insights. *Cell* 63:245–247.

Pietenpol, J.A., Stein, R.W., Moran, E., Yaciuk, P., Schlegel, R., Lyons, R.M., Pittelkow, M.R., Münger, K., Howley, P.M., and Moses, H.L. 1990. TGF-β1 inhibition of c-*myc* transcription and growth in keratinocytes is abrogated by viral transforming proteins with pRB binding domains. *Cell* 61:777–785.

Rauscher, F.J., III, Morris, J.F., Tournay, O.E., Cook, D.M., and Curran, T. 1990. Binding of the Wilms' tumor locus zinc finger protein to the EGR-1 consensus sequence. *Science* 250:1259–1262.

Raycroft, L., Wu, H., and Lozano, G. 1990. Transcriptional activation by wild-type but not transforming mutants of the *p53* anti-oncogene. *Science* 249:1049–1051.

Robbins, P.D., Horowitz, J.M., and Mulligan, R.C. 1990. Negative regulation of human c-*fos* expression by the retinoblastoma gene product. *Nature* 346:668–671.

Werness, B.A., Levine, A.J., and Howley, P.M. 1990. Association of human papillomavirus types 16 and 18 *E6* proteins with *p53*. *Science* 248:76–79.

ras, GAP, and the *NF1* Tumor Suppressor Gene

Ballester, R., Marchuk, D., Boguski, M., Saulino, A., Letcher, R., Wigler, M., and Collins, F. 1990. The *NF1* locus encodes a protein functionally related to mammalian GAP and yeast *IRA* proteins. *Cell* 63:851–859.

Martin, G.A., Viskochil, D., Bollag, G., McCabe, P.C., Crosier, W.J., Haubruck, H., Conroy, L., Clark, R., O'Connell, P., Cawthon, R.M., Innis, M.A., and McCormick, F. 1990. The GAP-related domain of the neurofibromatosis type 1 gene product interacts with *ras* p21. *Cell* 63:843–849.

Xu, G., Lin, B., Tanaka, K., Dunn, D., Wood, D., Gesteland, R., White, R., Weiss, R., and Tamanoi, F. 1990. The catalytic domain of the neurofibromatosis type 1 gene product stimulates *ras* GTPase and complements *ira* mutants of S. cerevisiae. *Cell* 63:835–841.

Xu, G., O'Connell, P., Viskochil, D., Cawthon, R., Robertson, M., Culver, M., Dunn, D., Stevens, J., Gesteland, R., White, R., and Weiss, R. 1990. The neurofibromatosis type 1 gene encodes a protein related to GAP. *Cell* 62:599–608.

The *DCC* Tumor Suppressor Gene and Cell Adhesion

Fearon, E.R., Cho, K.R., Nigro, J.M., Kern, S.E., Simons, J.W., Ruppert, J.M., Hamilton, S.R., Preisinger, A.C., Thomas, G., Kinzler, K.W., and Vogelstein, B. 1990. Identification of a chromosome 18q gene that is altered in colorectal cancers. *Science* 247:49–56.

Giancotti, F.G., and Ruoslahti, E. 1990. Elevated levels of the α5β1 fibronectin receptor suppress the transformed phenotype of Chinese hamster ovary cells. *Cell* 60:849–859.

The *bcl*-2 Oncogene and Programmed Cell Death

Hockenbery, D., Nuñez, G., Milliman, C., Schreiber, R.D., and Korsmeyer, S.J. 1990. *Bcl*-2 is an inner mitochondrial membrane protein that blocks programmed cell death. *Nature* 348:334–336.

PART III

CANCER PREVENTION AND TREATMENT

CAN CANCER BE ELIMINATED as a major killer in our society? The war on cancer can be fought on two levels, prevention and treatment, which are the subject of the third part of this book. Both the effectiveness and the limitations of these strategies for dealing with cancer can be understood in terms of the cellular and molecular alterations responsible for tumor development.

Chapter 12

Prospects for
Cancer Prevention

AS DISCUSSED IN CHAPTER 10, cancers develop as a result of accumulated damage to multiple oncogenes and tumor suppressor genes. Genetic damage occurs throughout life and in some cases is unavoidable. For example, mutations sometimes arise as a consequence of errors made during the normal replication of DNA in dividing cells. In addition, some chemical species (e.g., oxygen free radicals) that react with DNA and induce mutations are formed as a result of normal cellular metabolism. Thus, there is a background level of ongoing DNA damage that cannot be avoided and that presumably contributes to cancer development.

In addition, however, an individual's risk of developing cancer is affected by exposure to environmental agents. As discussed in chapter 3, it has been estimated that environmental carcinogens constitute risk factors for up to 80% of human cancers. In principle, then, much of the cancer burden could be eliminated by avoiding exposure to carcinogenic agents, which include chemicals, radiation, and viruses. This chapter will discuss the practical steps an individual can take, based on current knowledge, to lower his or her risk of developing cancer. Such steps include reducing or eliminating exposure to major known carcinogens and following good nutritional practices. The potentials and problems of other means of cancer prevention will also be considered. These include the possibility of developing medications that prevent cancer, as well as continuing efforts to identify and eliminate synthetic chemical carcinogens from the environment.

SMOKING

Not Smoking Is the Most Effective Means of Cancer Prevention

The major identified cause of cancer, discussed in detail in chapter 3, is unquestionably cigarette smoking. Smoking is responsible for almost all lung cancer, as well as contributing to the development of several other types of malignancy. Tobacco smoke contains a number of carcinogens, which probably act both to induce mutations and to stimulate cell proliferation. Mutations in the *ras*K oncogene and in the *p53* tumor suppressor gene, which frequently contribute to the development of lung cancers (see chapter 10), may well result from the mutagenic activity of these carcinogens. In total, nearly one-third of all cancer deaths in the United States can be attributed to smoking.

Clearly, then, the single most effective action an individual can take to prevent cancer is not to smoke. Smoking pipes or cigars, or using smokeless tobacco (chewing tobacco or snuff), is less risky than smoking cigarettes. However, all of these forms of tobacco use are associated with substantially increased cancer risk and, from the standpoint of cancer prevention, should be eliminated.

Smoking Is Still Prevalent in Our Society

The risk associated with smoking was first widely publicized more than 25 years ago, with the 1964 Report of the Surgeon General on Smoking and Health. It is now thoroughly established that smoking is not only a major cause of cancer, but also of heart disease, stroke, emphysema, and other respiratory diseases. The combined effect of all of these diseases is that one out of every three smokers will die from the habit. Such statistics make smoking the largest preventable cause of death in the United States. Nonetheless, over 50 million Americans, about one-third of the adult population, continue to smoke. Two to three million of these smokers are under the age of 18.

In the face of the striking risk associated with smoking, the continuing prevalence of smoking in our society seems surprising. Of course, individuals who have never smoked have the lowest risk of tobacco-related diseases, but the risk of disease for smokers of all ages is substantially reduced by quitting. For example, as discussed in chapter 3, the risk of lung cancer steadily declines for ex-smokers relative to those who continue smoking. Within about 20 years, the lung cancer risk for ex-smokers becomes close to that of nonsmokers. In terms of overall life expectancy, it has been estimated that individuals who quit smoking before age 50 have one-half the risk of dying before age 65 compared to continuing smokers. Although quitting smoking is hard, it has major health benefits—both immediate and long-term.

ALCOHOL

Excessive consumption of alcoholic beverages (more than four drinks a day) is associated with an increased risk of several cancers (see chapter 3). Cirrhosis, which can result from excess alcohol consumption, increases the risk of developing liver cancer, presumably as a result of chronic tissue damage leading to continuous cell proliferation. Alcohol is also a risk factor (particularly in combination with smoking) for cancers of the oral cavity, pharynx, larynx, and esophagus. In these cases, it is likely that the effect of alcohol is exerted primarily by potentiating the action of other carcinogens, such as those present in tobacco. It is generally recommended that alcohol be used only in moderation (two or fewer drinks a day). Because of the combined effects of alcohol and smoking, minimizing alcohol consumption would appear to be especially important for smokers.

RADIATION

Types of Carcinogenic Radiation

As discussed in chapter 3, both ultraviolet and ionizing radiation, which act to damage DNA directly, are risk factors for human cancer. Other kinds of radiation have also been discussed as potential carcinogens, including electromagnetic fields generated by power lines and electrical equipment. Although the possibility of a relationship between electromagnetic fields and cancer incidence is still being investigated, the evidence supporting this association is generally not considered convincing. Likewise, radiation from appliances such as televisions, computers, and microwave ovens has not been associated with increased cancer risk. Most exposure to the forms of radiation that do cause cancer is from natural sources, some of which cannot be avoided. However, exposure to some sources of potentially carcinogenic radiation can be minimized as part of an individual's cancer prevention effort. These include sunlight, medical and dental X-rays, and radon gas in the home.

Sunlight and Skin Cancer

Solar ultraviolet radiation is a major cause of skin cancer, including melanoma. Mutation of the *ras*N oncogene, discussed in chapter 10, may be one mechanism by which ultraviolet radiation can lead to melanoma development. Fair-skinned individuals are the most sensitive. It is advisable to avoid excessive exposure to the sun, if necessary by wearing protective clothing and using sunscreens.

Medical and Dental X-Rays

As discussed in chapter 3, about 80% of ionizing radiation exposure is from natural sources, such as cosmic rays and radioactive substances in the earth's crust. Medical and dental X-rays constitute most of the remainder (nearly 20%) of an individual's total exposure to ionizing radiation in the United States. These X-rays are now administered in such a way as to minimize radiation exposure to both the patient and physician. The risk associated with these procedures is small and is usually far outweighed by the resulting benefits. However, it is clearly advisable to avoid exposure to any X-rays that are not medically indicated.

Radon

Radon gas is a natural source of radiation (see chapter 3); it can be a significant source of radiation exposure in the home. The levels of radon vary widely in homes throughout the United States (Fig. 12.1), in some cases being high

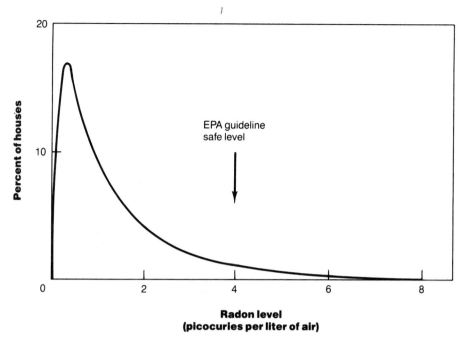

Figure 12.1   *Distribution of radon levels in United States homes. (From A.V. Nero, Sci. Amer. 258:42–48, 1988.)*

enough to impart an increased risk of lung cancer. The average level of radon in American homes is approximately 1.5 picocuries per liter of air. The Environmental Protection Agency's recommended guideline for a safe level of radon is 4 picocuries per liter. About 7% of American homes (approximately 4 million) have radon levels in excess of this guideline. A small percentage of homes (but still numbering in the tens of thousands) have indoor radon levels in excess of 25 picocuries per liter, which corresponds to about a fivefold increase in lung cancer risk. As with other carcinogens that contribute to lung cancer, the risk of radon exposure combines with the risk of smoking. Thus, particularly for smokers, high levels of radon in homes can constitute a significant hazard. It is therefore advisable for an individual to monitor indoor radon levels in his or her home and to take remedial action if indicated. Radon monitoring kits are obtainable in hardware stores. If necessary, the levels of radon in a home can be reduced by sealing cracks in basement walls and floors or by increasing ventilation.

Nuclear Power Plants

One area of public concern has been the potential increase in radiation exposure that might result from living in the vicinity of nuclear power plants. However, radiation pollution from these plants appears to make an insignificant contribution to the natural level of radiation exposure, and several studies have shown that residents in the vicinity of nuclear installations do not suffer an increased cancer risk. This assessment, of course, refers to the radiation pollution that results from normal day-to-day operation of these plants. Concerns over nuclear safety with respect to the likelihood of accidents, such as those that occurred at the Three Mile Island Power Station in 1979 or, more disastrously, at the Chernobyl Power Station in 1986, are obviously a different matter.

DIETARY FACTORS

General Dietary Recommendations

The significance of diet in carcinogenesis was discussed in detail in chapter 3. Although it is commonly thought that dietary factors contribute to the development of a significant percentage of cancers, attempts to pinpoint the role of individual dietary components have been largely inconclusive. Consequently, the possible effects of specific dietary practices on cancer incidence are not firmly established. Nonetheless, general dietary recommendations have been made by several organizations, including the National Academy of Sciences, the American Cancer Society, and the National Institutes of Health. In addition to being in line with good overall nutritional practice, these recommendations may also help to reduce cancer risk. The basic dietary recommendations for cancer prevention are (1) reduce consumption of high-fat and high-calorie foods, (2) increase consumption of fresh fruits, vegetables, and whole grain breads and cereals, and (3) consume cured, pickled, and smoked foods only in moderation.

Avoid Obesity

Obesity significantly increases the risk of cancer of the uterine endometrium and, to a lesser extent, that of breast cancer. It is therefore advisable to maintain normal body weight—if necessary, by restricting caloric intake. Recommended body weights for women are 100 pounds for 5 feet in height, plus an additional 5 pounds per inch. Body weights 40% or more in excess of these recommended levels are clearly associated with an increased risk of endometrial carcinoma, presumably resulting from excess stimulation of endometrial cell proliferation by estrogen produced by fat cells.

The Dangers of High-Fat Diets

High-fat diets have been linked to increased incidence of breast and, most convincingly, colon cancers. It is therefore advisable to reduce dietary fat intake. In the United States, fat constitutes approximately 37% of calories in the average diet and may be greater than 45% for some individuals. In contrast, fat constitutes a much smaller proportion of dietary intake in countries with substantially lower rates of breast and colon cancers, in some cases corresponding to less than

20% of total calories. It has been recommended by a National Academy of Sciences committee that Americans reduce fat consumption to 30% of total calorie intake, which might be expected to reduce the incidence of colon cancer by as much as twofold. The fat content of a variety of foods is summarized in Table 12.1. In general, a reduction in dietary fat can be achieved by eating more fruits and vegetables; utilizing lean meats, poultry, and fish; consuming low-fat dairy products; and eating fewer fried foods and bakery products.

Table 12.1   Fat Content of Foods

Food	Percent of Calories as Fat
Dairy products	
Whole milk	48
Lowfat milk (1%)	17
Butter	99
Cheddar cheese	70
Cottage cheese (lowfat)	18
Ice cream	47
Yogurt (lowfat)	8
Meats	
Ground beef (broiled)	64
Rib roast	53
Steak	56
Lamb (roasted)	39
Ham	61
Salami	68
Poultry	
Chicken (fried)	45
Chicken (roasted)	34
Turkey (roasted)	27
Seafood	
Flounder (baked with butter)	45
Flounder (baked without butter)	11
Shrimp (fried)	45
Shrimp (steamed)	9
Tuna (packed in oil)	38
Tuna (packed in water)	7
Baked Goods	
Cake	40
Cookies	54
Doughnuts	50
Fruits and Vegetables	nearly 0

The Benefits of Fruits and Vegetables

Fruits and vegetables are not only low in dietary fat but also serve as rich sources (together with whole grain breads and cereals) of several dietary components that appear to reduce cancer risk. These protective substances (discussed in chapter 3) include dietary fiber, β-carotene (a source of vitamin A), and vitamin C, in addition to other compounds that appear to interfere with the action of some carcinogens. Some of the mechanisms by which these protective agents may act are discussed later in this chapter (see The Possibility of Chemoprevention). It is important to note, however, that although diets rich in fruit and vegetables appear to be associated with a decreased risk of some cancers, the importance (and perhaps the identity) of the individual cancer-protective agents in these foods is not known. It is therefore recommended that individuals consume a variety of vitamin-rich vegetables and fruits, rather than relying on fiber, vitamin, or mineral supplements. In addition, some vitamins and minerals (including vitamin A and selenium) are toxic when taken in high amounts, so dietary supplementation with these agents can be dangerous. Whole-grain breads and cereals, as well as fruits and vegetables (especially beans and peas), are good sources of dietary fiber, which may serve to decrease the risk of colon cancer. Fruits and vegetables, especially green and yellow vegetables and citrus fruits, are rich sources of vitamins A and C. Cruciferous vegetables (broccoli, Brussels sprouts, cabbage, cauliflower, collards, kale, mustard greens, rutabagas, turnips, and turnip greens) also contain several additional compounds (flavones, indoles, and isothiocyanates) that may reduce cancer risk. Frequent consumption of a variety of these foods is generally recommended from the standpoints of both cancer prevention and good general nutrition.

Smoked, Cured, and Pickled Foods

Finally, excessive consumption of smoked, cured, and pickled foods has been linked to increased risk of stomach and esophageal cancers. These foods contain nitrites, which can be converted to carcinogenic N-nitroso compounds in the body. The activity of one such carcinogen (N-methyl-N-nitrosourea) in inducing mutations of *ras* oncogenes in rat mammary carcinomas was discussed in chapter 10, and it is possible that these carcinogens induce mutations of the *p53* tumor suppressor gene, which is frequently inactivated in esophageal cancers (see chapter 9). In any event, it is generally recommended that these foods be consumed only in moderation.

AVOIDANCE OF OCCUPATIONAL AND MEDICINAL CARCINOGENS

Carcinogens in the Workplace

Several occupations involve exposure to carcinogens in the workplace (Table 12.2). As industrial carcinogens have been identified, appropriate regulatory

Table 12.2 *Occupations Associated with Increased Cancer Risk*

Representative Occupations	Associated Cancer Risk
Chemical, dye, and rubber workers	Leukemia, bladder, lung, and liver cancer
Coal, gas, and petroleum workers	Bladder, lung, and skin cancer
Construction workers	Lung cancer
Furniture manufacturing workers	Nasal cancer
Leather workers	Nasal and bladder cancer
Metal workers	Lung cancer
Mustard gas workers	Lung, larynx, and nasal cancer
Nickel refining workers	Lung and nasal cancer
Underground miners	Lung cancer

Note: See Table 3.3 for identification of the carcinogens responsible for these occupational cancer risks.

actions have been taken to limit the exposure of workers to these agents. However, it is prudent for an individual to be aware of the potential carcinogenic hazards of his or her occupation, and to be sure that he or she follows recommended safety practices, such as wearing protective clothing or masks. Many industrial carcinogens, such as asbestos (see chapter 3), act in combination with smoking to impart a strikingly high risk of lung cancer. In many cases, therefore, simply not smoking substantially reduces the risk of industrial carcinogen exposure.

Some Medications Increase Cancer Risk

As discussed in chapter 3, many medicines are carcinogenic, but in most cases the benefit from treatment far outweighs the risk of cancer. For example, drugs used in organ transplant procedures are known to increase cancer incidence, presumably by suppressing the immune system, but the immediate treatment need far outweighs the risk that these medications will act as carcinogens. From the standpoint of cancer prevention, the major medical treatment currently associated with increased cancer risk is postmenopausal estrogen therapy. Long-term administration of estrogen alone clearly increases the risk of endometrial cancer. Thus, although postmenopausal estrogen therapy is beneficial to many women, the associated risk of cancer needs to be considered and discussed with a physician. It now appears that this risk is minimized by treatment with low doses of estrogen in combination with progesterone, which counteracts the effect of estrogen on endometrial cell proliferation.

ONCOGENIC VIRUSES

As discussed in chapter 4, viruses are important risk factors for some human cancers. The viruses implicated to date as direct-acting human carcinogens are hepatitis B virus (hepatocellular carcinoma), papillomaviruses (cervical and other anogenital carcinomas), Epstein-Barr virus (Burkitt's lymphoma and nasopharyngeal carcinoma), and human T-cell lymphotropic virus type 1 (adult T-cell leukemia). As discussed in chapter 8, the papillomaviruses, Epstein-Barr virus, and HTLV-1 contain viral oncogenes. Interestingly, the papillomavirus oncogenes function to inactivate the *RB* and *p53* tumor suppressor gene products, indicating a direct relationship between the products of viral oncogenes and cellular tumor suppressor genes (see chapter 11). Hepatitis B virus may also contain an oncogene, although it is alternatively possible that the induction of liver cancer by this virus results from chronic tissue damage leading to excess cell proliferation. In addition, patients with AIDS have a high frequency of developing certain cancers (particularly lymphomas and Kaposi's sarcoma) as a secondary consequence of immunodeficiency. The causative agent of AIDS, HIV, is thus an indirect cause of an increasing number of cancers throughout the world.

Vaccination Against Tumor Viruses

The identification of a virus as a cause of cancer presents the possibility of developing a vaccine to prevent virus infection. Vaccines have indeed been successful in combating many viruses that cause serious human diseases, such as polio and smallpox, and may ultimately prove effective in preventing virus-induced cancers. The greatest progress in this area so far has been the development of a safe and effective vaccine against hepatitis B virus. The efficacy of this vaccine in cancer prevention is currently being tested in The Gambia, a small African country with a high incidence of hepatitis B virus infection and hepatocellular carcinoma. Vaccination of children in The Gambia was initiated in 1986, and it appears that this program has been effective in reducing the incidence of virus infection. However, the effect of vaccination on cancer incidence will require long-term study. Since the lag time between virus infection and development of hepatocellular carcinoma is at least 20–30 years, the data needed to assess the effect of this vaccination program on cancer incidence will not be available until around the year 2010.

Minimizing the Risk of Infection

Although vaccination against other tumor viruses is not yet available, alternative steps can be taken to minimize the likelihood of virus infection. In the United States, the viruses of major concern are the papillomaviruses, hepatitis B virus, and HIV, since cancers induced by either Epstein-Barr virus or HTLV are rare in this country. Fortunately, none of these viruses is transmitted by casual contact, and there is no risk of acquiring cancer simply by exposure to an infected individual. Both the oncogenic papillomaviruses and HIV are sexually transmitted, so adherence to safe sexual practices is an important protective measure. Hepatitis B virus, as well as HIV, can be transmitted by contaminated blood products. Blood supplies used for transfusions are routinely screened to eliminate such viral contamination. However, it is critical for individuals to avoid other sources of contaminated blood, such as shared needles used for intravenous drug injection.

THE POSSIBILITY OF CHEMOPREVENTION

Identification of Chemopreventive Agents

As discussed above, several dietary constituents, such as fiber and certain vitamins, are thought to lower cancer risk. Evidence for the action of any of these factors is not definitive, so it is currently recommended that individuals eat a balanced diet rather than relying on specific dietary supplements. However, the possibility of identifying or developing specific medications that would serve to reduce cancer risk is an active area of cancer research, called **chemoprevention**.

Laboratory studies have identified hundreds of chemicals that appear to have some activity in reducing cancer risk in experimental animals. These include the dietary agents suspected of lowering human cancer incidence (e.g., vitamins A and C), as well as many other compounds. Such chemopreventive agents appear to act by interfering either with the action of carcinogens (**blocking agents**) or with the outgrowth of transformed cells (**suppressing agents**) (Fig. 12.2).

Blocking Agents

Many blocking agents affect the metabolism of potentially carcinogenic chemicals, either inhibiting the conversion of potential carcinogens to their active forms or accelerating the elimination of potential carcinogens from the body. Vitamins C and E, for example, can protect against stomach cancer by blocking the conversion of nitrites to carcinogenic N-nitroso compounds in the digestive tract. Additional examples of blocking agents are provided by the cancer protective compounds present in cruciferous vegetables, such as isothiocyanates. These compounds stimulate the activity of enzymes (in the liver and other tissues) that lead to the detoxification and elimination of a number of carcinogenic chemicals.

Other blocking agents may protect cells against the damage inflicted by carcinogens. Vitamin E and β-carotene, for example, protect cells against oxidative damage, which may contribute to carcinogenesis both by damaging DNA and by stimulating cell proliferation. In particular, vitamin E and β-carotene are efficient scavengers of **oxygen free radicals,** which are highly reactive compounds that induce mutations in DNA as well as cause damage to several other cellular constituents. As noted at the beginning of this chapter, oxygen free radicals are formed during normal cellular metabolism, as well as being generated within cells by chemical carcinogens and radiation. They may account for a substantial amount of natural and carcinogen-induced damage to DNA, so preventing this damage might significantly reduce rates of cancer development.

Figure 12.2 Actions of chemopreventive agents. Blocking agents interfere with the action of carcinogens, whereas suppressing agents inhibit the outgrowth of transformed cells.

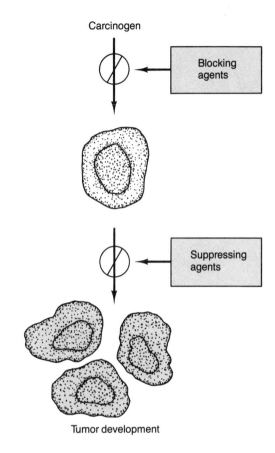

Suppressing Agents

The other class of chemopreventive agents, the suppressing agents, act to inhibit cell proliferation. As discussed in chapters 3 and 10, carcinogenesis requires excess proliferation of carcinogen-damaged (initiated) cells during the promotion phase of tumor development. Agents that suppress cell proliferation might, therefore, interfere with early stages of the outgrowth of cells that have already sustained carcinogen-induced mutations. The chemopreventive agents that act in this way include vitamin A and related compounds (retinoids), **tamoxifen** (an estrogen antagonist), and calcium. As discussed in chapter 11, retinoids induce the differentiation of a variety of cells, concomitant with a decrease in their proliferation. Tamoxifen blocks the stimulation of breast and endometrial cell proliferation normally induced by estrogen, thereby inhibiting proliferation of these cell types. Calcium likewise acts to induce differentiation, and inhibit proliferation, of colon cells. In each of these cases, such suppression of cell proliferation apparently serves to prevent, or at least slow down, cancer development.

Chemopreventive Activity of Retinoids in Human Trials

To date, the activities of these chemopreventive agents have been demonstrated in experimental animals. However, studies are currently underway to determine the possible efficacy of several promising chemopreventive agents, including β-carotene, vitamin A, vitamin C, vitamin E, and calcium, in humans. Moreover, encouraging results have recently been obtained to support a significant activity of retinoids in prevention of cancers of the oral cavity, pharynx, and larynx. The primary risk factor for these cancers, like lung cancer, is the use of tobacco. Because of their generally heavy exposure to tobacco carcinogens, patients who have been successfully treated for one such cancer are at high risk of developing a second. This risk was found to be substantially reduced (about fivefold) by administration of retinoids, although the high doses used in these studies also produced significant toxic side effects. A great deal of further work in this area is needed, but it is possible that in the future, chemopreventive agents might be successfully employed to lower cancer risk.

THE QUESTIONABLE IMPORTANCE OF SYNTHETIC CHEMICALS IN THE ENVIRONMENT

Only Low Levels of Carcinogens Are Present as Environmental Pollutants

Another strategy in cancer prevention has been the adoption of regulatory measures to eliminate carcinogens from the environment. The focus of this approach has been the identification and subsequent elimination of man-made synthetic chemicals that cause cancer. At first sight, this makes good sense. Individuals can do little to protect themselves against industrial pollution, so the possibility that carcinogens are being released into the environment clearly seems an appropriate area for government intervention and regulation. Indeed, such regulation has been of major importance in controlling the exposure of many workers to dangerous levels of occupational carcinogens. Environmental pollutants, however, are present in minute quantities compared to occupational carcinogens in the workplace, so the hazards resulting from environmental pollution are less clear. The question of whether pollution with synthetic chemicals contributes significantly to cancer incidence is, in fact, highly controversial, as is the efficacy of attempting to identify and eliminate such low-level environmental carcinogens.

As discussed in chapter 3, many known occupational carcinogens are also present as pollutants in the general environment. However, the levels of these carcinogens in the environment are extremely low, so it is unlikely that these pollutants make any significant contribution to overall cancer incidence. The level of asbestos generally present in city air, for example, is more than 1,000-fold less than the level of asbestos currently considered safe for occupational exposure. Even cases of chemical contamination of toxic waste sites, such as in the Love Canal area, have not been associated with any apparent increase in cancer incidence.

Nonetheless, it can be argued that any exposure to a carcinogen might cause cancer and therefore should be prevented. The problem thus becomes one of risk assessment. How can carcinogenic pollutants be identified, and how can one determine the degree of hazard associated with such potential carcinogens?

Testing Potential Carcinogens in Rodents Is Controversial

The potential carcinogenicity of suspected chemicals is generally tested in rodents. In order to minimize the cost and the number of animals required for such tests, common practice is to administer a large amount of the suspected carcinogen. In fact, carcinogens are usually tested at doses near the highest doses that are tolerated without severe toxicity, the maximum tolerated dose. The significance of identifying carcinogens by this procedure has been seriously challenged by some scientists. In the first place, these doses are sometimes hundreds of thousands of times higher than those to which humans are exposed. Moreover, more than half of either the synthetic or natural chemicals investigated have been found to cause cancer in rodents when administered at the maximum tolerated dose. This is a surprisingly high fraction of chemicals to be carcinogens, and critics of the tests consider this to be a misleading result of the test procedure. The concern is that since tested agents are administered at very high, near-toxic doses, cancers may result from increased cell proliferation required to repair chronic tissue damage caused by the test substance. Since such excess cell proliferation can clearly contribute to tumor development (for example, in liver cancer associated with cirrhosis), it is possible that this accounts for much of the carcinogenicity observed in these assays. If so, the induction of cancer by these chemicals may be an artifact of high-dose toxicity, which would then be meaningless in terms of the lower doses to which humans are exposed. On the other hand, supporters of these assays argue that the mechanism of carcinogenicity of the tested agents is not known. Therefore, it is unwarranted to assume that positive results in rodent tests are misleading.

Evaluating the Risk Associated with Exposure to Low Doses of Carcinogens

Even granted that high-dose carcinogenicity in rodents can be extrapolated to humans, is it likely that the much lower doses to which humans are exposed contribute significantly to cancer? Does it make sense to eliminate chemicals from the environment on the basis of high-dose rodent carcinogenicity? One way to address these questions is to estimate what fraction of carcinogen exposure comes from the synthetic chemicals that are the targets of screening and elimination programs. An illustrative example is concern over contamination of vegetables with synthetic pesticides, including DDT, and other industrial pollutants, such as polychlorinated biphenyls (PCBs). A basis for comparison is the fact that all plants contain a number of natural pesticides, including glucosino-lates, indoles, isothiocyanates, cyanide compounds, terpenes, and phenols, to protect themselves from insects. High doses of about half of these natural pesti-cides, like the synthetic pesticides, are carcinogenic in rodents. Importantly, the amount of natural plant pesticides consumed by Americans has been estimated to be more than 1,000 times higher than the amount of synthetic pesticides consumed as a result of residual contamination. Thus, the synthetic pesticides appear to make an insignificant contribution to overall carcinogen intake. More-over, in spite of their relatively high content of natural pesticides, dietary studies clearly indicate that consumption of vegetables lowers, rather than increases, cancer incidence. Indeed, as discussed in the preceding section, some of the natural pesticides, such as the isothiocyanates, actually serve to block the activity of other carcinogens by stimulating the activity of detoxifying enzymes. Thus, neither the relatively high levels of natural carcinogens in vegetables nor the much lower levels of contaminating synthetic carcinogens pose a risk from the standpoint of human cancer.

The Need for Balanced Risk Assessment

Finally, one must consider the relative risk/benefit associated with a suspected synthetic carcinogen. In many cases, the minimal risk associated with synthetic chemicals is dwarfed by the benefits they provide. This consideration applies to all levels at which our society benefits from synthetic chemicals, but the point can be made by just focusing on risk/benefit in terms of cancer. As discussed above, the cancer risk associated with the contamination of fruits and vegetables by synthetic pesticides is minimal. However, if pesticides were not used, fruits and vegetables would be more difficult to produce and consequently more expen-sive and less widely available. Since consumption of fresh fruits and vegetables clearly lowers cancer risk, it seems likely that a ban on pesticides would be highly counterproductive from the standpoint of cancer prevention.

Another example is provided by the artificial sweetener saccharin. As dis-cussed in chapter 3, high doses of saccharin, 100 to 1,000 times higher than the amounts consumed by humans, cause bladder cancer in rats. However, epidemi-ological studies have found no evidence that human consumption of saccharin is linked to increased bladder cancer risk. On the basis of the animal data, saccha-rin was temporarily banned, although this ban has since been lifted. One pre-sumed consequence of banning saccharin would be the use of sugar in its place, which in some cases would be expected to contribute to obesity—an established risk factor for endometrial carcinoma.

The identification and elimination of carcinogens from the environment is clearly a potentially important undertaking. However, caution must be exer-cised, in order to focus efforts on those agents that make a significant contribu-tion to human cancer incidence. Attempts to eliminate potential carcinogens that are present at only minute levels in the environment are not likely to reduce cancer risk substantially. On the other hand, the cost of such efforts may divert resources from more profitable avenues of cancer prevention, such as elimination of smoking. In addition, as discussed above, attempts to remove all synthetic chemicals that are potential carcinogens may ultimately prove harmful rather than beneficial to public health. A balanced risk assessment policy that considers the levels of human hazard associated with low-level exposure to potential carcinogens is clearly needed, but how to arrive at such a policy remains a subject of disagreement among experts.

SUMMARY

An individual can take several steps to reduce cancer risk. Foremost among these is avoiding the use of tobacco, which would eliminate about 30% of all cancer deaths in the United States. In addition, it is possible to minimize exposure to other known carcinogens, such as alcoholic beverages, some sources of radiation, occupational and medicinal carcinogens, and oncogenic viruses. A number of dietary factors may also influence cancer risk, and although the effect of these factors is not known, it is prudent to follow generally recommended dietary practices. In total, it is estimated that these steps might prevent up to 40–50% of cancers in the United States. However, risk factors (and therefore preventive measures) for many cancers remain unknown. The possibility of chemoprevention may hold promise for future developments. On the other hand, the identification and elimination of potential carcinogens present at low levels in the environment does not seem likely to reduce cancer incidence significantly.

KEY TERMS

smoking
alcohol
ultraviolet radiation
ionizing radiation
radon
radiation pollution
obesity
high-fat diets
β-carotene
vitamin A
dietary fiber
cruciferous vegetables
nitrites
N-nitroso compounds
industrial carcinogens
medicinal carcinogens

immune suppression

postmenopausal estrogen therapy

vaccination

chemoprevention

blocking agent

suppressing agent

oxygen free radical

retinoids

tamoxifen

environmental pollution

carcinogen testing

maximum tolerated dose

chronic tissue damage

synthetic pesticides

natural pesticides

saccharin

REFERENCES AND FURTHER READING

General References

American Cancer Society. 1990. *Cancer facts and figures—1990.* American Cancer Society, Atlanta.

Loeb, L.A. 1989. Endogenous carcinogenesis: molecular oncology into the twenty-first century. *Cancer Res.* 49:5489–5496.

National Cancer Institute. 1984. *Cancer prevention.* National Institutes of Health, Bethesda.

Smoking

U.S. Department of Health and Human Services. 1990. *The health benefits of smoking cessation.* A report of the Surgeon General.

Radiation

Cook-Mozaffari, P., Darby, S., and Doll, R. 1989. Cancer near potential sites of nuclear installations. *Lancet* 2:1145–1147.

Evans, H.J. 1990. Leukaemia and nuclear facilities: data in search of explanation. *Nature* 347:712–713.

Forman, D., Cook-Mozaffari, P., Darby, S., Davey, G., Stratton, I., Doll, R., and Pike, M. 1987. Cancer near nuclear installations. *Nature* 329:499–505.

Guimond, R.J. 1991. Radon risk and EPA. *Science* 251:724–725.

Nero, A.V., Jr. 1988. Controlling indoor air pollution. *Sci. Amer.* 258:42–48.

Pool, R. 1991. EMF–cancer link still murky. *Nature* 349:554.

Dietary Factors

National Cancer Institute. 1987. *Diet, nutrition, and cancer prevention: a guide to food choices.* National Institutes of Health, Bethesda.

National Research Council. 1989. *Diet and health: implications for reducing chronic disease risk.* National Academic Press, Washington.

Willett, W.C., Stampfer, M.J., Colditz, G.A., Rosner, B.A., and Speizer, F.E. 1990. Relation of meat, fat, and fiber intake to the risk of colon cancer in a prospective study among women. *N. Engl. J. Med.* 323:1664–1672.

Avoidance of Occupational and Medicinal Carcinogens

Henderson, B.E., Ross, R., and Bernstein, L. 1988. Estrogens as a cause of human cancer: The Richard and Hinda Rosenthal Foundation Award Lecture. *Cancer Res.* 48:246–253.

IARC Working Group. 1980. An evaluation of chemicals and industrial processes associated with cancer in humans based on human and animal data: IARC monographs volumes 1 to 20. *Cancer Res.* 40:1–12.

Oncogenic Viruses

The Gambia Hepatitis Study Group. 1987. The Gambia hepatitis intervention study. *Cancer Res.* 47:5782–5787.

Whittle, H.C., Inskip, H., Hall, A.J., Mendy, M., Downes, R., and Hoare, S. 1991. Vaccination against hepatitis B and protection against chronic viral carriage in The Gambia. *Lancet* 337:747–750.

Zuckerman, A.J. 1989. Prevention of primary liver cancer by immunization. *Cancer Detection and Prevention* 14:309–315.

The Possibility of Chemoprevention

Bertram, J.S., Kolonel, L.N., and Meyskens, F.L., Jr. 1987. Rationale and strategies for chemoprevention of cancer in humans. *Cancer Res.* 47:3012–3031.

Boone, C.W., Kelloff, G.J., and Malone, W.E. 1990. Identification of candidate cancer chemopreventive agents and their evaluation in animal models and human clinical trials: a review. *Cancer Res.* 50:2–9.

Hong, W.K., Lippman, S.M., Itri, L.M., Karp, D.D., Lee, J.S., Byers, R.M., Schantz, S.P., Kramer, A.M., Lotan, R., Peters, L.J., Dimery, I.W., Brown, B.W., and Goepfert, H. 1990. Prevention of second primary tumors with isotretinoin in squamous-cell carcinoma of the head and neck. *N. Engl. J. Med.* 323:795–801.

Knabbe, C., Lippman, M.E., Wakefield, L.M., Flanders, K.C., Kasid, A., Derynck, R., and Dickson, R.B. 1987. Evidence that transforming growth factor-β is a hormonally regulated negative growth factor in human breast cancer cells. *Cell* 48:417–428.

Lipkin, M., and Newmark, H. 1985. Effect of added dietary calcium on colonic epithelial-cell proliferation in subjects at high risk for familial colonic cancer. *N. Engl. J. Med.* 313:1381–1384.

Meyskens, F.L., Jr. 1990. Coming of age—the chemoprevention of cancer. *N. Engl. J. Med.* 323:825–827.

Richter, C., Park, J.-W., and Ames, B.N. 1988. Normal oxidative damage to mitochondrial and nuclear DNA is extensive. *Proc. Natl. Acad. Sci. USA* 85:6465–6467.

Sporn, M.B., and Roberts, A.B. 1991. Interactions of retinoids and transforming growth factor-β in regulation of cell differentiation and proliferation. *Mol. Endo.* 5:3–7.

Wattenberg, L.W. 1985. Chemoprevention of cancer. *Cancer Res.* 45:1–8.

The Questionable Importance of Synthetic Chemicals in the Environment

Ames, B.N., and Gold, L.S. 1990. Chemical carcinogenesis: too many rodent carcinogens. *Proc. Natl. Acad. Sci. USA* 87:7772–7776.

Ames, B.N., and Gold, L.S. 1990. Too many rodent carcinogens: mitogenesis increases mutagenesis. *Science* 249:970–971.

Ames, B.N., Profet, M., and Gold, L.S. 1990. Dietary pesticides (99.99% all natural). *Proc. Natl. Acad. Sci. USA* 87:7777–7781.

Ames, B.N., Profet, M., and Gold, L.S. 1990. Nature's chemicals and synthetic chemicals: comparative toxicology. *Proc. Natl. Acad. Sci. USA* 87:7782–7786.

Heath, C.W., Jr., Nadel, M.R., Zack, M.M., Jr., Chen, A.T.L., Bender, M.A., and Preston, J. 1984. Cytogenetic findings in persons living near the Love Canal. *J. Amer. Med. Assoc.* 251:1437–1440.

Higginson, J. 1988. Changing concepts in cancer prevention: limitations and implications for future research in environmental carcinogenesis. *Cancer Res.* 48:1381–1389.

Janerich, D.T., Burnett, W.S., Feck, G., Hoff, M., Nasca, P., Polednak, A.P., Greenwald, P., and Vianna, N. 1981. Cancer incidence in the Love Canal area. *Science* 212:1404–1407.

Koshland, D.E., Jr. 1989. Scare of the week. *Science* 244:9.

Marx, J. 1990. Animal carcinogen testing challenged. *Science* 250:743–745.

Perera, F.P. 1990. Carcinogens and human health. *Science* 250:1644–1645.

Weinstein, I.B. 1991. Mitogenesis is only one factor in carcinogenesis. *Science* 251:387–388.

Chapter 13

Early Detection and Diagnosis

THE NEXT BEST THING to prevention of cancer is early detection. As discussed in chapters 2 and 10, cancers do not arise as fully developed malignancies. Instead, they develop gradually, apparently as the result of the accumulation of mutations in multiple oncogenes and tumor suppressor genes. As these abnormalities accumulate, tumor cells progressively acquire the characteristics of malignancy: increased proliferative capacity, invasiveness, and metastatic potential.

The importance of early detection arises from the progressive nature of tumor development. Prior to metastasis, most cancers can be cured by localized treatments, such as surgery or radiotherapy. Premalignant tumors (e.g., colon adenomas) and cancers that have not yet invaded surrounding normal tissue (carcinomas *in situ*) are usually completely curable, frequently by relatively minor procedures. More extensive surgery, perhaps in combination with other therapeutic modalities such as radiation, is required for invasive cancers, but such localized treatments are still effective as long as extensive spread of the cancer has not occurred. Once a cancer has metastasized to distant body sites, however, localized treatments are no longer sufficient, and a cure becomes much less likely.

The early detection of cancer is thus critical to the outcome of the disease. If the earliest stages of cancer (e.g., carcinoma *in situ*) could be reliably detected, lethality would be prevented, often by comparatively minor treatment. Steps taken to detect such early stages of tumor development are referred to as **secondary prevention**. For some kinds of cancer, routine **screening** of healthy individuals, before any symptoms are evident, is an effective way to detect the earliest stages of tumor development and reduce mortality from the disease. Other kinds of cancer, however, cannot usually be diagnosed until later stages, when symptoms of the disease have appeared.

THE PAP SMEAR AND CERVICAL CARCINOMA

The **Pap smear,** developed in the 1930s, is the most outstanding illustration of an effective screening program. It is estimated that regular screening by Pap smear would prevent over 90% of deaths from cervical carcinoma, and such screening is responsible for the fact that mortality from this disease has decreased 75% since 1940 (see chapter 1). As an early screening test, the Pap smear possesses several desirable features, which account for its effectiveness.

Advantages of the Pap Test as a Screening Program

First, the Pap smear is a safe, reliable, and inexpensive procedure that involves minimal discomfort. A physician simply scrapes a sample of cells from the cervix with a cotton swab or wooden spatula, as part of a routine pelvic examination. The procedure is painless and free of risk. The sample is then smeared onto a microscope slide, fixed, stained, and examined for the presence of abnormal cells (Fig. 13.1). Carcinoma *in situ*, as well as even earlier **preneoplastic** stages of disease (**dysplasias**), can be reliably detected by this analysis, which costs less than $20.

Not only can early stages of cervical carcinoma be detected by the Pap smear, but the course of the disease makes such early detection highly beneficial to the patient. Cervical dysplasia and carcinoma *in situ* usually persist for several years before progressing to invasive carcinoma. Thus, regular screening by the Pap smear is quite likely to detect early stages of the disease before it becomes life-threatening. Moreover, dysplasias and carcinoma *in situ* can be easily treated by several methods, including minor local surgery, which are virtually 100% effective.

Impact of the Pap Test on Cervical Cancer Mortality

There is little question that the Pap smear provides highly effective protection against cervical carcinoma. In the United States in 1990, about 50,000 cases of cervical carcinoma were detected as *in situ* tumors, compared to 13,500 cases that were not detected until they had reached the invasive stage. Moreover, it is thought that regular Pap smears could have prevented the vast majority of the remaining invasive carcinoma cases, which still account for about 6,000 American deaths annually (approximately 2.5% of cancer deaths in women). The American Cancer Society recommends annual Pap smears starting at 18 years of age.

EARLY DETECTION OF BREAST CANCER

Breast cancer is the most frequent cancer in women, occurring with an incidence of 150,000 cases a year in the United States. It accounts for about 30% of all female cancers, striking nearly one in every ten women. The mortality from breast cancer is about 44,000 deaths annually, making it second only to lung cancer (50,000 deaths annually) as a leading cause of cancer death in American women.

Importance of Early Detection

The prognosis for breast cancer strongly depends on early detection of the disease (Fig. 13.2). Five-year survival rates are nearly 100% for carcinoma *in situ*, 90% for locally invasive carcinoma, 68% for carcinoma that has spread regionally, and only 18% for those cancers that have metastasized to distant body sites. Thus, it is apparent that screening procedures to detect early stages of breast cancer could have major public health benefits.

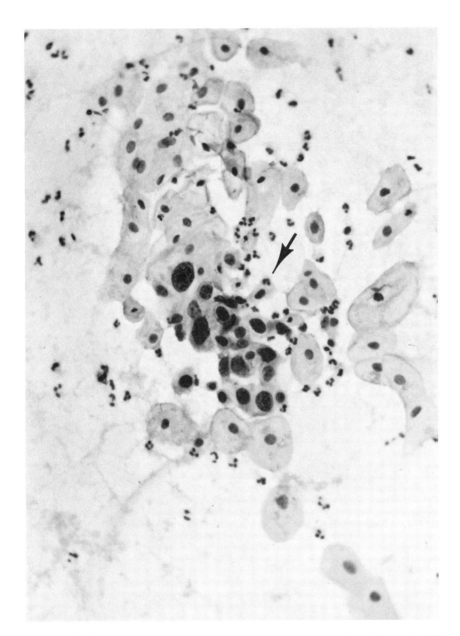

Figure 13.1   Detection of cervical carcinoma by the Pap smear. A cluster of abnormal cells arising from a cervical carcinoma *in situ* is present in the center of this photomicrograph (arrow). The carcinoma cells can be readily distinguished from adjacent normal cells by their large nuclei. (From L.V. Crowley, *Introduction to Human Disease,* 1988.)

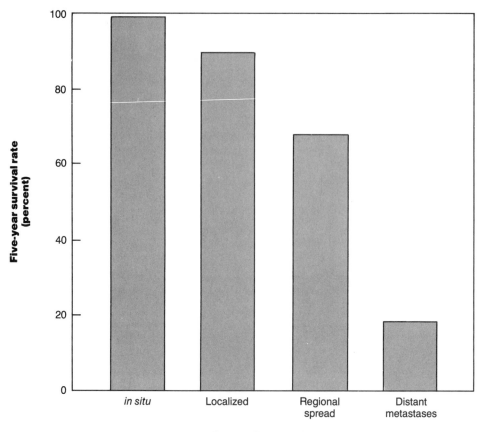

Figure 13.2 Survival rates following detection of different stages of breast cancer. Five-year survival rates for patients diagnosed with carcinoma *in situ*, with invasive carcinoma that is still localized to its site of origin, with carcinoma that has spread to regional lymph nodes, and with carcinoma that has already metastasized to distant body sites. (From American Cancer Society, *Cancer Facts and Figures,* 1990.)

Three approaches to early detection of breast cancer are recommended by the American Cancer Society: (1) monthly breast self-examination, (2) an annual physical examination for women above age 40, and (3) mammography, as discussed below. Changes and lumps in breast tissue can be detected by palpation, either during self-examination or by a physician. Any such abnormalities detected in self-examination should then be investigated by a physician to determine whether or not they represent development of a tumor. Even smaller breast lesions, which cannot yet be felt, can be detected by **mammography,** a low-dose X-ray of the breast. Thus, mammography has the potential of detecting breast cancer at the earliest stages of its development. As a cancer screening test, however, mammography has not achieved the wide success of the Pap smear; in fact, it still remains a subject of some controversy.

Mammography Reduces Breast Cancer Mortality by About 30%

Unfortunately, the clinical efficacy of mammography is not as great as that of the Pap smear, although there is convincing data to indicate that regular mammography can lower breast cancer mortality. Several studies have compared breast cancer mortality rates among women who receive regular mammography and those who do not. These studies all indicate that breast cancer death rates are reduced by about 25–30% as a result of screening with mammography and physical examination (Fig. 13.3). This is a significantly lower benefit than that derived from the Pap smear, which is thought to reduce cervical carcinoma death rates by over 90%. Nonetheless, given the high frequency of breast cancer, a 25% reduction in mortality would correspond to saving the lives of over 10,000 American women each year. From the standpoint of an individual woman in the United States, the lifetime risk of dying from breast cancer is approximately 4%. Regular mammography might reduce this risk to about 3%.

Drawbacks to Mammography

Thus, although mammography is not likely to abolish breast cancer mortality, it does appear to confer significant benefits. These benefits must be weighed

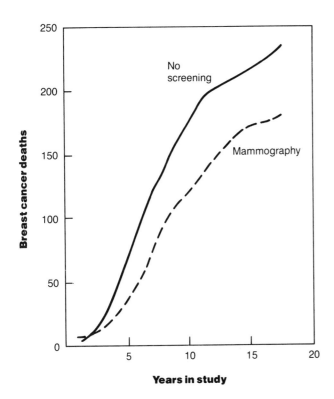

Figure 13.3   Effect of mammography on breast cancer mortality. The number of breast cancer deaths among women who participated in a study designed to evaluate the efficacy of screening by mammography. One group of women received regular mammography (designated *mammography*), while a comparable group of women had no regular screening tests (*no screening*). (From S. Shapiro, *World J. Surg.* 13:9–18, 1989.)

against potential risk, expense, and discomfort, all of which appear to be more significant drawbacks to mammography than to the Pap smear.

One concern that has been raised about mammography is the potential carcinogenic effect of regular X-irradiation of breast tissue. However, the X-ray doses used for mammography are now quite low, and the risk of developing mammography-induced breast cancer is negligible. For example, it has been estimated that mammography might induce (at most) 1–5 breast cancers among a group of 10,000 women screened yearly starting at age 40. However, nearly 1,000 of these women would be expected to develop breast cancer in the natural course of events. The increased risk of breast cancer resulting from mammography is less than 1%, which is clearly outweighed by a 25–30% reduction in breast cancer mortality as a result of early detection.

The other negative features of mammography are cost, discomfort, and the comparatively high frequency of false-positive test results. There is some discomfort from breast compression during mammography, but surveys indicate that most women find such discomfort only minor. The cost of mammography is relatively high, generally around $100 to $150. The frequency of false positives (suspected lesions that are biopsied but ultimately do not turn out to be cancer) is about 80%. That is, only about 1 out of 5 biopsies performed following mammography reveal cancer. The problem arising from these false positives, of course, is the need for unnecessary biopsies, with concomitant expense and anxiety for the patients.

Mammography Is Underutilized

Nonetheless, it is generally felt that the demonstrated benefits of screening for breast cancer by mammography far outweigh its disadvantages. Consequently, the use of mammography as a regular program of breast cancer early detection, together with physical examination, is recommended by a number of organizations, including the American Cancer Society, the National Cancer Institute, and the American Medical Association. These consensus guidelines for mammography screening are for women to have a baseline mammogram between ages 35 and 40, mammograms every 1–2 years from ages 40 to 50, and yearly mammograms at age 50 and above. Unfortunately, many physicians fail to recommend mammograms, which appears to contribute significantly to their underutilization.

SCREENING FOR COLORECTAL CANCER

Potential Benefits of Screening

Colorectal cancer is the third type of cancer for which early detection by screening is commonly recommended. Like breast cancer, colorectal cancer is extremely common, accounting for 155,000 cases and 61,000 deaths in the United States in 1990. In addition, as discussed extensively in chapters 2 and 10, colorectal cancers clearly develop in a gradual, progressive manner, and many of the steps in colorectal carcinogenesis have been identified. Furthermore, the benefits of treatment are much greater for patients with less advanced disease. For example, survival rates for patients with localized colon and rectal carcinomas are close to 90% and 80%, respectively. Survival rates drop to about 50%, however, for cancers that have spread to adjacent organs and lymph nodes, and survival rates for patients with distant metastases are only about 6%. Detection and treatment of precancerous lesions (adenomas) and early carcinomas could thus be expected to yield significant benefits.

Three approaches to the early detection of colorectal cancer are recommended by the American Cancer Society and other organizations: (1) digital rectal examination, (2) sigmoidoscopy, and (3) fecal occult blood testing. Each of these tests has different strengths and weaknesses. Their recommendation is further complicated by the fact that, in contrast to the situation with cervical and breast cancers, there is as yet no clear data to demonstrate that screening for colorectal cancers actually reduces mortality from the disease. Such studies to determine the efficacy of screening are ongoing, however, and it is apparent that early detection can yield important benefits in the effectiveness of treatment. Screening for colorectal cancer is therefore currently recommended even in the absence of clear evidence for its presumed effectiveness in reducing mortality.

Digital Rectal Examination

Digital rectal examination is a simple test, which is easily performed as part of a routine physical exam. However, it is relatively insensitive as a screening test, because only about 10% of colorectal tumors develop within the limited portion of the rectum that can be reached by an examining finger. Therefore, although it should certainly be included in regular physical examinations, the effectiveness of the digital rectal exam for early detection of colorectal cancer is limited.

Sigmoidoscopy

A significantly larger fraction of colorectal tumors can be detected by **sigmoidoscopy,** which is examination of the rectum and lower part of the colon with a thin lighted tube. Using the most sensitive of these instruments, the flexible sigmoidoscope, it is possible to examine far enough up in the colon to detect about 50% of colorectal tumors. This is a significant improvement over the digital exam, but the drawback is a considerable degree of discomfort for the patient.

Fecal Occult Blood Testing

Screening for colorectal tumors by **fecal occult blood testing** has the potential advantage of detecting tumors in any part of the colon. The principle behind this test is that developing tumors cause minor bleeding, resulting in the presence of small amounts of blood in the stool. To test for such small amounts of blood (occult blood), a sample of stool is smeared on a slide impregnated with a chemical, usually guaiac, that changes color in the presence of hemoglobin. The stool sample can be obtained either at home or during a digital rectal exam, and the test is simple and easy to perform. Unfortunately, however, the occult blood test suffers from a high frequency of both false-negative and false-positive results. Many colorectal tumors do not release large enough amounts of blood to be detected. Rectal bleeding is often intermittent, so it is common practice to test multiple stool samples obtained over several days. Even when this is done, however, occult blood tests on 20 to 30% of patients with colon cancer are negative. The sensitivity of detection of premalignant adenomas and early cancers is still lower, probably about 50%. Thus, a high proportion of early colon cancers are not detected by this method. Conversely, occult blood tests are frequently positive in the absence of colon tumors. Such false positives can be the result of bleeding from ulcers, fissures, or hemorrhoids. An additional source of false positives is consumption of foods, such as red meat, that contain hemoglobin or other substances that react in the test. In any case, such incorrect results currently account for over 80% of positive occult blood tests. Nonetheless, any positive results must be followed up by further evaluations, which may include examination of the entire colon by **colonoscopy** or X-irradiation after a barium enema. Since these further tests entail a considerable degree of discomfort and expense, the high rate of false positives is a significant problem in occult blood testing.

Despite these drawbacks, screening for colorectal cancer could have significant payoffs. As noted above, the actual reductions in mortality that result from current screening programs are not known, but such mortality reductions have been estimated to be about 30%. This would be similar to the effect of screening on breast cancer mortality. Since colorectal cancers are responsible for over 60,000 deaths per year in the United States, such a reduction in mortality from the disease would correspond to saving the lives of 20,000 Americans. In the absence of conclusive data, the American Cancer Society recommends (1) yearly digital rectal exams starting at age 40, (2) yearly fecal occult blood tests starting at age 50, and (3) sigmoidoscopy every three to five years starting at age 50.

EARLY DETECTION OF OTHER CANCERS

The Importance of Annual Physical Exams

Several other cancers can be detected at early stages of disease during routine physical examinations. Consequently, an annual physical exam is recommended starting at age 40. Such annual physicals should include exams for cancer of the lymph nodes, oral cavity, skin, prostate, testes, ovaries, and thyroid, as well as the tests for breast, cervical, and colorectal cancers discussed above. Prostate cancer can be detected by a digital rectal exam—another reason to recommend such an exam, in addition to the possible detection of some rectal carcinomas.

Additional procedures that are being evaluated for early detection of prostate cancer are ultrasound examination (described in the section of this chapter on Diagnosis and Staging) and blood tests for **prostate-specific antigen,** an enzyme secreted into the serum by prostate epithelial cells. For women, pelvic exams are important for detection of uterine and ovarian cancers. Unfortunately, most ovarian cancers have already reached an advanced stage before they are detected in pelvic exams, so the possible use of more sensitive early detection methods, such as ultrasound exams, is currently being explored. Early stages of oral cancers can frequently be detected visually during a medical or dental examination. Periodic self-examination is important for detection of skin cancers, including melanoma. Testicular cancers can also be detected by self-examination, as well as by a physician.

Some Cancers Cannot Be Detected at Early Stages

For many cancers, however, there are no available means to detect early stages of the disease, before symptoms are evident. Unfortunately, this group includes lung cancer, for which there is no effective early detection strategy. Periodic screening by chest X-ray is not recommended. Most lung cancers have already spread by the time they are large enough to be seen in X-rays, so their detection by this method does not confer any significant advantage to the patient.

Early detection of cancer by screening prior to the appearance of symptoms is therefore an important, but not universally effective, step to reduce cancer mortality. In total, it is estimated that early detection could have prevented up to 50,000 American cancer deaths in 1990—approximately a 10% decrease in total cancer mortality.

SYMPTOMS OF CANCER

Those cancers that are not detected by screening prior to the development of symptoms still need to be diagnosed as early as possible to maximize the likelihood of a cure and the benefits of treatment. The American Cancer Society has called attention to seven early warning signals of cancer, which spell out the word *CAUTION.* Such symptoms do not necessarily indicate the presence of a cancer, but they should be investigated by a physician. Unfortunately, these symptoms are often not detected until the cancer has progressed to a relatively advanced stage.

1. Change in bowel or bladder habits.
2. A sore that does not heal.
3. Unusual bleeding or discharge.
4. Thickening or lump in breast or elsewhere.
5. Indigestion or difficulty in swallowing.
6. Obvious change in wart or mole.
7. Nagging cough or hoarseness.

DIAGNOSIS AND STAGING

Analysis of Blood and Urine

When cancer is suspected, either because of positive results in a screening test or possible symptoms, further tests are undertaken to diagnose the disease definitively. The first step is a complete physical examination. In addition to palpation for abnormal tissue masses, the physical exam will include laboratory analysis of blood and urine. Microscopic examination of the cells present in blood can indicate the presence of leukemia (Fig. 13.4), and other laboratory tests can provide clues to the presence of other cancers. For example, blood in the urine may be an indication of bladder cancer, just as blood in the stool may suggest a colon tumor. Prostate cancers frequently produce enzymes, prostatic acid phosphatase, and prostate-specific antigen (discussed above), which can be detected in blood. The presence of abnormal immunoglobulins in blood or urine may signal **multiple myeloma,** a cancer of immunoglobulin-producing lymphocytes. High levels of certain hormones in blood may be indicative of cancers of hormone-producing cells, such as testicular cancers.

Tumor Markers

Other substances in blood may also serve as useful markers for some cancers, although they are not sufficiently specific or sensitive to form the basis of a definitive diagnosis. An example of such a **tumor marker** is **carcinoembryonic antigen (CEA),** which is a protein expressed on the surface of some cancer cells and some embryonic cell types. CEA is frequently secreted by colon and rectum carcinomas, but it is also produced by other cancers, including cancers of the breast, lung, and pancreas. Since CEA is usually only detectable in patients with advanced tumors, it is not a useful early diagnostic test. However, CEA is frequently used to follow the course of patients after treatment. For example, the detection of high levels of CEA following surgical removal of a colon cancer might indicate recurrence of the tumor or development of a metastatic lesion.

X-Rays and Other Imaging Techniques

The next step in diagnosis, following a physical exam and blood tests, frequently involves delineation of any suspected tumor by X-rays or other imaging techniques. **Imaging** refers to a variety of noninvasive methods that can be used to view the inside of the body. In addition to conventional X-rays, a number of sophisticated imaging techniques are used for cancer diagnosis. **Computed tomography** (**CT** or **CAT scan**) employs computer analysis of scanning X-ray images to generate cross-sectional pictures that portray a tumor's size and location much more accurately than conventional X-rays (Fig. 13.5). This method is particularly useful for identifying and precisely locating abnormalities of internal organs, such as tumors within the abdomen. **Nuclear magnetic resonance** or **magnetic resonance imaging** (**MRI**) is another technologically advanced imaging

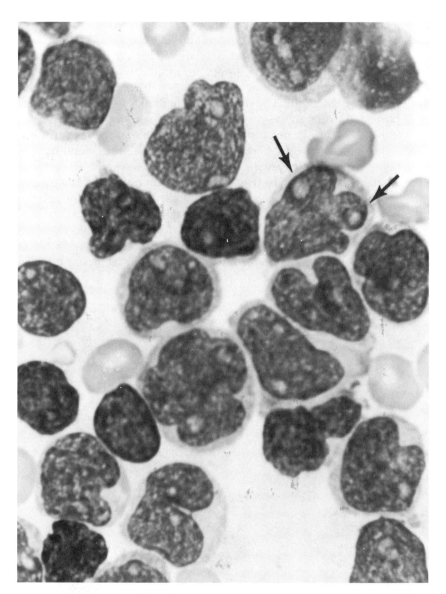

Figure 13.4 Blood smear of a patient with leukemia. Multiple abnormal white cells (example indicated by arrows) are present. (From L.V. Crowley, *Introduction to Human Disease*, 1988.)

technique; coupled with computer analysis, it can provide extremely sensitive and precise information. It has the advantage of not using X-rays, so it is not associated with exposure to carcinogenic radiation. MRI is particularly useful for analysis of tumors in tissues surrounded by bone, such as tumors of the brain or spinal cord. **Angiography** is X-ray examination of the blood vessels, which may reveal distortions or the formation of new blood vessels indicative of the presence of a tumor. **Radioisotope scanning** methods permit detection of some lesions, including tumors in liver, bone, brain, and the thyroid gland, that are not revealed by X-rays. In this method, radioactive isotopes that localize in the target tissues are administered to the patient. Scanning for radioactivity can then detect abnormal patterns of isotope accumulation, allowing delineation of a tumor mass. **Ultrasound** is a technique in which high frequency sound waves are directed at a part of the body. The "echoes" of these sound waves are then detected, revealing the size, shape, and location of tissue masses. Ultrasound, like MRI, does not involve exposure to radiation. In addition, ultrasound has the advantage of being much less expensive than either CT scan or MRI, both of which are costly procedures. It can be used for delineation of tumors in a variety of sites, such as the stomach, pancreas, kidney, uterus, and ovary, although the image produced by ultrasound is not as clear as that obtained from a CT scan. Often, several different imaging methods will be used in combination, to evaluate both a primary tumor and the possible extent of metastasis. These imaging methods are important, not only in detecting an abnormal tissue mass, but also in precisely locating a possibly malignant lesion for biopsy.

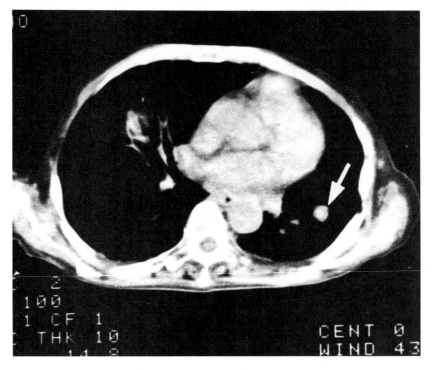

Figure 13.5   Detection of a lung tumor in a CT scan. The heart and mediastinum appear white in the center of the scan, whereas the lungs on both sides are less dense. The lung tumor, indicated by the arrow, is denser than normal lung tissue and appears as a white nodule. (From L.V. Crowley, *Introduction to Human Disease*, 1988.)

Endoscopy

In addition to imaging, many tumors can be viewed directly by **endoscopy**, the use of a flexible lighted tube to examine internal body cavities. **Sigmoidoscopy,** examination of the lower part of the colon, was discussed above as a colon cancer screening procedure. Examination of the entire colon, **colonoscopy,** is not practical for screening asymptomatic individuals, but colonoscopy is performed diagnostically, for example, to determine the cause of positive results in a fecal occult blood test. Other internal organs that can be inspected directly by endoscopy include the esophagus, stomach, bladder, larynx, pharynx, bronchi, uterus, and ovaries. These procedures are of considerable value in the diagnosis of cancer, since they allow examination of internal organs without major surgery.

Biopsy and Clinical Staging

A definitive diagnosis ultimately requires a biopsy, so that cells in the suspected lesion can be examined directly by a pathologist. For leukemias, diagnosis involves examination of both blood samples and bone marrow biopsies. For solid tumors, a tissue sample is obtained by one of a number of techniques, depending on the type and location of the suspected lesion. Biopsy procedures range from removing a small sample of tissue with a needle (aspiration and needle biopsies) to excising the entire abnormal tissue mass. Many potential tumors (e.g., breast tumors) can be biopsied externally, whereas biopsies of other tumors (e.g., colon tumors) can be obtained by endoscopy. Needle biopsies coupled with sophisticated imaging techniques are of considerable importance in cancer diagnosis, since these procedures allow tissue samples to be obtained from many suspected lesions without the need for major surgery. Tissue masses deep within the chest or abdomen, for example, can be successfully biopsied in this way.

Analysis of biopsy specimens will determine if a tumor is present and, if so, whether the tumor is benign or malignant (Fig. 13.6). If a malignant tumor (cancer) is diagnosed, it is important to determine how far the malignancy has progressed. Particularly critical information in this regard are (1) the extent to which the tumor has invaded surrounding normal tissue, (2) whether the malignancy has spread to lymph nodes in the region of the tumor, and (3) whether the tumor has metastasized to distant body sites. As discussed in chapter 2, these factors determine the **clinical stage** of the tumor, which is of major importance in choosing an appropriate treatment strategy.

Tumor Grading

In addition to clinical staging, a number of other histologic and laboratory analyses may provide information that is useful in predicting the course of disease and response to therapy. **Tumor grading** is based on cell morphology and on the fraction of tumor cells undergoing mitosis. In general, more malignant cancers are characterized by abnormal cell morphologies and large numbers of dividing cells.

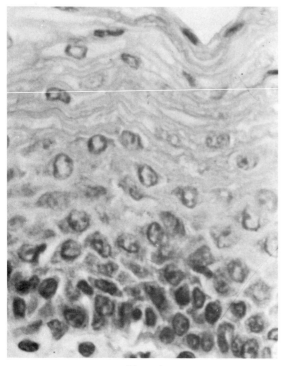
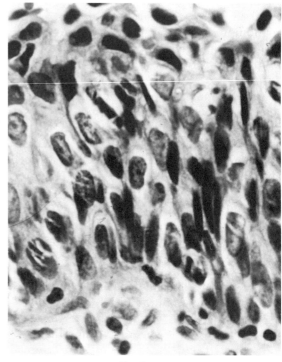

Normal   Carcinoma

Figure 13.6 Biopsy of a cervical carcinoma. A photomicrograph of normal cervical epithelium (left panel) compared to cervical carcinoma (right panel). The tumor *cells have not invaded underlying connective tissue and therefore represent carcinoma in situ. (From L.V. Crowley, Introduction to Human Disease, 1988.)*

The proliferative rate of tumor cells can also be analyzed by measuring the fraction of cells engaged in DNA synthesis (i.e., in S phase) by assaying the incorporation of precursor molecules, such as radioactive thymidine, into DNA.

Other Diagnostic Markers

It is also frequently useful to determine the DNA content and chromosomal composition of tumor cells. Abnormalities in amounts of cellular DNA and chromosome numbers usually indicate a less favorable prognosis, perhaps because they signal the accumulation of genetic abnormalities, such as oncogene amplification and loss of tumor suppressor genes, which occur during tumor progression (see chapter 10). In addition, specific chromosome translocations are diagnostic of some cancers. The prototype example is chronic myelogenous leukemia, in which the *abl* oncogene is translocated from chromosome 9 to chromosome 22, forming the Philadelphia chromosome (see chapter 8). The expression of particular proteins may also provide a useful prognostic indicator for some types of cancer. For breast cancers, for example, expression of estrogen and progesterone receptors generally indicates a more favorable prognosis.

The Use of Oncogenes in Diagnosis

Analysis of oncogene abnormalities is also beginning to form part of the diagnostic picture for some cancers. For example, amplification of the *erb*B-2 oncogene in breast cancers and of the N-*myc* oncogene in neuroblastoma are indicative of more rapidly progressing tumors (see chapter 10) and may dictate more aggressive treatments. Molecular analysis of oncogene abnormalities, such as the translocation of *abl* in chronic myelogenous leukemia, can also provide sensitive methods for monitoring the course of disease following treatment.

SUMMARY

Because of the progressive nature of the development of tumors, early detection and diagnosis is critical to the outcome of cancer. For some cancers, specific screening tests are recommended for healthy individuals, in order to detect the earliest possible stages of tumor development. The Pap smear, which is highly effective in detecting cervical carcinoma at a readily treatable stage, is an ideal example of such a screening method. Early screening tests are also recommended for detection of breast and colorectal cancers, although the screening tests for these diseases (including mammography and fecal occult blood testing) are less effective than the Pap smear. Other cancers can be detected by self-examination, during routine physical exams, or by recognition of early disease symptoms. A number of further procedures are then used to investigate a suspected lesion, ultimately including a biopsy to determine definitively whether cancer is present. If a diagnosis of cancer is made, clinical staging of the disease, together with other diagnostic assays, provides information critical to the choice of an appropriate treatment.

KEY TERMS

secondary prevention
screening
Pap smear
carcinoma *in situ*
preneoplastic
dysplasia
mammography
digital rectal exam

sigmoidoscopy

fecal occult blood test

colonoscopy

prostate-specific antigen

tumor marker

carcinoembryonic antigen (CEA)

imaging

computed tomography (CT scan)

nuclear magnetic resonance

magnetic resonance imaging (MRI)

angiography

radioisotope scanning

ultrasound

endoscopy

biopsy

clinical staging

tumor grading

DNA content

chromosome number

chromosome translocations

estrogen receptor

progesterone receptor

oncogene abnormalities

REFERENCES AND FURTHER READING

General References

American Cancer Society. 1990. *Cancer facts and figures—1990.* American Cancer Society, Atlanta.

Crowley, L.V. 1988. *Introduction to human disease.* 2nd ed. Jones and Bartlett, Boston.

DeVita, V.T., Jr., Hellman, S., and Rosenberg, S.A., eds. 1989. *Cancer: principles and practice of oncology.* 3rd ed. J.B. Lippincott, Philadelphia.

Miller, A.B., ed. 1985. *Screening for cancer.* Academic Press, Orlando.

The Pap Smear and Cervical Carcinoma

Eddy, D.M. 1990. Screening for cervical cancer. *Ann. Intern. Med.* 113:214–226.

Early Detection of Breast Cancer

Bassett, L.W., Manjikian, V., III, and Gold, R.H. 1990. Mammography and breast cancer screening. *Surg. Clin. N. Amer.* 70:775–800.

Cooper, R.A. 1989. Mammography. *Clin. Obs. Gynecol.* 32:768–785.

Council on Scientific Affairs. 1989. Mammographic screening in asymptomatic women aged 40 years and older. *J. Amer. Med. Assoc.* 261:2535–2542.

Moskowitz, M. 1988. Breast cancer screening: all's well that ends well, or much ado about nothing? *Amer. J. Radiol.* 151:659–665.

Seidman, H., Gelb, S.K., Silverberg, E., LaVerda, N., and Lubera, J.A. 1987. Survival experience in the Breast Cancer Detection Demonstration Project. *CA—A Cancer Journal for Clinicians* 37:258–290.

Shapiro, S. 1989. The status of breast cancer screening: a quarter of a century of research. *World J. Surg.* 13:9–18.

Screening for Colorectal Cancer

Eddy, D.M. 1990. Screening for colorectal cancer. *Ann. Intern. Med.* 113:373–384.

Hardcastle, J.D., Armitage, N.C., Chamberlain, J., Amar, S.S., James, P.D., and Balfour, T.W. 1986. Fecal occult blood screening for colorectal cancer in the general population. *Cancer* 58:397–403.

Khubchandani, I.T., Karamchandani, M.C., Kleckner, F.S., Sheets, J.A., Stasik, J.J., Rosen, L., and Riether, R.D. 1989. Mass screening for colorectal cancer. *Dis. Colon Rectum* 32:754–758.

Kolata, G. 1985. Debate over colon cancer screening. *Science* 229:636–637.

Lieberman, D.A. 1990. Colon cancer screening: the dilemma of positive screening tests. *Arch. Intern. Med.* 150:740–744.

Simon, J.B. 1985. Occult blood screening for colorectal carcinoma: a critical review. *Gastroenterology* 88:820–837.

Early Detection of Other Cancers

Catalona, W.J., Smith, D.S., Ratliff, T.L., Dodds, K.M., Coplen, D.E., Yuan, J.J.J., Petros, J.A., and Andriole, G.L. 1991. Measurement of prostate-specific antigen in serum as a screening test for prostate cancer. *N. Engl. J. Med.* 324:1156–1161.

Miller, A.B. 1986. Screening for cancer: issues and future directions. *J. Chron. Dis.* 39:1067–1077.

Smart, C.R. 1990. Screening and early cancer detection. *Sem. Oncol.* 17:456–462.

Symptoms of Cancer

American Cancer Society. 1990. *Cancer facts and figures—1990.* American Cancer Society, Atlanta.

Diagnosis and Staging

American Joint Committee on Cancer. 1987. *Manual for staging of cancer.* 3rd ed. J.B. Lippincott, Philadelphia.

International Union Against Cancer. 1990. *TNM Atlas.* 3rd ed. Springer-Verlag, New York.

Jandl, J.H. 1991. *Blood: pathophysiology.* Blackwell Scientific Publications, Boston.

Look, A.T., Hayes, F.A., Shuster, J.J., Douglass, E.C., Castleberry, R.P., Bowman, L.C., Smith, E.I., and Brodeur, G.M. 1991. Clinical relevance of tumor cell ploidy and N-

myc gene amplification in childhood neuroblastoma: a pediatric oncology group study. *J. Clin. Oncol.* 9:581–591.

McGuire, W.L., Tandon, A.K., Allred, D.C., Chamness, G.C., and Clark, G.M. 1990. How to use prognostic factors in axillary node-negative breast cancer patients. *J. Natl. Cancer Inst.* 82:1006–1015.

Rubin, E., and Farber, J.L., eds. 1990. *Essential pathology.* J.B. Lippincott, Philadelphia.

Sawyers, C.L., Timson, L., Kawasaki, E.S., Clark, S.S., Witte, O.N., and Champlin, R. 1990. Molecular relapse in chronic myelogenous leukemia patients after bone marrow transplantation detected by polymerase chain reaction. *Proc. Natl. Acad. Sci. USA* 87:563–567.

Squire, L.F., and Novelline, R.A. 1988. *Fundamentals of radiology.* 4th ed. Harvard University Press, Cambridge.

Chapter 14

Treatment of Cancer

ONCE CANCER IS DIAGNOSED, a variety of possible treatment options are considered. The choice of treatment is determined by the type of cancer and the extent to which the disease has progressed. Substantial advances have been made in the treatment of some cancers, particularly childhood leukemias and lymphomas—which, in most cases, are now curable diseases. For most other types of cancer, however, the likelihood of a cure is highly dependent on detection at early stages of the disease, before the cancer has spread from its site of origin. This chapter considers the current methods of cancer treatment, as well as some new, still experimental, treatment strategies.

SURGERY

The Effectiveness of Surgery Is Limited by Invasion and Metastasis

Surgery is the first line of attack against most cancers. For benign tumors, surgical removal of the tumor mass results in a complete cure. The only benign tumors that are life-threatening are those that are inoperable because of their location, such as some brain tumors, or those that go untreated and become infected or compress vital structures as they increase in size. For malignant tumors, however, the success of surgery depends on removing all of the cancer cells. If this is not accomplished, the remaining cells will continue to grow and metastasize even after removal of the bulk of the tumor. This is the reason that early diagnosis is so important. Early stage tumors (e.g., carcinomas *in situ*) that have not yet invaded surrounding normal tissue can be completely removed and are virtually 100% curable. Once invasion of adjacent normal tissues has begun, however, it becomes difficult to be sure that all of the cancer cells are eliminated by surgical removal of the tumor mass and immediately adjacent tissues. The extent of surgical removal of normal tissue surrounding an invasive cancer is therefore highly dependent on the type of tumor and its extent of progression. In some cases, such as most skin cancers, it is sufficient to remove the tumor together with a small margin of surrounding normal tissue. In other cases, it is necessary to remove larger amounts of adjacent tissue, perhaps including regional lymph nodes to which the cancer may have spread, in order to eliminate all of the cancer cells.

Unfortunately, about 70% of cancers have already metastasized by the time of diagnosis and therefore cannot be cured by surgery alone. Nonetheless, surgical removal of the primary tumor plays an important role in treatment, in combination with other therapeutic modalities discussed in subsequent sections of this chapter. First, the specimens of tumor and surrounding normal tissues, such as lymph nodes, obtained in surgery are of critical importance in determining the extent of tumor progression (clinical staging) and thereby designing an optimum strategy for further treatment. Even when a tumor has already metastasized, surgical removal of the primary tumor mass still, in some instances, forms an integral part of treatment, in combination with radiation and chemotherapy, for dealing with both local spread and distant metastasis. In addition to removal of a primary tumor mass, surgery can sometimes be used effectively to eliminate isolated metastatic tumors. Even in advanced cancer cases, appropriate surgical procedures, although not curative, may significantly reduce pain and other disease symptoms.

The role of surgery in cancer treatment is thus not confined to those cases in which the disease can be cured solely by removal of the primary tumor. However, surgery is ultimately a local treatment, the effectiveness of which is frequently defeated by the spread of tumors to regions that are too extensive for surgical removal (for example, throughout entire organs), as well as by metastasis to distant body sites.

Complications of Surgery and Patient Support Groups

Even when successful, surgical removal of cancers can result in disfigurement or disablement. Such consequences of surgery often have profound emotional effects on both the patient and family members. Examples include radical head and neck surgery, loss of a breast from mastectomy, loss of bowel control from colostomy, amputation of a limb, loss of speech from laryngectomy, and impotence resulting from prostatectomy. Patients can often be assisted in dealing with these and other complications in a variety of ways, including breast prostheses, reconstructive surgery, artificial limbs, and the use of mechanical or esophageal speech. A number of support groups, such as the Reach to Recovery, Laryngectomy Rehabilitation, and Ostomy Rehabilitation programs of the American Cancer Society, have been formed to help patients cope with these problems.

RADIATION THERAPY

Use of Radiation in Cancer Treatment

Radiation therapy, like surgery, is primarily a treatment for localized cancers, but one that can attack cancer cells that have spread through normal tissue beyond the scope of surgical removal. Thus, localized tumors are sometimes treated with radiation instead of surgery, and radiation is frequently used in combination with surgery to eliminate cancer cells that have invaded normal tissues surrounding the primary tumor site. In addition, some cancers are particularly sensitive to radiation, making it a preferential form of treatment for these diseases.

Previous chapters have discussed radiation as a cause of cancer, because it induces mutations by damaging DNA. Such damage to DNA can also kill cells, which is the basis for the use of radiation in cancer therapy. However, the effectiveness of radiation is limited by the fact that radiation is not specific for cancer cells. It also kills normal cells and consequently can be highly toxic to the patient. This problem of toxicity, resulting from an inability of a treatment to distinguish between cancer cells and normal cells, similarly limits the effectiveness of chemotherapy, as discussed in the next section of this chapter.

A variety of types of radiation are used in cancer therapy, including X-rays, radiation produced by the decay of radioactive elements (e.g., cobalt), and beams of particles (e.g., electrons) produced by linear accelerators. The radiation source is usually located outside of the body, and a beam of radiation is directed at the tumor. In some cases, however, radiation is administered from internal sources, such as an implant of radioactive material placed directly in the vicinity of a tumor. Cervical carcinomas, for example, can be treated by implanting a capsule of radioactive material (e.g., radium) within the cervix and upper vagina for a two- or three-day period.

Toxicity to Rapidly Dividing Normal Cells Limits the Effectiveness of Radiotherapy

The primary action of radiation is to damage DNA, either directly or by the generation of reactive chemical species (such as oxygen free radicals) within cells. Such damage to the genetic material is most lethal to cells that are rapidly dividing, so radiation selectively kills actively proliferating cells. Unfortunately, this includes normal cells as well as cancer cells. As discussed in previous chapters, several kinds of cells normally continue to divide throughout adult life. These include the blood-forming cells of the bone marrow, the epithelial cells that line the intestine, skin cells, the cells that form hair, and cells of the reproductive organs. The sensitivity of these normal cells to radiation is responsible for a number of side effects, which may include anemia, nausea, vomiting, diarrhea, skin damage, hair loss, and sterility. The extent of these side effects depends on the amount of radiation delivered and on the area of the body that is irradiated. In some cases, radiation can be directed precisely at a tumor, with little damage to surrounding normal cells and minimal side effects. In other cases, sensitive normal tissues must be irradiated at the same time, limiting the amount of radiation that an individual can receive during treatment.

Some Tumors Are Preferentially Treated by Radiotherapy

Radiotherapy is sometimes a preferred alternative to surgery for curative treatment of localized tumors. An example is the use of radiotherapy for laryngeal cancer, where radiation provides effective treatment without the loss of speech that would result from surgical removal of the vocal cords. Similarly, radiation is used to treat some cancers in locations that are hard to treat by surgery. For example, skin cancers at sites such as the eyelid or the tip of the nose may be treated by radiotherapy. Radiation may also be used instead of surgery for treatment of some cancers of the cervix, esophagus, and oral cavity. Hodgkin's disease, a form of lymphoma, is also effectively treated by radiation of the lymph nodes.

Radiation in Combination with Surgery

In addition to the use of radiotherapy as a primary mode of treatment, radiation is frequently used in combination with surgery to kill any tumor cells that remain after surgical removal of a primary tumor. In some cases, this is particularly useful in limiting the extent of surgery that needs to be performed. Early stages of breast cancer, for example, need no longer be treated by mastectomy. Instead, the generally recommended course of treatment is surgical removal of only the primary tumor and nearby lymph nodes, followed by radiation to kill remaining tumor cells. Some testicular cancers (seminomas) are also particularly sensitive to radiation and are generally treated by limited surgery followed by radiotherapy of regional lymph nodes.

Radiation can thus be viewed as a localized cancer therapy that covers a broader area than surgery. In this context, radiation can extend the effective range of treatment of localized cancers by eliminating cancer cells that have spread from a primary tumor into surrounding normal tissues that cannot be completely removed by surgery. Tumors that have already metastasized to distant body sites, however, can no longer be treated by localized means (either surgery or radiotherapy) alone. Instead, the treatment of these cancers requires the use of chemotherapy to reach cancer cells that have become disseminated throughout the body.

CHEMOTHERAPY

Although localized cancers can be effectively treated by surgery or radiotherapy, the success of these treatments is frequently limited by metastasis of cancer cells to distant body sites. Metastatic growths are often present by the time of diagnosis, although they may be too small to be detected. The existence of such micrometastases limits the success of localized treatment, necessitating the administration of chemotherapeutic drugs to attempt to kill tumor cells throughout the body.

The Action and Toxicity of Chemotherapeutic Drugs

Unfortunately, the drugs available for use in chemotherapy are not specific for cancer cells. Most chemotherapeutic agents act either by damaging DNA or by interfering with DNA synthesis. Thus, like radiation, they kill rapidly dividing cells—not only cancer cells but also the normal cells in the body that are undergoing cell division, particularly cells of the intestinal epithelium, blood-forming cells (**hematopoietic cells**) of the bone marrow, and cells that form hair. As in the case of radiation, the common toxic effects of chemotherapy result from the effects of chemotherapeutic drugs on these normal cell populations. Damage to the intestinal epithelium leads to nausea, vomiting, and diarrhea. The bone marrow is a site of major toxicity, since the loss of blood-forming cells causes anemia, ineffective blood clotting, and suppression of the immune system. Hair loss is also a common side effect of chemotherapy. Unlike radiation, chemotherapeutic drugs circulate throughout the body, so these toxic manifestations cannot be avoided. The effectiveness of chemotherapeutic drugs is thus limited by their toxicity, and the potential success of chemotherapy is determined by the relative sensitivities of cancer cells and normal cells to the drug being used. The goal is to kill all of the cancer cells while allowing sufficient numbers of normal cells to survive. The battle is fought by the administration of carefully regulated doses of the available drugs, and the physician attempts to walk a fine line between cancer cell death and normal cell survival.

Antimetabolites Interfere with DNA Synthesis

A number of different drugs are used in cancer chemotherapy (Table 14.1), and these drugs act in several ways to inhibit cell division. One class of drugs, called **antimetabolites**, interfere with one or more steps in the synthesis of DNA. Since the genetic material must be duplicated each time a cell divides, these drugs block cell division, resulting in the death of actively proliferating cells. The antimetabolites used in cancer chemotherapy include compounds such as **methotrexate**, **fluorouracil**, **cytosine arabinoside**, **mercaptopurine**, **thioguanine**, and **hydroxyurea**. These agents either inhibit cellular enzymes responsible for the synthesis of one or more of the nucleotides that are the precursors of DNA, or they inhibit the synthesis of DNA itself (Fig. 14.1). Fluorouracil, for example, blocks the synthesis of thymine (TMP) by inhibiting an enzyme called **thymidylate synthetase.** Because fluorouracil inhibits this enzyme, TMP is not made, and DNA cannot be replicated. Consequently, cell division is blocked, and cells that are attempting to divide eventually die. Methotrexate, hydroxyurea, thioguanine, and mercaptopurine also appear to act by inhibiting the synthesis of DNA precursors, whereas cytosine arabinoside probably acts mainly as a consequence of being incorporated into DNA in place of cytosine, resulting in defects in further DNA replication.

Table 14.1   *Chemotherapeutic Drugs*

Drug	Mechanism of Action
Actinomycin D	Inhibition of RNA synthesis
Asparaginase	Degrades the amino acid asparagine in blood
Bischloroethylnitrosourea	Damages DNA
Bleomycin	Damages DNA
Chlorambucil	Damages DNA
Cisplatin	Damages DNA
Cyclohexylchloroethylnitrosourea	Damages DNA
Cyclophosphamide	Damages DNA
Cytosine arabinoside	Inhibition of DNA synthesis
Daunomycin	Inhibition of topoisomerase II
Doxorubicin	Inhibition of topoisomerase II
Etoposide	Inhibition of topoisomerase II
Fluorouracil	Inhibition of thymidylate synthetase (TMP synthesis)
Hydroxyurea	Inhibition of ribonucleotide reductase (synthesis of deoxyribonucleotides)
Melphalan	Damages DNA
Mercaptopurine	Inhibition of AMP and GMP synthesis
Methotrexate	Inhibition of dihydrofolate reductase (blocks synthesis of AMP, GMP, and TMP)
Mitomycin C	Damages DNA
Nitrogen mustard	Damages DNA
Procarbazine	Damages DNA
Teniposide	Inhibition of topoisomerase II
Thioguanine	Inhibition of AMP and GMP synthesis
Thiotepa	Damages DNA
Vinblastine	Inhibition of mitosis
Vincristine	Inhibition of mitosis

Some Drugs Act by Damaging DNA

Other chemotherapeutic drugs act, like radiation, by damaging DNA directly. Also like radiation, the DNA damage induced by these drugs can sometimes contribute to the development of secondary cancers (particularly leukemias) in treated patients (see chapter 3). However, successful treatment of a patient's primary cancer is usually a much higher priority.

The largest group of DNA-damaging drugs are the **alkylating agents**, which react chemically with DNA molecules (Fig. 14.2). Some of the alkylating agents used in cancer chemotherapy include **mechlorethamine (nitrogen mustard)**, **cyclophosphamide, melphalan, bischloroethylnitrosourea (BCNU), cyclohexyl-chloroethylnitrosourea (CCNU), thiotepa, chlorambucil,** and **procarbazine**.

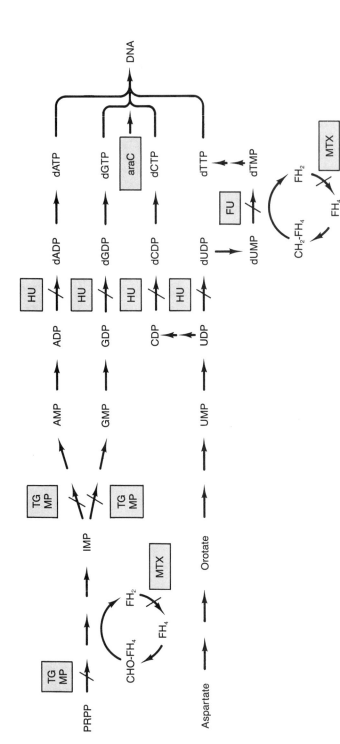

Figure 14.1 Inhibition of DNA synthesis by antimetabolites. DNA synthesis proceeds by incorporation of four deoxyribonucleoside triphosphates into the growing DNA molecule: dATP (deoxyadenosine triphosphate), dGTP (deoxyguanosine triphosphate), dCTP (deoxycytidine triphosphate), and dTTP (deoxythymidine triphosphate). The figure illustrates some of the steps leading to synthesis of these deoxyribonucleoside triphosphates, as well as the sites of action of antimetabolites used in cancer chemotherapy.

Cytosine arabinoside (araC) is converted to its nucleoside triphosphate (araCTP) and incorporated into DNA in place of dCTP, inhibiting further DNA synthesis. *Fluorouracil* (FU) is converted to its deoxyribonucleoside monophosphate (dFUMP), which blocks synthesis of dTTP by inhibiting conversion of deoxyuridine monophosphate (dUMP) to deoxythymidine monophosphate

(dTMP). This reaction also requires the participation of folic acid and is blocked by *methotrexate* (MTX), which inhibits the conversion of dihydrofolate (FH_2) to tetrahydrofolate (FH_4). Folic acid is also required for early steps in the synthesis of dATP and dGTP, which are similarly inhibited by MTX. *Hydroxyurea* (HU) blocks synthesis of all four deoxyribonucleoside triphosphates by inhibiting the conversion of ribonucleoside diphosphates (ADP, GDP, CDP, and UDP) to the corresponding deoxyribonucleoside diphosphates (dADP, dGDP, dCDP, and dUDP). *Thioguanine* (TG) and *mercaptopurine* (MP) inhibit early steps in the synthesis of dATP and dGTP, including the initial step in conversion of phosphoribosyl pyrophosphate (PRPP) to inosine monophosphate (IMP), as well as the subsequent conversion of IMP to both AMP and GMP.

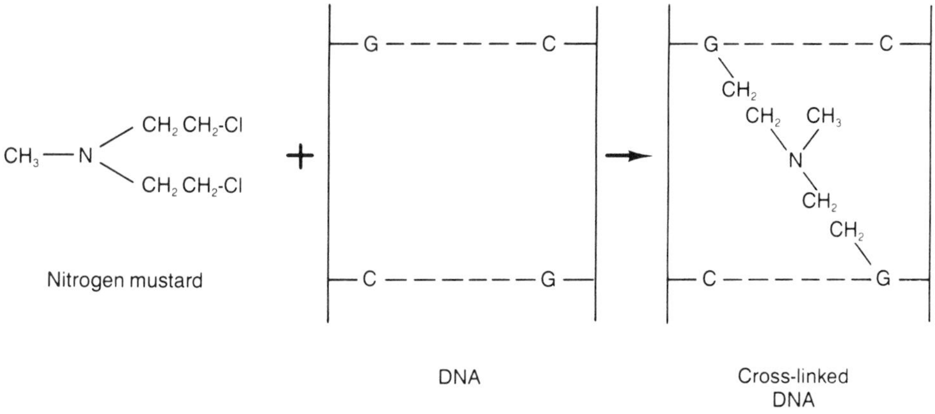

Figure 14.2 Cross-linking of DNA by alkylating agents. Nitrogen mustard (mechlorethamine) reacts with guanine residues on opposite strands of DNA.

These agents induce several types of DNA damage, including DNA breakage and the formation of cross-links between opposite strands of DNA, thereby blocking further DNA replication. Other chemotherapeutic drugs, including **bleomycin, cisplatin,** and **mitomycin C** also react directly to damage the DNA molecule. Still other drugs, including **daunomycin, doxorubicin, etoposide (VP-16),** and **teniposide,** lead to DNA damage by interacting with the enzyme **topoisomerase II** (Fig. 14.3). Normally, topoisomerase II acts to break and then rejoin DNA strands within the cell. Such breakage-rejoining reactions are required for a number of cell processes, including DNA replication. Etoposide and teniposide form a complex with DNA and topoisomerase, inhibiting the rejoining function of the enzyme, so that the DNA breaks are not repaired and the DNA molecule is cleaved. Daunomycin and doxorubicin bind to DNA and also result in the formation of topoisomerase II-induced breaks that are not repaired.

Other Chemotherapeutic Agents

A few chemotherapeutic drugs interfere with cell division by inhibiting other cellular processes. **Actinomycin D** binds to DNA, but it appears to act primarily by blocking subsequent RNA synthesis. The agents **vincristine** and **vinblastine** block the process of mitosis by binding to the microtubule protein tubulin. This disrupts the mitotic spindle and blocks cell division at the M phase of the cell cycle (see chapter 6). The enzyme **asparaginase** is used for therapy of acute leukemias. This enzyme degrades the amino acid asparagine, which is essential for protein synthesis. Most cells synthesize their own asparagine, but many leukemic cells do not and are therefore dependent on obtaining asparagine as an essential nutrient from the blood supply. Asparaginase destroys asparagine in the blood and is therefore effective in blocking the growth of these tumors.

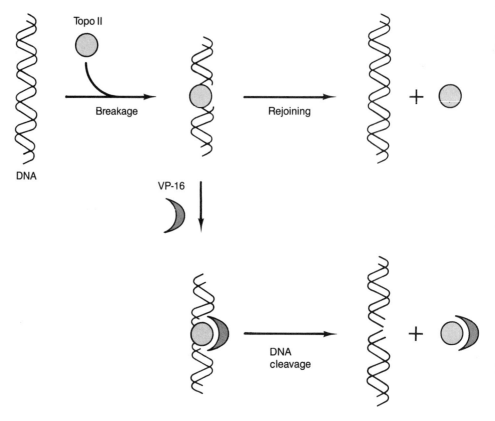

Figure 14.3 Inhibition of topoisomerase II by etoposide. Topoisomerase II (Topo II) acts to break transiently and then rejoin opposite strands of DNA. Etoposide (VP-16) inhibits the rejoining reaction, leading to cleavage of the DNA molecule.

Chemotherapy Is Successful Against Some Types of Cancer

The chemotherapeutic agents discussed above are representative of those in use, but they make up only a small fraction of the compounds that have been and are being tested for anticancer activity. Nonetheless, it is apparent that in spite of the investigation of a large number of cancer chemotherapeutic agents, the cellular targets against which these drugs act are limited and not very specific for cancer cells. Most chemotherapeutic drugs either inhibit DNA synthesis, induce DNA damage, or otherwise block cell division. Therefore, the limiting factor in chemotherapy is the relative drug sensitivity of the cancer cells compared to the rapidly proliferating normal cells of the body. Some tumors are particularly drug sensitive and can therefore be successfully treated by chemotherapy. Often, these tumors are rapidly growing and therefore most sensitive to DNA damage or inhibition of DNA synthesis, but the basis of selective toxicity to tumor cells versus normal cells is not well understood. Notable examples of successes with chemotherapy include the treatment of Burkitt's lymphoma, Hodgkin's disease, acute lymphocytic leukemia, choriocarcinoma, and testicular cancer. However, many other forms of cancer, including most carcinomas of adults, are less responsive to chemotherapeutic drugs, which then become ineffective because of the overwhelming toxic side effects that result from attempts to administer drug doses high enough to kill the cancer cells.

DRUG RESISTANCE AND COMBINATION CHEMOTHERAPY

Cancer Cells Frequently Become Resistant to Chemotherapeutic Drugs

As discussed above, the problem faced in chemotherapy is how to eradicate cancer cells while minimizing toxicity to normal cells of the patient. Since most chemotherapeutic agents have limited target specificity, the doses that can be employed must always represent a compromise between these two considerations, so complete elimination of tumor cells is a goal not easily attained. One of the major problems is that some cells in a cancer are frequently resistant to the action of a given drug, so they survive treatment. Even if most of the cancer cells are killed, a small number of such drug-resistant survivors will grow back to form a tumor that no longer responds to the initial treatment. The phenomenon of **drug resistance** thus poses a major problem to successful chemotherapy.

There are several ways in which cancer cells can become resistant to chemotherapeutic drugs, and the generation of drug-resistant cancer cells occurs frequently, in part as a result of the characteristic genetic instability of cancer cells (see chapter 6). One common mechanism of drug resistance is gene amplification, which occurs at particularly high frequencies in cancer cells. Tumor cells become resistant to methotrexate, for example, by amplification of the gene encoding **dihydrofolate reductase,** the target enzyme that methotrexate inhibits (see Fig. 14.1). This results in elevated levels of the enzyme, so that inhibition by methotrexate is no longer effective, and DNA synthesis is able to proceed in the tumor cells. Tumor cells can also become resistant to antimetabolites as a result of structural mutations in their target enzymes—for example, a mutation altering dihydrofolate reductase so that methotrexate no longer inhibits the mutant enzyme. In addition, cancer cells sometimes become drug-resistant as a result of mutations that make the tumor cells unable to take up chemotherapeutic drugs or convert them to their active forms inside of cells.

Multidrug Resistance

In addition to mutations that render cancer cells resistant to single chemotherapeutic agents, cancer cells sometimes become resistant to multiple different drugs at the same time. This phenomenon, known as **multidrug resistance**, is obviously particularly troublesome from the standpoint of chemotherapy. One form of multidrug resistance results from expression of the gene encoding a plasma membrane protein (called **P-glycoprotein**) that acts to "pump" a variety of chemotherapeutic drugs out of the cell before they can act. Normally, the P-glycoprotein serves to protect cells against potentially toxic foreign compounds. Overexpression of the P-glycoprotein gene, however, renders cancer cells simultaneously resistant to several different chemotherapeutic drugs, including actinomycin D, daunomycin, doxorubicin, vinblastine, vincristine, mitomycin C, etoposide, and teniposide. Because of the importance of this gene in drug resistance in human cancers, agents that inhibit activity of the P-glycoprotein are being tested as drugs that might be used to overcome this form of drug resistance. Another form of multidrug resistance results from alterations in topoisomerase II which result in inability of the drugs daunomycin, doxorubicin, etoposide, and teniposide to interact with this enzyme and generate DNA breaks. Increased activity of cellular enzymes that repair damage to DNA can also lead to cross-resistance to multiple alkylating agents.

The Advantages of Drug Combinations

An immediate approach to overcoming the problem posed by drug resistance is **combination chemotherapy**—the administration of combinations of multiple chemotherapeutic agents. The rationale is that although many cancer cells may be resistant to one or even two drugs in such a combination, it is unlikely that any one cell will be resistant to all of the drugs being administered. Consequently, an appropriate combination of drugs has the theoretical potential of eradicating all cells in a tumor. Given the phenomenon of multidrug resistance, it is obviously important to choose a spectrum of drugs such that cancer cells do not become resistant to all drugs in a combination simultaneously.

An additional advantage is attained by using combinations of drugs with nonoverlapping patterns of toxicity to normal cells. An example is the use of doxorubicin, which is primarily toxic to the bone marrow, in combination with vincristine, which causes relatively little bone marrow toxicity. Such combinations allow the physician to maximize the antitumor effects of each drug, while remaining within the limits of toxicity tolerated by the patient.

The use of drug combinations is responsible for most of the progress that has been made in cancer chemotherapy over the last several decades. Single chemotherapeutic agents have shown notable activity only in the treatment of Burkitt's lymphoma (cyclophosphamide) and choriocarcinoma (methotrexate). However, combinations of drugs have resulted in successful treatment of several other kinds of cancer, including acute lymphocytic leukemia, Hodgkin's disease, non-Hodgkin's lymphoma, and testicular carcinoma. Testicular carcinoma, for example, is treated by combination chemotherapy with cisplatin, bleomycin, and etoposide. Unfortunately, however, curative chemotherapy for most common adult malignancies (e.g., breast, colon, and lung carcinomas) remains elusive.

BONE MARROW TRANSPLANTATION

Bone Marrow Transplantation Bypasses Toxicity to the Blood-Forming Cells

As discussed above, the hematopoietic (blood-forming) stem cells of the bone marrow are one of the rapidly proliferating normal cell types that limit the effectiveness of chemotherapy, because of toxicity. **Bone marrow transplantation** is an approach to bypassing this toxicity, thereby allowing the use of much higher doses of drugs or radiation to eliminate tumor cells. In this procedure, the patient is treated with intensive high doses of radiation or chemotherapy, which would normally be intolerable due to destruction of hematopoietic stem cells of the bone marrow. The patient is rescued following the conclusion of treatment by receiving a transplant of new healthy marrow. The stem cells contained within this donor marrow are able to repopulate and restore the hematopoietic system of the patient, which therefore ultimately regains normal function.

Bone marrow transplantation is most frequently used for treatment of leukemias and lymphomas, although it is also being evaluated as a possible therapy for some solid tumors. Usually, a normal tissue-matched donor provides the marrow that the patient receives after radiation or chemotherapy. In some cases, however, the patient's own bone marrow may be harvested before chemotherapy, stored, and then returned to the patient after chemotherapy has been completed. In these cases, the marrow must either be free of cancer cells at the time of harvest or treated to remove cancer cells before it is reintroduced into the patient.

Severe Toxicity and Complications

Because of the high-dose chemotherapy used in this procedure, toxic effects, including nausea and vomiting, are severe. In addition, the patient's normal hematopoietic functions, including those of the immune system, are suppressed until the donor bone marrow becomes fully functional. Marrow function is usually restored in a period of two to four weeks, but it generally takes several months before the patient's immune system recovers full function. Because of the resulting high susceptibility to infection, patients are sometimes maintained in special rooms that are kept as free from infectious agents as possible. Other immunological complications, such as **graft versus host disease,** may also occur, and patients are intensively monitored both during hospitalization and for several months after being discharged.

HORMONE THERAPY

A hallmark of cancer cells is their failure to respond to the mechanisms that regulate normal cell proliferation. In many cancers, however, this lack of regulation is not complete, and proliferation of the cancer cells can still be modulated by some of the factors that control normal cell division. In particular, some tumor cells are still responsive to the steroid hormones that regulate proliferation of their normal counterparts. In these cases, manipulation of hormone levels can be an effective means of cancer treatment.

Treatment of Breast Cancer with Tamoxifen

As discussed in previous chapters, estrogen stimulates the proliferation of epithelial cells of both the breast and the uterine endometrium. Indeed, excess estrogen stimulation is associated with an increased risk of cancer at these sites. Conversely, interference with the hormonal pathways that drive normal cell proliferation can sometimes inhibit the proliferation of cancer cells. For example, breast cancers can be treated by interfering with estrogen-stimulated cell proliferation by administration of the antiestrogen **tamoxifen**. Tamoxifen is structurally similar to estrogen (Fig. 14.4) and apparently binds in place of estrogen to the estrogen receptor. However, the binding of tamoxifen fails to activate the receptor appropriately, so expression of estrogen-responsive target genes is not normally induced. Although tamoxifen does exert some weak estrogenic effects, it primarily serves to antagonize estrogen-induced processes, thereby inhibiting the growth of estrogen-dependent breast cancers. About 50% of patients with breast cancer (usually those patients whose tumors express detectable levels of estrogen receptor) respond to such treatment. Although tamoxifen induces some side effects, they are much less severe than those resulting from chemotherapeutic drugs that interfere more generally with cell proliferation.

Progesterone and Endometrial Cancer

The proliferation of endometrial cells during the menstrual cycle is stimulated by estrogen and inhibited by progesterone. The most common hormone therapy for

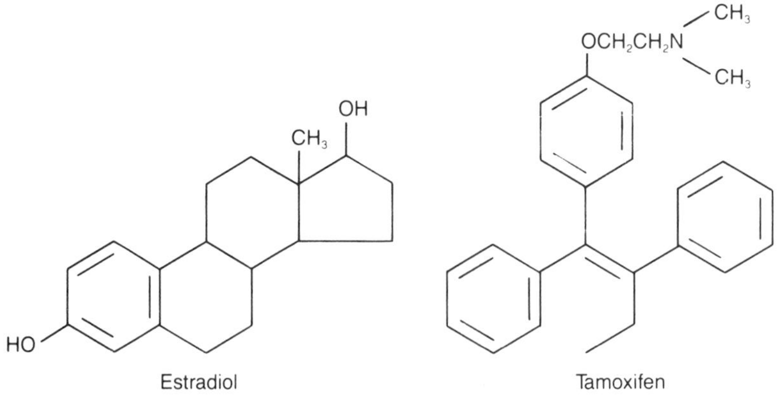

Figure 14.4 Structures of estradiol (an estrogen) and tamoxifen (an estrogen antagonist).

endometrial cancer is administration of synthetic progesterones to inhibit cell growth, a strategy that is successful in about one-third of patients. Anti-estrogens, such as tamoxifen, may also be active against some endometrial cancers.

Prostate Cancer

Prostate cancers are also hormone-responsive, in this case being stimulated by **androgens,** such as testosterone. Hormone therapy is standard treatment to slow the growth of advanced prostate cancers that have spread through the body. Different strategies are employed, all with the goal of blocking androgen stimulation of the cancer cells (Fig. 14.5). First, since the testes are the primary source of androgens in the body, removal of the testicles (orchiectomy) may be recommended. Alternatively, the production of androgens can be inhibited by administration of other hormones. Androgen production in the testis is triggered by peptide hormones (**gonadotropins**) that are produced by the pituitary gland. Estrogens, such as **diethylstilbestrol (DES)**, suppress the production of pituitary gonadotropins and consequently inhibit androgen synthesis in the testis. A simi-lar effect is achieved by administration of analogs of **gonadotropin-releasing hormone (GnRH)**, such as the drug **luprolide**. GnRH is a hormone produced by the hypothalamus which acts on the pituitary to signal gonadotropin release. Continuous administration of GnRH analogs, however, suppresses gonadotro-pin release, consequently inhibiting androgen synthesis. Finally, antiandrogens can act at the level of the prostate cancer cell by blocking the binding of andro-gens to their receptor protein. Such hormone therapies are not curative, but they do significantly inhibit tumor progression, relieve pain, and prolong the lives of many patients.

Steroid Hormones in the Treatment of Leukemias and Lymphomas

Steroid hormones produced by the adrenal gland (adrenal glucocorticoids) in-hibit the proliferation of lymphoid cells. Consequently, glucocorticoids (usually **prednisone**) are commonly employed in the treatment of leukemias and lympho-mas, including acute lymphocytic leukemia, Hodgkin's disease, non-Hodgkin's lymphomas, and myeloma. A standard therapeutic regimen for Hodgkin's dis-ease, for example, is the combination of two alkylating agents (mechlorethamine and procarbazine), vincristine, and prednisone.

Retinoic Acid, the *RAR* Oncogene, and Acute Promyelocytic Leukemia

A recent example of hormone therapy is the treatment of acute promyelocytic leukemia with retinoic acid (vitamin A). Retinoic acid appears to induce these leukemic cells to differentiate, concomitant with cessation of their prolifera-tion. As discussed in chapter 11, retinoic acid acts, like the steroid hormones, through a nuclear protein receptor that regulates gene expression, frequently

Figure 14.5 Hormone therapy of prostate carcinoma. Proliferation of prostate cancer cells is stimulated by androgens, which are produced by the testes. Androgen production is triggered by pituitary gonadotropins, whose release is signaled by gonadotropin-releasing hormone (GnRH) produced by the hypothalamus. Strategies for hormone therapy include inhibition of gonadotropin release by administration of estrogen or GnRH analogs, orchiectomy (removal of the testicles), and administration of antiandrogens.

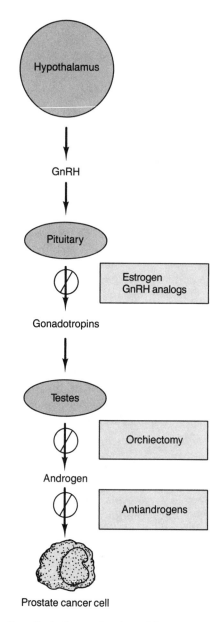

Prostate cancer cell

promoting cell differentiation. It is noteworthy that the retinoic acid receptor apparently acts as an oncogene in this leukemia, being translocated from chromosome 15 to chromosome 17 (see chapter 8). In some way, then, malfunction of a retinoic acid pathway leading to cell differentiation contributes to the development of acute promyelocytic leukemia, and further manipulation of this pathway appears to reverse the leukemic cell phenotype. This is the first possible example of a direct relationship between an oncogene and a therapeutic regimen, although (as will be discussed in chapter 18) the possibility of directing therapies to specific oncogene targets is an area of active consideration in current cancer research.

IMMUNOTHERAPY

Helping the Body's Natural Defense Against Cancer

Immunotherapy is the attempt to use the body's natural defense mechanism, the immune system, to fight cancer. The notion that the immune system does provide a defense against cancer is supported by the fact that individuals with inherited or acquired immunodeficiencies suffer a high incidence of certain tumors (see chapters 3–5). It is noteworthy, however, that such immunodeficiencies particularly increase susceptibility to a few specific kinds of cancers, such as lymphomas, rather than to all cancers in general. Interestingly, the cancers that occur most frequently in immunodeficient individuals are those associated with tumor viruses (e.g., Epstein-Barr virus-induced lymphomas, discussed in chapter 4), suggesting that a normal immune response is particularly important as a defense against virus-induced tumors. Immune reactions against a variety of other kinds of tumors also occur, however, so the immune system may provide a general defense against tumor development.

The goal of immunotherapy is to bolster the effectiveness of a patient's immune system in an attempt to eliminate cancer cells. Two general approaches are used: (1) nonspecific stimulation of the patient's own immune response and (2) the administration of specific kinds of immune cells or secreted cellular products that act against the patient's cancer cells.

Levamisole and Colon Cancer

Nonspecific stimulation of the immune response has generally been ineffective as a therapeutic strategy against most cancers in either humans or experimental animals. However, this approach has recently been found to yield significant benefits in the treatment of human colon carcinoma. The drug employed as a nonspecific immune stimulant in this case is **levamisole**, which stimulates a variety of immune responses and has been found to inhibit the growth of some tumors in experimental animals. Based on these studies, combination treatment with levamisole and fluorouracil (an antimetabolite discussed above) was tried as a postsurgical treatment for patients with colon carcinomas that had spread to regional lymph nodes. The five-year survival rate for these patients after surgery alone is less than 50%, due to metastases that are not eliminated by surgical removal of the primary tumor mass. Neither fluorouracil nor levamisole alone has a significant effect on this postsurgical prognosis. However, these two drugs in combination resulted in a reduction of about one-third in the rates of tumor recurrence and mortality following surgery. Although modest, this represents a clear benefit of postsurgical treatment with levamisole plus fluorouracil over surgery alone. Consequently, this treatment modality has now been accepted as standard therapy to be offered to patients with lymph node-positive colon carcinoma.

Tumor-Infiltrating Lymphocytes and Interleukin-2

In contrast to levamisole and other drugs that nonspecifically stimulate the patient's immune system, alternative modes of immunotherapy are specifically directed against the patient's tumor. An example of such a therapy currently under clinical evaluation is treatment with **tumor-infiltrating lymphocytes**. In this approach, large numbers of lymphocytes with anti-tumor reactivity are administered to the patient. These lymphocytes are obtained from surgically removed tumor specimens, grown to large cell numbers in culture, and then transferred back to the patient together with a growth factor (**interleukin-2**) to further stimulate their proliferation and antitumor activity. The hope is that sufficiently large numbers of these antitumor lymphocytes will effectively react against the cancer. To date, such treatments have been most active in patients with advanced renal cancers and melanoma, with positive responses being achieved in about 20% of such patients. Although far from a complete success, such a response rate is favorable compared to other treatments of these two tumor types. The results of these trials are therefore encouraging, and further studies are continuing.

Monoclonal Antibodies

Another variety of specific immunotherapy is the administration of **monoclonal antibodies**, which are antibodies produced by single clones of B lymphocytes against a specific antigen. Lymphocytes producing monoclonal antibodies can be generated and propagated in the laboratory, so large quantities of these highly specific antibodies can be obtained. A number of monoclonal antibodies that react against proteins on the surface of cancer cells have been tried as therapeutic reagents, but this approach has not proven very successful. An alternative application, however, is the use of monoclonal antibodies to deliver cytotoxic drugs or sources of radiation specifically to tumor cells. For example, an antitumor antibody coupled to a radioactive isotope might serve to direct this source of radiation specifically to cancer cells, resulting in their selective eradication. Such applications of monoclonal antibodies in drug targeting are being actively studied.

Cytokines

Yet another variation in immunotherapy is the administration of regulatory factors secreted by lymphocytes, rather than administration of lymphocytes or antibodies *per se*. Such factors (called **cytokines**) include **interleukin-2**, the **interferons**, and **tumor necrosis factor**. They may have both antiproliferative effects on tumor cells and immune stimulatory effects on the host. Interleukin-2 stimulates the activity of lymphocytes with antitumor activity. As noted above, it is utilized together with the transfer of tumor-infiltrating lymphocytes to maximize their antitumor response. Interferon, the most thoroughly studied of the antitumor cytokines, appears both to stimulate the immune system of the host and to have direct effects on tumor cells. So far, interferon has shown activity against only a few human cancers (particularly certain leukemias), but further studies of the combined effects of interferon with other chemotherapeutic drugs are continuing. Tumor necrosis factor is another cytokine with direct effects against tumor cells. Unfortunately, however, treatment with tumor necrosis factor also results in considerable toxic side effects, limiting its therapeutic effectiveness.

Thus, although immunotherapy offers the promise of a natural, and therefore potentially nontoxic, approach to cancer treatment, this promise has not yet been translated into widespread practice. Nonetheless, this is an active area of cancer research, and it is possible that further studies will provide more effective means of manipulating the immune system to target tumor cells for selective elimination.

CLINICAL TRIALS VERSUS UNPROVEN TREATMENTS

Several of the treatment modalities discussed in this chapter represent therapies that are still under investigation, rather than established, standard cancer treatments. Indeed, since cancer treatments are obviously less than entirely satisfactory, new drugs and treatment protocols are continuously being developed.

Drug Development Is a Long Process

The process of drug development is a painstaking one, which involves extensive laboratory and experimental animal studies of any new drug prior to its use in humans. On the average, it takes 12 years and $240 million for a pharmaceutical company to develop a successful new drug from conception to general use in medical practice. The process is closely regulated by the Food and Drug Administration (FDA), which must approve human use of any new drug based on its satisfactory performance in a series of required tests in animals. Only when a drug has shown significant antitumor activity in experimental animals, with tolerable side effects, is it finally ready to be tested in humans.

The Design and Conduct of Clinical Trials

Once a new drug or treatment regimen is ready for human testing, known as **clinical trial**, it undergoes evaluation via a three-stage process. At each stage, patients who volunteer to participate as subjects are informed of the possible risks and benefits. Phase I involves administration of the drug to a small number of advanced cancer patients (usually about 20), in order to determine how well the drug is tolerated in humans. If no prohibitive toxicity is encountered, the drug is tested in phase II trials on a larger group of advanced cancer patients (perhaps up to 100), with an attempt to determine efficacy in different types of tumors. Patients asked to participate in either phase I or II trials are advanced cancer patients whose malignancies have failed to respond to standard treatments, so their participation in these tests does not interfere with conventional treatment and has the possibility of beneficial effects. If a drug appears effective in phase II trials, it proceeds to phase III testing. These studies involve much larger numbers of patients (up to 1,000 or more) who have earlier stages of a particular malignancy. In these trials, the test treatment is compared to the current standard treatment for the type of cancer being investigated; subjects are usually assigned randomly to receive the standard or the experimental treatment. An example is provided by the trial of levamisole plus fluorouracil in colon cancer (discussed above). Approximately 1,200 patients participated in this trial. They all received surgery (the standard treatment), and some then received follow-up treatment with levamisole plus fluorouracil (the experimental treatment). Once the efficacy of this drug combination became apparent, postsurgical treatment with levamisole plus fluorouracil was accepted as standard for future practice.

Many patients hesitate to participate in clinical trials for fear of being used as "guinea pigs." Clinical trials are indeed experimental, but without them no progress in the development of new drugs could be made. It is important to realize that a considerable amount of drug testing and review has already taken place by the clinical trial stage. Moreover, patients participating in such trials are not denied any treatment known to be effective for their tumors. On the other hand, participants may benefit from the treatment being evaluated, as in the case of the levamisole plus fluorouracil testing.

The Difference Between Clinical Trials and Unproven Treatments

However, clinical trials must be distinguished from what are generously called "unconventional" or "alternative" cancer treatments and less kindly referred to as "quackery." These treatments lie outside of the mainstream of medical practice and have not passed the scrutiny of rigorous animal testing and clinical trial to establish their validity. In the end, the only way to determine whether a treatment works or not is to compare patients who have received the treatment in question with those who have not. This is the outcome of a phase III clinical trial, which is undertaken only on the basis of positive results in animals as well as in phase I and II trials in humans. Only if patients receiving the treatment in a phase III trial respond favorably compared to controls is the treatment considered beneficial. "Unconventional cancer treatments" do not undergo such objective evaluation but are offered to the public on the basis of limited experience, which is not broadly accepted in science or medicine.

Laetrile

Laetrile, a naturally occurring substance extracted from the pits of apricots and other fruits, is a good example of such an unconventional cancer treatment. Laetrile was first used to treat cancer in the 1950s. It became popular in the 1970s, when several members of the John Birch Society founded the "Committee for the Freedom of Choice in Cancer Therapy," the goal of which was to promote the right of cancer patients to use laetrile. Approximately 70,000 people used laetrile as an anticancer drug in this period, in spite of the fact that a variety of studies in experimental animals failed to detect any antitumor activity. Nonetheless, because of its prominence as an unconventional treatment, the National Cancer Institute undertook phase I and II trials of laetrile. These clinical trials, completed in the early 1980s, again failed to support the claims of laetrile's antitumor activity.

Laetrile, like other unconventional treatments, thus failed to meet objective criteria for activity as an anticancer drug. Nonetheless, laetrile is still offered to the public in Mexican clinics as an effective treatment. There are several dangers to cancer patients who go the route of such unconventional therapies. Not only may they be denying themselves the potential benefits of established therapies, but the unconventional treatments may be dangerous *per se*. Laetrile, for example, breaks down in the body to form cyanide, which has led to toxic effects in many patients. There are many other examples of such dangerous supposed treatments. Their effectiveness is highly touted by their proponents, notwithstanding the lack of objective evidence. It is important for patients and family members to be aware of this problem and to make sure that any proposed treatment has a recognized scientific and medical basis.

SUMMARY

The limiting factor in cancer treatment is metastasis. Localized cancers can usually be effectively treated by surgery or radiotherapy, but over half of cancers have already metastasized to distant body sites by the time of diagnosis. A variety of chemotherapeutic drugs are then used to attempt to kill cancer cells throughout the body. These drugs are usually administered in combinations, in order to maximize their effects and to eliminate cancer cells that may become resistant to any one of the drugs being used. Unfortunately, the chemotherapeutic drugs currently available are not specific for cancer cells. Rather, they interfere generally with cell division and therefore also act against those normal cells in the body that are undergoing active proliferation—particularly the blood-forming cells of the bone marrow, the cells that line the intestine, and the cells that form hair. Toxicity to these normal cell populations limits the doses of chemotherapeutic drugs that can be tolerated by the patient, thus limiting the effectiveness of chemotherapy. Thus, while some types of cancer are particularly drug-sensitive and can be successfully treated, many cancers are not responsive to current chemotherapy protocols. In some cases, hormones can be used to inhibit the growth of cancer cells, but these treatments usually slow tumor progression rather than eliminating the cancer. Immunotherapy, the possibility of manipulating the immune system to fight cancer more effectively, is an area of active research that has shown promise as a treatment for some cancers but has not yet achieved broad clinical application. It is clear that significant strides in the treatment of cancer have been made, and over 50% of cancer patients can now be cured. However, treatment of the most common kinds of cancer is ineffective once metastasis has occurred, so nearly half of all cancer patients still die of their disease. The still unmet challenge in cancer treatment is the development of drugs that act specifically against cancer cells.

KEY TERMS

surgery
radiation therapy
radiation implant
chemotherapy
hematopoietic cells
antimetabolites
methotrexate
fluorouracil
cytosine arabinoside
mercaptopurine
thioguanine
hydroxyurea
alkylating agents
mechlorethamine
nitrogen mustard
cyclophosphamide
melphalan
bischloroethylnitrosourea (BCNU)
cyclohexylchloroethylnitrosourea (CCNU)
thiotepa
chlorambucil
procarbazine
bleomycin
cisplatin
mitomycin C
daunomycin
doxorubicin
etoposide (VP-16)
teniposide
topoisomerase II
actinomycin D
vincristine
vinblastine
asparaginase
combination chemotherapy
drug resistance
multidrug resistance

P-glycoprotein

bone marrow transplantation

hormone therapy

tamoxifen

androgen

gonadotropin

gonadotropin-releasing hormone

prednisone

retinoic acid

immunotherapy

levamisole

tumor-infiltrating lymphocyte

interleukin-2

monoclonal antibody

cytokine

interferon

tumor necrosis factor

clinical trial

unproven cancer treatment

laetrile

REFERENCES AND FURTHER READING

General References

American Cancer Society. 1990. *Cancer facts and figures—1990.* American Cancer Society, Atlanta.

DeVita, V.T., Jr., Hellman, S., and Rosenberg, S.A., eds. 1989. *Cancer: principles and practice of oncology.* 3rd ed. J.B. Lippincott, Philadelphia.

National Cancer Institute. 1988. *What you need to know about cancer.* National Institutes of Health, Bethesda.

Surgery

Rosenberg, S.A. 1989. Principles of surgical oncology. In *Cancer: principles and practice of oncology,* ed. DeVita, V.T., Jr., Hellman, S., and Rosenberg, S.A. 3rd ed. J.B. Lippincott, Philadelphia. pp. 236–246.

Radiation Therapy

Friedberg, E.C. 1985. *DNA repair.* W.H. Freeman, New York.

Goffman, T.E., Raubitschek, A., Mitchell, J.B., and Glatstein, E. 1990. The emerging biology of modern radiation oncology. *Cancer Res.* 50:7735–7744.

Hellman, S. 1989. Principles of radiation therapy. In *Cancer: principles and practice of oncology,* ed. DeVita, V.T., Jr., Hellman, S., and Rosenberg, S.A. 3rd ed. J.B. Lippincott, Philadelphia. pp. 247–275.

Chemotherapy

Chabner, B.A., and Collins, J.M, eds. 1990. *Cancer chemotherapy: principles and practice.* J.B. Lippincott, Philadelphia.

DeVita, V.T., Jr. 1989. Principles of chemotherapy. In *Cancer: principles and practice of oncology,* ed. DeVita, V.T., Jr., Hellman, S., and Rosenberg, S.A. 3rd ed. J.B. Lippincott, Philadelphia. pp. 276–300.

Heidelberger, C. 1970. Chemical carcinogenesis, chemotherapy: cancer's continuing core challenges—G.H.A. Clowes Memorial Lecture. *Cancer Res.* 30:1549–1569.

Schweitzer, B.I., Dicker, A.P., and Bertino, J.R. 1990. Dihydrofolate reductase as a therapeutic target. *FASEB J.* 4:2441–2452.

Stryer, L. 1988. *Biochemistry.* 3rd ed. W.H. Freeman, New York.

Drug Resistance and Combination Chemotherapy

Chabner, B.A., and Collins, J.M, eds. 1990. *Cancer chemotherapy: principles and practice.* J.B. Lippincott, Philadelphia.

DeVita, V.T., Jr. 1989. Principles of chemotherapy. In *Cancer: principles and practice of oncology,* ed. DeVita, V.T., Jr., Hellman, S., and Rosenberg, S.A. 3rd ed. J.B. Lippincott, Philadelphia. pp. 276–300.

Goldstein, L.J., Galski, H., Fojo, A., Willingham, M., Lai, S.-L., Gazdar, A., Pirker, R., Green, A., Crist, W., Brodeur, G.M., Lieber, M., Cossman, J., Gottesman, M.M., and Pastan, I. 1989. Expression of a multidrug resistance gene in human cancers. *J. Natl. Cancer Inst.* 81:116–124.

Pastan, I., and Gottesman, M. 1987. Multiple-drug resistance in human cancer. *N. Engl. J. Med.* 316:1388–1393.

Schimke, R.T. 1984. Gene amplification in cultured animal cells. *Cell* 37:705–713.

Bone Marrow Transplantation

Butturini, A., Keating, A., Goldman, J., and Gale, R.P. 1990. Autotransplants in chronic myelogenous leukaemia: strategies and results. *Lancet* 335:1255–1258.

National Cancer Institute. 1990. *Bone marrow transplantation.* National Institutes of Health, Bethesda.

Storb, R. 1989. Bone marrow transplantation. In *Cancer: principles and practice of oncology,* ed. DeVita, V.T., Jr., Hellman, S., and Rosenberg, S.A. 3rd ed. J.B. Lippincott, Philadelphia. pp. 2474–2489.

Thomas, E.D. 1987. Bone marrow transplantation. *CA—A Cancer Journal for Clinicians* 37:291–301.

Hormone Therapy

Castaigne, S., Chomienne, C., Daniel, M.T., Ballerini, P., Berger, R., Fenaux, P., and Degos, L. 1990. All-*trans* retinoic acid as a differentiation therapy for acute promyelocytic leukemia. I. Clinical results. *Blood* 76:1704–1709.

Chomienne, C., Ballerini, P., Balitrand, N., Daniel, M.T., Fenaux, P., Castaigne, S., and Degos, L. 1990. All-*trans* retinoic acid in acute promyelocytic leukemias. II. *In vitro* studies: structure-function relationship. *Blood* 76:1710–1717.

Huang, M., Ye, Y., Chen, S., Chai, J., Lu, J.-X., Zhoa, L., Gu, L., and Wang, Z. 1988. Use of all-*trans* retinoic acid in the treatment of acute promyelocytic leukemia. *Blood* 72:567–572.

Lerner, L.J., and Jordan, V.C. 1990. Development of antiestrogens and their use in breast cancer: Eighth Cain Memorial Award Lecture. *Cancer Res.* 50:4177–4189.

Swain, S.M., and Lippman, M.E. 1990. Endocrine therapies of cancer. In *Cancer chemotherapy: principles and practice,* ed. Chabner, B.A., and Collins, J.M. J.B. Lippincott, Philadelphia. pp. 59–109.

Yen, S.S.C., and Jaffe, R.B., eds. 1986. *Reproductive endocrinology.* 2nd ed. W.B. Saunders, Philadelphia.

Immunotherapy

Abbas, A.K., Lichtman, A.H., and Pober, J.S. 1991. *Cellular and molecular immunology.* W.B. Saunders, Philadelphia.

Borden, E.C., and Sondel, P.M. 1990. Lymphokines and cytokines as cancer treatment. *Cancer* 65:800–814.

Grem, J.L. 1990. Levamisole as a therapeutic agent for colorectal carcinoma. *Cancer Cells* 2:131–137.

Itoh, K., Tilden, A.B., and Balch, C.M. 1986. Interleukin 2 activation of cytotoxic T-lymphocytes infiltrating into human metastatic melanomas. *Cancer Res.* 46:3011–3017.

Moertel, C.G., Fleming, T.R., MacDonald, J.S., Haller, D.G., Laurie, J.A., Goodman, P.J., Underleider, J.S., Emerson, W.A., Tormey, D.C., Glick, J.H., Veeder, M.H., and Mailliard, J.A. 1990. Levamisole and fluorouracil for adjuvant therapy of resected colon carcinoma. *N. Engl. J. Med.* 322:352–358.

NIH Consensus Conference. 1990. Adjuvant therapy for patients with colon and rectal cancer. *J. Amer. Med. Assoc.* 264:1444–1450.

Rosenberg, S.A., Longo, D.L., and Lotze, M.T. 1989. Principles and applications of biologic therapy. In *Cancer: principles and practice of oncology,* ed. DeVita, V.T., Jr., Hellman, S., and Rosenberg, S.A. 3rd ed. J.B. Lippincott, Philadelphia. pp. 301–347.

Rosenberg, S.A., Packard, B.S., Aebersold, P.M., Solomon, D., Topalian, S.L., Toy, S.T., Simon, P., Lotze, M.T., Yang, J.C., Seipp, C.A., Simpson, C., Carter, C., Bock, S., Schwartzentruber, D., Wei, J.P., and White, D.E. 1988. Use of tumor-infiltrating lymphocytes and interleukin-2 in the immunotherapy of patients with metastatic melanoma. *N. Engl. J. Med.* 319:1676–1680.

Rosenberg, S.A., Spiess, P., and Lafreniere, R. 1986. A new approach to the adoptive immunotherapy of cancer with tumor-infiltrating lymphocytes. *Science* 233:1318–1321.

Schiller, J.H., Storer, B.E., Witt, P.L., Alberti, D., Tombes, M.B., Arzoomanian, R., Proctor, R.A., McCarthy, D., Brown, R.R., Voss, S.D., Remick, S.C., Grem, J.L., Borden, E.C., and Trump, D.L. 1991. Biological and clinical effects of intravenous tumor necrosis factor-α administered three times weekly. *Cancer Res.* 51:1651–1658.

Wadler, S., and Schwartz, E.L. 1990. Antineoplastic activity of the combination of interferon and cytotoxic agents against experimental and human malignancies: a review. *Cancer Res.* 50:3473–3486.

Clinical Trials Versus Unproven Treatments

Simon, R.M. 1989. Design and conduct of clinical trials. In *Cancer: principles and practice of oncology,* ed. DeVita, V.T., Jr., Hellman, S., and Rosenberg, S.A. 3rd ed. J.B. Lippincott, Philadelphia. pp. 396–420.

U.S. Congress, Office of Technology Assessment. 1990. *Unconventional cancer treatments.* U.S. Government Printing Office, Washington.

PART IV
OVERVIEW OF MAJOR TYPES OF CANCER

THE PRECEDING PARTS OF THIS BOOK have considered human cancer from the standpoint of an overall understanding of its causes, molecular and cellular basis, prevention, and treatment. This part of the book takes a different approach. The major types of cancer are discussed as distinct topics, in order to provide an integrated overview of the multiple factors relevant to each individual disease.

Chapter 15

Leukemias and Lymphomas

MULTIPLE MYELOMA
SUMMARY
KEY TERMS
REFERENCES AND FURTHER READING

LEUKEMIAS AND LYMPHOMAS are cancers of the blood and lymph systems. Together, they account for approximately 8% of United States cancer incidence—an estimated 82,000 cases annually. In addition, leukemias and lymphomas are the most common cancers in children, representing about half of all childhood malignancies (approximately 3,800 cases annually).

ORIGIN AND CLASSIFICATION

Formation of Blood Cells and Lymphocytes

There are many different types of leukemias and lymphomas, classified according to the type of cell involved and how rapidly the disease progresses. As discussed in chapter 6, all of the different cells in blood and lymph are derived from a common precursor cell—the pluripotent stem cell, in bone marrow. Leukemias and lymphomas result from the continuous proliferation of cells that are blocked at various stages of their normal development (Fig. 15.1). The first step in blood cell differentiation gives rise to **myeloid** and **lymphoid** progenitor cells. Cells derived from the myeloid lineage include blood **platelets,** red blood cells (**erythrocytes**), **granulocytes** (basophils, eosinophils, and neutrophils), **monocytes,** and **macrophages.** These cells function in blood coagulation (platelets), oxygen transport (red blood cells), inflammatory reactions (granulocytes, monocytes, and macrophages), and the immune response (macrophages). The lymphoid lineage gives rise to **B** and **T lymphocytes,** which are responsible for antibody secretion and cell-mediated immunity, respectively. B lymphocytes develop within the bone marrow, whereas the precursors of T lymphocytes migrate to the thymus and develop there. Precursors of both B and T lymphocytes then migrate to the **lymphatic system,** which includes the lymph nodes and spleen in addition to the thymus, tonsils, adenoids, and aggregates of lymphatic tissue in the bone marrow and intestine. In addition to lymphocytes, lymphatic tissues contain macrophages, which also function in the immune response.

Types of Leukemias and Lymphomas

Leukemias arise in the blood-forming cells of the bone marrow and can result from abnormal growth and differentiation of any of the different kinds of cells within either the myeloid or the lymphoid lineages. **Lymphomas** develop from lymphocytes or macrophages in lymphatic tissues. In children, leukemias are

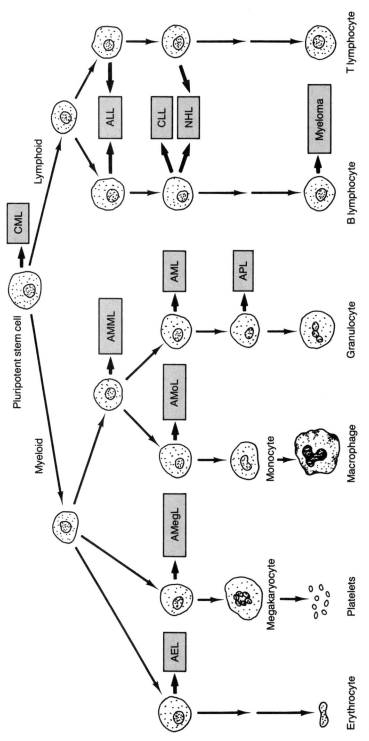

Figure 15.1 Cellular origins of leukemias and lymphomas. Different kinds of leukemias and lymphomas most frequently result from continued proliferation of cells blocked at the indicated stages of their normal differentiation. CML, chronic myelogenous leukemia; AEL, acute erythroid leukemia; AMegL, acute megakaryocytic leukemia; AMoL, acute monocytic leukemia; AMML, acute myelomonocytic leukemia; AML, acute myelocytic leukemia; APL, acute promyelocytic leukemia; ALL, acute lymphocytic leukemia; CLL, chronic lymphocytic leukemia; NHL, non-Hodgkin's lymphomas.

Table 15.1 Leukemias and Lymphomas

Type of Cancer	Approximate Cases per Year (United States Population)	
	Children	Adults
Acute lymphocytic leukemia	1,900	3,000
Acute nonlymphocytic leukemia	600	8,000
Chronic lymphocytic leukemia	—	8,000
Chronic myelogenous leukemia	100	5,000
Hodgkin's disease	500	7,000
Non-Hodgkin's lymphomas	700	35,000
Myeloma	—	12,000
Total	3,800	78,000

about twice as frequent as lymphomas, whereas lymphomas are more common in adults. The most prevalent of these diseases (Table 15.1) are discussed in this chapter.

ACUTE LYMPHOCYTIC LEUKEMIA

Acute lymphocytic leukemia (ALL) is the most common leukemia in children, accounting for approximately 1,900 cases of childhood leukemia annually in the United States (approximately 75% of childhood leukemias and 25% of all cancers in children). It is less common in adults, where it represents about 12% of all leukemias. Acute lymphocytic leukemia is a rapidly progressing disease (*acute*) characterized by an abnormal increase in immature lymphocytes (*lymphocytic*) in the blood and bone marrow (Fig. 15.1). About 20% of these leukemias arise from **T-cells,** the remainder being of **B-cell** origin.

Risk Factors and Oncogenes

The causes of most cases of ALL (and of other leukemias) are unknown, although the risk of ALL is increased by exposure to radiation. In addition, ALL occurs more frequently in individuals with certain inherited diseases that are associated with instability of the genetic material, including Down's syndrome, Fanconi's anemia, Bloom's syndrome, and ataxia telangiectasia (see chapter 5). Consistent with this general association of DNA damage with increased leukemia risk, a number of alterations in oncogenes and tumor suppressor genes are frequently observed in ALL. These include point mutation of the *ras*N oncogene, chromosome translocation of the *abl*, c-*myc*, *E2A*, and interleukin-3 oncogenes, and loss or inactivation of the *p53* tumor suppressor gene. However, the precise roles of these oncogenes and tumor suppressor genes in the development of leukemia are not yet known.

Diagnosis and Treatment

Early detection of ALL, like other leukemias, is difficult. Initial symptoms are usually nonspecific, resembling those associated with common infectious diseases, such as fever and fatigue. Additional signs of disease include frequent infections, a tendency to bleed or bruise easily, paleness, and weight loss. When leukemia is suspected, diagnosis is made by microscopic examination of the blood and bone marrow.

Fortunately, ALL, particularly in children, is one of the cancers that responds well to chemotherapy. Indeed, childhood ALL was the first leukemia to be successfully treated with chemotherapy, and it has served as a model for the development of current concepts of cancer treatment. The disease is currently treated using combinations of drugs, including vincristine, prednisone, asparaginase, doxorubicin, methotrexate, and mercaptopurine. Such combination chemotherapy is curative for more than half of children with this disease. It should be noted, however, that treatment of ALL (like that of all cancers) is a complicated undertaking, best performed by specialized teams with appropriate training and experience. The effectiveness of therapy for childhood ALL indeed varies between different institutions, with cure rates as high as 75% in some medical centers. Adult patients with ALL generally respond less well than children, but chemotherapy is still an effective treatment.

ACUTE NONLYMPHOCYTIC LEUKEMIAS

Types, Risk Factors, and Oncogenes

The acute nonlymphocytic leukemias affect blood cell types other than lymphocytes (Fig. 15.1). The most common is **acute myelocytic leukemia (AML),** which accounts for the majority of acute leukemias in adults. Other types of acute nonlymphocytic leukemias include **acute promyelocytic leukemia, acute myelomonocytic leukemia, acute monocytic leukemia,** and **erythroleukemia.** Risk factors and symptoms for these diseases are similar to those discussed for ALL above. Exposure to some occupational carcinogens, such as benzene (see chapter 3), also increases the risk of AML. In addition, some of the drugs used in cancer chemotherapy—particularly the alkylating agents, which induce DNA damage (see chapter 14)—can lead to the development of secondary leukemias. Point mutations of the *ras*N oncogene frequently occur in AML, often as an early event in the disease process. Interestingly, acute promyelocytic leukemia is characterized by chromosome translocation of the retinoic acid receptor gene, which appears to prevent normal differentiation of these leukemic cells.

Treatment

The acute nonlymphocytic leukemias are treated by chemotherapy with combinations of drugs, frequently including daunomycin and cytosine arabinoside. Chemotherapy of the acute nonlymphocytic leukemias is less successful than that of ALL, however, and treatment of the majority of these patients fails to eliminate the leukemic cells completely. The currently used combination chemotherapy protocols are therefore curative for only about 20% of AML patients. Given the role of the retinoic acid receptor as an oncogene in acute promyelocytic leukemia, it is noteworthy that retinoic acid appears to be of therapeutic benefit in this disease by inducing differentiation, and inhibiting proliferation, of the leukemic cells.

CHRONIC LYMPHOCYTIC LEUKEMIA

Chronic lymphocytic leukemia (CLL) is rare in children but accounts for about one-third of all leukemias in adults. It occurs most frequently in patients over 60 years of age. Usually the leukemic cells are immature B lymphocytes.

CLL is a much more slowly progressing disease than ALL. In the early stages, which may last for many years, patients have no symptoms of disease, and diagnosis is only made by the observation of abnormal numbers of blood lymphocytes. The average survival time is nearly ten years from diagnosis at this stage. Because of its indolent course, treatment of CLL is generally not initiated until symptoms appear. Once symptoms do become evident, patients are treated with low dose chemotherapy, usually with alkylating agents. Such treatment is not curative but is usually effective in slowing the course of disease. Normal function of the immune system is gradually lost as the disease progresses, and most patients with CLL die of infection or other illnesses not directly related to the leukemia.

CHRONIC MYELOGENOUS LEUKEMIA

Cellular Origin and the *abl* Oncogene

Chronic myelogenous leukemia (CML) is also a slowly progressing disease; it is rare in children but accounts for approximately 20% of adult leukemias. It originates in the pluripotent stem cell in the bone marrow—the cell that gives rise to all of the different cell types of the hematopoietic system, including cells of both the lymphoid and myeloid lineages (see Fig. 15.1 and chapter 6). It is characterized predominantly by the accumulation of abnormal numbers of granulocytes in the blood and bone marrow. CML always involves activation of the *abl* oncogene by translocation from chromosome 9 to chromosome 22—a chromosomal rearrangement known as the Philadelphia translocation, after the city in which it was discovered. Because this oncogene translocation is uniformly present in the leukemic cells, it is an important diagnostic marker for following the status of disease in CML patients.

Chronic Phase and Blast Crisis

The course of CML is divided into two stages: chronic phase and blast crisis. The chronic phase of the disease is associated with minimal symptoms and may persist for years. Eventually, however, patients progress to an accelerated stage of the disease, which culminates in the acute, life-threatening phase known as blast crisis. Blast crisis resembles an acute leukemia and is characterized by the accumulation of large numbers of rapidly proliferating leukemic cells, called blasts. The blast cells are myeloid in about two-thirds of cases, and lymphoid in the remainder. One of the events involved in progression from the chronic phase of CML to blast crisis may be loss or inactivation of the *p53* tumor suppressor gene.

Treatment

CML patients in blast crisis are treated with chemotherapy regimens similar to those used in ALL or AML, depending on whether the blast cells are lymphoid or myeloid. If successful, such therapy may induce remission to the chronic phase. Chemotherapy is also used during the chronic phase of the disease, but it does not usually succeed in totally eradicating the leukemic cells. Bone marrow transplantation may be curative for up to 50% of chronic phase CML patients, although this procedure is associated with significant risks due to its severe toxic side effects (see chapter 14). These treatment risks are lower for patients early in the course of disease, and the cure rate from bone marrow transplantation may be as high as 80% if performed within the first year after diagnosis.

HODGKIN'S DISEASE

Hodgkin's disease is the most frequent type of lymphoma, accounting for about 7,400 cases annually in the United States. It frequently affects young adults, with most cases occurring between the ages of 15 and 35. No known causes or risk factors for Hodgkin's disease have been identified, although there is some speculation that it may be associated with an infectious agent.

Origin and Spread Through the Lymphatic System

The site of origin of Hodgkin's disease is a lymph node, frequently in the neck. It then spreads through the lymphatic system (Fig. 15.2) and, in advanced stages, affects other organs, including the spleen, liver, lungs, and bone marrow. The initial symptom is usually painless lymph node enlargement, sometimes accompanied by fever, night sweats, itching of the skin, or weight loss. Diagnosis is made by microscopic examination of a lymph node biopsy.

Figure 15.2 The lymphatic system.

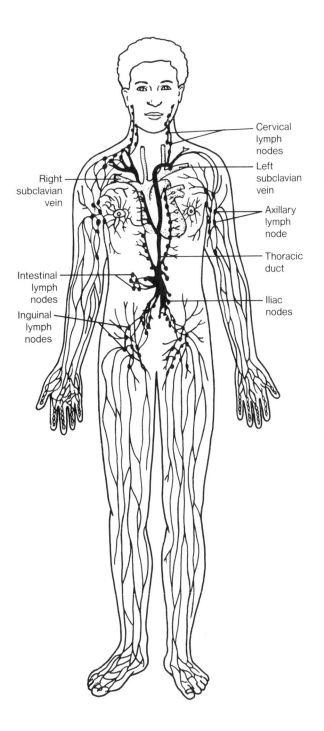

Cervical lymph nodes

Left subclavian vein

Right subclavian vein

Axillary lymph node

Thoracic duct

Intestinal lymph nodes

Inguinal lymph nodes

Iliac nodes

The Reed-Sternberg Cell

Hodgkin's disease is distinguished by the presence of a unique type of cell, called the **Reed-Sternberg cell,** in affected lymph nodes. Unlike most cancers, most of the tumor mass in Hodgkin's disease is composed of normal lymphocytes and connective tissue, rather than of tumor cells. The Reed-Sternberg cell, which is probably of lymphoid origin, appears to be the actual malignant cell in the disease. Activation of the *bcl*-2 oncogene by chromosome translocation may be involved in the development of these lymphomas.

Treatment

Hodgkin's disease can be successfully treated by both radiation and chemotherapy. Early stages of the disease, in which only one lymph node is involved or in which only limited spread of the tumor to regional lymph nodes has occurred, can be cured by intensive radiotherapy in nearly 90% of cases. Chemotherapy is also effective for both early and advanced stages of Hodgkin's disease. A number of different drug combinations are employed, one example being the regimen called MOPP, which consists of mechlorethamine, vincristine (Oncovin), procarbazine, and prednisone. Such combination chemotherapy is curative for over half of patients—even those with advanced stages of Hodgkin's disease. Indeed, treatment of Hodgkin's disease has become sufficiently successful that long-term side effects are important concerns, particularly since the disease frequently affects young people. Such treatment side effects include sterility and an increased incidence of secondary leukemias induced by the chemotherapeutic drugs.

NON-HODGKIN'S LYMPHOMAS

Origin and Classification

Lymphomas other than Hodgkin's disease account for about 36,000 cases per year in the United States. The majority (about 80%) of **non-Hodgkin's lymphomas** are malignancies of B lymphocytes; about 15% arise from T lymphocytes and the remaining 5% from macrophages. Several kinds of lymphomas, which differ considerably in their causes, prognosis, and treatment, are grouped together under this classification.

A number of different systems have been used to classify the non-Hodgkin's lymphomas, according to the type of malignant cell involved and the aggressiveness of the disease. In fact, six different systems are used throughout the world, and in spite of attempts to achieve a uniform classification, no single system has yet been broadly accepted. One of the most commonly used classifications is the Rappaport system (Table 15.2). In this scheme, lymphomas are first classified by growth pattern, either nodular or diffuse. They are then subclassified by cell type: lymphocytic if the cells are small and resemble lymphocytes; histiocytic if

Table 15.2 Non-Hodgkin's Lymphomas

Rappaport Classification	Type of Cell	Percent of Cases
Low Grade		
Diffuse well-differentiated lymphocytic	B	10–15
Nodular poorly differentiated lymphocytic	B	15–20
Nodular mixed lymphocytic histiocytic	B	5–10
Intermediate Grade		
Nodular histiocytic	B	<5
Diffuse poorly differentiated lymphocytic	B or T	15–20
Diffuse mixed lymphocytic histiocytic	B or T	5–10
Diffuse histiocytic	B or T	30
High Grade		
Diffuse histiocytic	B or T	5–10
Diffuse lymphoblastic	T	<5
Diffuse undifferentiated	B	<5

the cells are large and resemble macrophages or histiocytes; and mixed if the cells are of both sizes. Finally, the lymphoma cells are further described as well differentiated or poorly differentiated, depending on their extent of morphological similarity to normal cells. The main drawback of this system is that the classification of cell types is a description of morphology that does not correctly identify the actual origin of the malignant cell. In particular, the so-called histiocytic lymphomas are most frequently malignancies of B lymphocytes, not of histiocytes (macrophages). Nonetheless, since the Rappaport system correctly predicts biological behavior and clinical response, it remains one of the most widely used classifications of lymphomas. It will therefore form the basis of the present discussion.

Aggressive and Indolent Lymphomas

Several types of lymphomas (the intermediate- and high-grade lymphomas) are aggressive malignancies that grow rapidly and are quickly fatal if not treated. These include nodular and diffuse histiocytic lymphoma, diffuse lymphoblastic lymphoma, diffuse poorly differentiated lymphocytic lymphoma, diffuse undifferentiated lymphoma, and diffuse mixed lymphocytic-histiocytic lymphoma. About half of patients with non-Hodgkin's lymphomas have one of these aggressive diseases, and they are the most common lymphomas in children.

Other lymphomas (low grade) are referred to as indolent diseases because, like the chronic leukemias, they grow slowly and may persist for years with minimal symptoms. These tumors, which usually occur in adults, include nodular poorly differentiated lymphocytic lymphoma, diffuse well-differentiated lymphocytic lymphoma, and nodular mixed lymphocytic-histiocytic lymphoma. In time, however, these indolent lymphomas evolve to more aggressive diseases, accompanied by symptoms such as fever, night sweats, and weight loss.

Viruses, Immunosuppression, and Oncogenes

Epstein-Barr virus (EBV) is a causative agent of Burkitt's lymphoma (a diffuse undifferentiated lymphoma in the Rappaport system), which occurs with a high incidence in some regions of Africa (see chapter 4). In addition, EBV is associated with the high frequency of lymphomas in immune deficient patients. As discussed in chapters 3, 4, and 5, patients with inherited or acquired immunodeficiencies (resulting from immunosuppressive drugs or AIDS) develop EBV-associated B-cell lymphomas about 100 times more frequently than the general population.

Burkitt's and other aggressive B-cell lymphomas also regularly involve activation of the c-*myc* oncogene by chromosome translocation. In contrast, the indolent follicular (nodular) B-cell lymphomas involve activation of a different oncogene, *bcl*-2. As discussed in chapter 11, *bcl*-2 acts to prolong cell survival rather than promoting active cell proliferation—an activity that seems consistent with the slow progression of these indolent diseases. The further progression of these indolent lymphomas to more aggressive neoplasms may involve translocation of c-*myc* as a second event, resulting in increased tumor cell proliferation and more rapid disease progression.

Treatment

The prognosis for patients with these lymphomas seems paradoxical, in that the more aggressive lymphomas are more curable. In particular, the aggressive lymphomas respond well to combination chemotherapy, which is curative in over 50% of cases. Commonly used drugs include cyclophosphamide, vincristine, procarbazine, and prednisone. The indolent lymphomas are also treated by chemotherapy and radiation, but not usually cured. In many cases, symptom-free patients with these lymphomas are closely monitored but do not receive treatment unless the disease shows signs of progressing to a more aggressive form.

MULTIPLE MYELOMA

Multiple myeloma is a neoplasm of plasma cells, which are mature antibody-secreting B lymphocytes. There are about 12,000 cases annually in the United States. Multiple myeloma usually develops in older adults, with the average age of onset being around 70. Initial symptoms usually include bone pain, anemia, and fatigue. In addition to microscopic examination of blood and bone marrow, diagnosis includes analysis of blood and urine for antibodies secreted by the myeloma cells. There are no established causes or reproducible genetic alterations associated with multiple myeloma in humans, although the c-*myc* oncogene is involved in a closely related cancer (plasmacytoma) in mice. The disease is usually treated by chemotherapy, with combinations of drugs including alkylating agents (melphalan or cyclophosphamide) and prednisone. Such treatment is frequently effective in controlling disease progression for a period of two to four years, but it is not generally curative.

SUMMARY

Leukemias and lymphomas are cancers of the blood and lymph systems. They account for about half of all childhood cancers and about 8% of cancers in adults. In most cases, the causes are unknown, although infection with Epstein-Barr virus in combination with immunodeficiency results in a high risk of lymphoma development. Different types of leukemias and lymphomas involve alterations in the *abl*, *bcl*-2, *E2A*, c-*myc*, and *ras*N oncogenes, the retinoic acid receptor, interleukin-3, and the *p53* tumor suppressor gene. Some of these genes appear to affect cell differentiation (retinoic acid receptor) and cell survival (*bcl*-2) as well as cell proliferation. Leukemias are treated by chemotherapy, and lymphomas by both chemotherapy and radiation. Such treatments are effective for several of the most virulent of these diseases, including acute lymphocytic leukemia (the most common childhood cancer), Hodgkin's disease, and aggressive non-Hodgkin's lymphomas.

KEY TERMS

myeloid
lymphoid
platelet
erythrocyte
granulocyte
monocyte
macrophage
B lymphocyte
T lymphocyte
lymphatic system
acute lymphocytic leukemia
acute nonlymphocytic leukemias
acute myelocytic leukemia
acute promyelocytic leukemia
acute myelomonocytic leukemia
acute monocytic leukemia
erythroleukemia
chronic lymphocytic leukemia
chronic myelogenous leukemia
chronic phase

blast crisis
Hodgkin's disease
Reed-Sternberg cell
non-Hodgkin's lymphomas
Rappaport classification system
aggressive lymphomas
indolent lymphomas
multiple myeloma

REFERENCES AND FURTHER READING

General References

American Cancer Society. 1990. *Cancer facts and figures—1990.* American Cancer Society, Atlanta.

Bonner, H. 1990. The blood and the lymphoid organs. In *Essential pathology,* ed. Rubin, E., and Farber, J.L. J.B. Lippincott, Philadelphia. pp. 550–613.

Cooper, G.M. 1990. *Oncogenes.* Jones and Bartlett, Boston.

Crowley, L.V. 1988. *Introduction to human disease.* 2nd ed. Jones and Bartlett, Boston.

DeVita, V.T., Jr., Hellman, S., and Rosenberg, S.A., eds. 1989. *Cancer: principles and practice of oncology.* 3rd ed. J.B. Lippincott, Philadelphia.

National Cancer Institute. 1991. *Physician data query.* BRS Colleague, New York. (*Note:* This is a computer database that contains frequently updated information on the diagnosis, staging, and treatment of each different type of cancer. It is available at most medical libraries and through several software vendors in addition to BRS Colleague.)

National Cancer Institute. 1988. *Research report: Hodgkin's disease and the non-Hodgkin's lymphomas.* National Institutes of Health, Bethesda.

National Cancer Institute. 1988. *Research report: leukemia.* National Institutes of Health, Bethesda.

Page, H.S., and Asire, A.J. 1985. *Cancer rates and risks.* 3rd ed. National Institutes of Health, Bethesda.

Origin and Classification

Abbas, A.K., Lichtman, A.H., and Pober, J.S. 1991. *Cellular and molecular immunology.* W.B. Saunders, Philadelphia.

Jandl, J.H. 1991. *Blood: pathophysiology.* Blackwell Scientific Publications, Boston.

Sawyers, C.L., Denny, C.T., and Witte, O.N. 1991. Leukemia and the disruption of normal hematopoiesis. *Cell* 64:337–350.

Acute Lymphocytic Leukemia

Ahuja, H.G., Foti, A., Bar-Eli, M., and Cline, M.J. 1990. The pattern of mutational involvement of *RAS* genes in human hematologic malignancies determined by DNA amplification and direct sequencing. *Blood* 75:1684–1690.

Rivera, G.K., Raimondi, S.C., Hancock, M.L., Behm, F.G., Pui, C.-H., Abromowitch, M., Mirro, J., Jr., Ochs, J.S., Look, A.T., Williams, D.L., Murphy, S.B., Dahl, G.V., Kalwinsky, D.K., Evans, W.E., Kun, L.E., Simone, J.V., and Crist, W.M. 1991. Improved outcome in childhood acute lymphoblastic leukaemia with reinforced early treatment and rotational combination chemotherapy. *Lancet* 337:61–66.

Sugimoto, K., Toyoshima, H., Sakai, R., Miyagawa, K., Hagiwara, K., Hirai, H., Ishikawa, F., and Takaku, F. 1991. Mutations of the *p53* gene in lymphoid leukemia. *Blood* 77:1153–1156.

Acute Nonlymphocytic Leukemias

Borrow, J., Goddard, A.D., Sheer, D., and Solomon, E. 1990. Molecular analysis of acute promyelocytic leukemia breakpoint cluster region on chromosome 17. *Science* 249: 1577–1580.

Castaigne, S., Chomienne, C., Daniel, M.T., Ballerini, P., Berger, R., Fenaux, P., and Degos, L. 1990. All-*trans* retinoic acid as a differentiation therapy for acute promyelocytic leukemia. I. Clinical results. *Blood* 76:1704–1709.

Chomienne, C., Ballerini, P., Balitrand, N., Daniel, M.T., Fenaux, P., Castaigne, S., and Degos, L. 1990. All-*trans* retinoic acid in acute promyelocytic leukemias. II. *In vitro* studies: structure-function relationship. *Blood* 76:1710–1717.

de Thé, H., Chomienne, C., Lanotte, M., Degos, L., and Dejean, A. 1990. The t(15;17) translocation of acute promyelocytic leukaemia fuses the retinoic acid receptor α gene to a novel transcribed locus. *Nature* 347:558–561.

Huang, M., Ye, Y., Chen, S., Chai, J., Lu, J.-X., Zhoa, L., Gu, L., and Wang, Z. 1988. Use of all-*trans* retinoic acid in the treatment of acute promyelocytic leukemia. *Blood* 72:567–572.

Slingerland, J.M., Minden, M.D., and Benchimol, S. 1991. Mutation of the *p53* gene in human acute myelogenous leukemia. *Blood* 77:1500–1507.

Stone, R.M., and Mayer, R.J. 1990. The unique aspects of acute promyelocytic leukemia. *J. Clin. Oncol.* 8:1913–1921.

Vogelstein, B., Civin, C.I., Preisinger, A.C., Krischer, J.P., Steuber, P., Ravindranath, Y., Weinstein, H., Elfferich, P., and Bos, J. 1990. *RAS* gene mutations in childhood acute myeloid leukemia: a pediatric oncology group study. *Genes, Chromosomes and Cancer* 2:159–162.

Chronic Myelogenous Leukemia

Ahuja, H., Bar-Eli, M., Advani, S.H., Benchimol, S., and Cline, M. J. 1989. Alterations in the *p53* gene and the clonal evolution of the blast crisis of chronic myelocytic leukemia. *Proc. Natl. Acad. Sci. USA* 86:6783–6787.

Butturini, A., Keating, A., Goldman, J., and Gale, R.P. 1990. Autotransplants in chronic myelogenous leukaemia: strategies and results. *Lancet* 335:1255–1258.

Sawyers, C.L., Timson, L., Kawasaki, E.S., Clark, S.S., Witte, O.N., and Champlin, R. 1990. Molecular relapse in chronic myelogenous leukemia patients after bone marrow transplantation detected by polymerase chain reaction. *Proc. Natl. Acad. Sci. USA* 87:563–567.

Hodgkin's Disease

Cleary, M., and Rosenberg, S.A. 1990. The *bcl*-2 gene, follicular lymphoma, and Hodgkin's disease. *J. Natl. Cancer Inst.* 82:808–809.

Stetler-Stevenson, M., Crush-Stanton, S., and Cossman, J. 1990. Involvement of the *bcl*-2 gene in Hodgkin's disease. *J. Natl. Cancer Inst*. 82:855–858.

Non-Hodgkin's Lymphomas

Gauwerky, C.E., Haluska, F.G., Tsujimoto, Y., Nowell, P.C., and Croce, C.M. 1988. Evolution of B-cell malignancy: pre-B-cell leukemia resulting from *MYC* activation in a B-cell neoplasm with a rearranged *BCL*2 gene. *Proc. Natl. Acad. Sci. USA* 85:8548–8552.

Link, M.P., Donaldson, S.S., Berard, C.W., Shuster, J.J., and Murphy, S.B. 1990. Results of treatment of childhood localized non-Hodgkin's lymphoma with combination chemotherapy with or without radiotherapy. *N. Engl. J. Med*. 322:1169–1174.

Multiple Myeloma

Kyle, R.A. 1990. Newer approaches to the therapy of multiple myeloma. *Blood* 76:1678–1679.

Chapter 16
Childhood Solid Tumors

CHILDHOOD CANCER IS RARE, accounting for approximately 7,600 cases annually in the United States (less than 1% of the total cancer incidence). On the other hand, cancer is second only to accidents as the leading cause of death in children under age 15. The common adult cancers, carcinomas, occur very infrequently in children. Instead, as discussed in the preceding chapter, about half of childhood cancers are leukemias and lymphomas. The remainder are varieties of solid tumors (Table 16.1) that are rare in adults. Many of these tumors are evident soon after birth and probably arise during early embryonic development. In general, childhood tumors are rapidly growing cancers, which are often more responsive to chemotherapy than the common solid tumors of adults.

BRAIN TUMORS

Types of Childhood Brain Tumors

Brain tumors are the most common solid tumors of childhood, accounting for about 20% of all cancers in children. There are several different types of brain tumors, the most common being **astrocytomas, medulloblastomas,** and **ependymomas.** Astrocytomas, which account for nearly two-thirds of childhood brain tumors, occur in children of all ages. Ependymomas and medulloblastomas are most frequent in children under age 5 and between the ages of 5 and 10, respectively. The causes and risk factors for the development of childhood brain tumors are not known. In some cases, inactivation of the *p53* tumor suppressor gene may be involved.

Diagnosis and Treatment

Symptoms of brain tumors include headaches, dizziness, blurred vision, and problems with coordination. Diagnosis involves a variety of imaging techniques, particularly CT scan and MRI, as well as recording of the brain's electrical activity by electroencephalography (EEG). Surgery is the primary therapy, the goal being to remove the entire tumor. This is not always possible, however, and brain tumors are unusual in that even benign tumors can be life-threatening if their location precludes surgical removal. Tumors that are not accessible to surgery or that cannot be completely removed are treated by radiation and chemotherapy (for example, with drugs such as vincristine and actinomycin D).

The overall cure rate for brain tumors in children is greater than 50%, but this varies considerably according to the tumor type. Five-year survival rates for children with ependymoma or medulloblastoma are 70–80%, but the prognosis for children with astrocytomas is less favorable. Low-grade astrocytomas can sometimes be cured by surgery, and five-year survival rates for children with these tumors are around 50%, depending on the location of the tumor. However, the more aggressive astrocytomas, anaplastic astrocytomas and **glioblastomas,**

Table 16.1   *Childhood Cancers*

Type of Cancer	Approximate Cases per Year (United States Population)	
Leukemias and lymphomas	3,800	(50%)
Brain tumors	1,500	(20%)
Neuroblastoma	600	(8%)
Wilms' tumor	400	(5%)
Bone tumors	400	(5%)
Soft tissue sarcomas	400	(5%)
Retinoblastoma	200	(3%)
Total	7,300	(96%)

Percentages refer to total estimated incidence of all cancers in children under 15.

are considerably more malignant tumors, and average survival times for children with these diseases are 2–3 years and about 1 year, respectively.

NEUROBLASTOMA

Progression and the N-*myc* Oncogene

Neuroblastoma is the next most common childhood cancer, accounting for approximately 8% of all childhood malignancies. It is a neoplasm of embryonic neural cells and usually occurs by age 2. Rare cases are inherited (see chapter 5), but causes for the majority of cases are unknown. Progression of neuroblastomas to more aggressively growing advanced stages of disease is associated with amplification of the N-*myc* oncogene and loss of a tumor suppressor gene located on chromosome 1 (see chapter 10). Once neuroblastomas have progressed to these advanced stages (stages III and IV) they often respond poorly to treatment, so amplification of N-*myc* and loss of the chromosome 1 tumor suppressor are important indicators of prognosis.

Tumor Stage, Prognosis, and Treatment

Neuroblastomas most frequently originate in the abdomen; they are usually detected as a swelling or abnormal tissue mass. The disease is treated with surgery, radiation, and chemotherapy using combinations of drugs, frequently including cyclophosphamide and doxorubicin. The overall survival rate is about 50%, but this is highly dependent on the age of the patient and on the extent of disease progression at the time of diagnosis (Fig. 16.1). In children under 1 year of age, all stages of neuroblastoma can be effectively treated, with cure rates of about 90%. In older children, however, only the early stages of disease respond well to therapy. Thus, while 80–90% of children older than 1 year are cured of

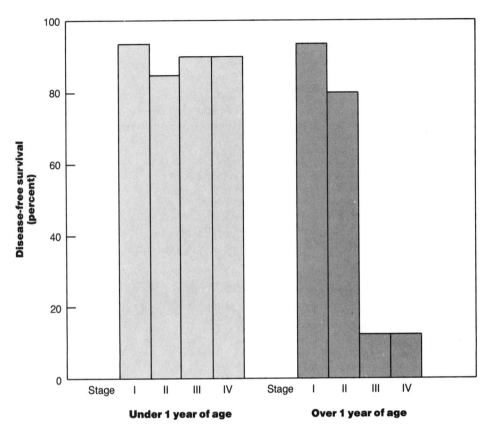

Figure 16.1   *Prognosis for children with neuroblastoma. Disease-free survival for children either under or over 1 year of age is plotted as a function of the stage of tumor at diagnosis. The staging system for neuroblastoma is described in Table 10.1. (Data from P.A. Pizzo et al., 1989.)*

stage I and II neuroblastomas, survival rates fall to less than 20% for children of this age having more advanced stage III and IV disease.

RETINOBLASTOMA

The *RB* Tumor Suppressor Gene

Retinoblastoma is an eye tumor, arising from embryonic retinal cells, which accounts for about 3% of childhood cancers. It usually occurs by age 3. Despite its rarity, retinoblastoma has been important as the prototype example of the involvement of a tumor suppressor gene in inherited cancer (see chapters 5 and 9). Both inherited and noninherited retinoblastomas involve inactivation or loss of the *RB* tumor suppressor gene. About 40% of retinoblastoma cases are hereditary, arising from inheritance of a defective *RB* allele from one parent.

Children with inherited retinoblastoma frequently develop multiple tumors in both eyes.

Diagnosis and Treatment

Retinoblastoma is usually detected by a change in the appearance of the pupil, particularly the occurrence of a white reflection (Fig. 16.2). If diagnosed early, retinoblastoma can be cured by surgery or radiotherapy without loss of vision.

WILMS' TUMOR

Inheritance and Tumor Suppressor Genes

Wilms' tumor, a tumor of embryonal kidney cells, accounts for about 5% of childhood cancer incidence. It usually occurs by age 5. Like retinoblastoma, Wilms'

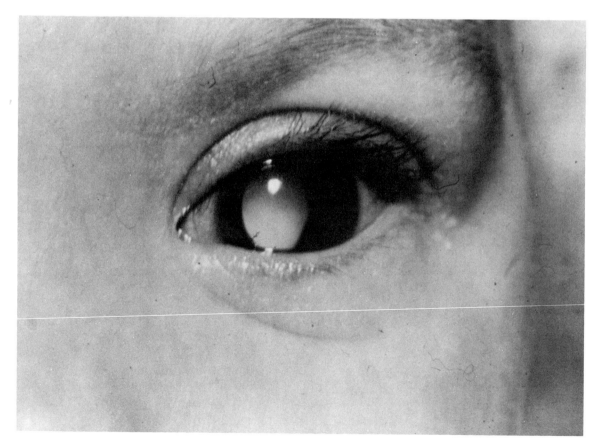

Figure 16.2 The eye of a child with retinoblastoma, showing the white mass of tumor through the pupil. (From L.V. Crowley, *Introduction to Human Disease*, 1988.)

tumor develops as a result of loss or inactivation of a tumor suppressor gene (in this case either *WT1* or a distinct tumor suppressor gene on chromosome 11). Some cases of Wilms' tumor are inherited, and these are likely to involve the development of multiple tumors in both kidneys. Wilms' tumors are also frequently associated with other congenital abnormalities, including aniridia (absence of the iris), genitourinary defects, and mental retardation (the WAGR syndrome).

Diagnosis and Treatment

Wilms' tumors are usually detected as a swelling or mass in the abdomen and diagnosed by X-rays and other imaging techniques. Treatment of Wilms' tumor generally involves a combination of surgery, radiation, and chemotherapy, particularly using the drugs actinomycin D and vincristine. The response of Wilms' tumor patients is good, and cures are obtained in over 80% of cases.

BONE TUMORS

Types and Tumor Suppressor Genes

Two types of bone cancer, **osteosarcoma** and **Ewing's sarcoma,** constitute about 5% of childhood malignancies. Both usually occur in children between the ages of 10 and 18. Osteosarcomas involve loss or inactivation of the *RB* tumor suppressor gene, and they often occur as secondary tumors in patients with inherited retinoblastoma. The *p53* tumor suppressor gene is also frequently lost or inactivated in osteosarcomas.

Diagnosis and Treatment

The primary symptom of both osteosarcomas and Ewing's sarcomas is pain and swelling in the area of the tumor. Diagnosis is then made by X-rays and biopsy. Both of these bone cancers respond well to treatment. Osteosarcomas are usually treated by surgical removal, coupled with chemotherapy using methotrexate, doxorubicin, and cisplatin. In many cases, it is possible to remove only a section of bone to which the tumor is localized, rather than amputating an entire limb. Ewing's sarcoma is highly sensitive to radiation. It is usually treated by radiation plus chemotherapy, with drug combinations including vincristine, cyclophosphamide, actinomycin D, and doxorubicin. Cure rates for these bone cancers are about 50%.

SOFT TISSUE SARCOMAS

Rhabdomyosarcoma

Rhabdomyosarcoma, a cancer of skeletal muscle cells, is the most frequent soft tissue sarcoma in children, accounting for approximately 4% of all childhood cancers. It occurs most frequently in children in two age groups: 2–6 or 14–18. In some cases, rhabdomyosarcomas are inherited—for example, in the Li-Fraumeni cancer family syndrome (see chapter 5). Loss or inactivation of the _RB_ and _p53_ tumor suppressor genes frequently occurs in these tumors.

The initial sign of rhabdomyosarcoma is usually a painless lump or mass, which is then diagnosed by biopsy. Surgery is the primary treatment, combined with radiation and chemotherapy, depending on the extent of tumor spread. Chemotherapy is effective against rhabdomyosarcoma, usually employing drug combinations that include vincristine, actinomycin D, and cyclophosphamide. The overall survival rate is approximately 70%.

Other Soft Tissue Sarcomas

Other soft tissue sarcomas together account for about 1% of childhood cancers. They include fibrosarcomas (cancers of fibrous connective tissue), liposarcomas (cancers of fat cells), hemangiopericytomas (cancers of cells that surround the blood vessels), and synovial sarcomas (cancers of cells that line joint cavities and tendon sheaths). The primary treatment for these diseases is surgery, possibly in combination with radiation and chemotherapy.

SUMMARY

Together with the leukemias and lymphomas discussed in chapter 15, the tumors reviewed above account for about 95% of childhood cancers. The same tumors occur only rarely in adults, where the vast majority of cancers are carcinomas of other sites. Most cancers of children are rapidly growing malignancies, which frequently arise from embryonal cell types. There are both inherited and noninherited forms of a number of childhood cancers, most notably retinoblastoma and Wilms' tumor. Hereditary cases of these cancers result from inheritance of a defective copy of a tumor suppressor gene, _RB_ or _WT1_, respectively. Loss or inactivation of the _RB_ and _p53_ tumor suppressor genes is also important in bone tumors and soft tissue sarcomas. In addition, loss of a tumor suppressor gene on chromosome 1 and amplification of the N-_myc_ oncogene are important events in the progression of neuroblastomas to advanced stages of disease. Many of the childhood cancers are responsive to radiation and chemotherapy, and overall cure rates for these diseases range from 50% to 90%.

KEY TERMS

astrocytoma
medulloblastoma
ependymoma

glioblastoma

neuroblastoma

retinoblastoma

Wilms' tumor

osteosarcoma

Ewing's sarcoma

rhabdomyosarcoma

REFERENCES AND FURTHER READING

General References

American Cancer Society. 1990. *Cancer facts and figures—1990.* American Cancer Society, Atlanta.

Cooper, G.M. 1990. *Oncogenes.* Jones and Bartlett, Boston.

Crist, W.M., and Kun, L.E. 1991. Common solid tumors of childhood. *N. Engl. J. Med.* 324:461–471.

Crowley, L.V. 1988. *Introduction to human disease.* 2nd ed. Jones and Bartlett, Boston.

Knudson, A.G., Jr. 1986. Genetics of human cancer. *Ann. Rev. Genet.* 20:231–251.

National Cancer Institute. 1991. *Physician data query.* BRS Colleague, New York. (Note: This is a computer database that contains frequently updated information on the diagnosis, staging, and treatment of each different type of cancer. It is available at most medical libraries and through several software vendors in addition to BRS Colleague.)

Page, H.S., and Asire, A.J. 1985. *Cancer rates and risks.* 3rd ed. National Institutes of Health, Bethesda.

Pizzo, P.A., Horowitz, M.E., Poplack, D.G., Hays, D.M., and Kun, L.E. 1989. Solid tumors of childhood. In *Cancer: principles and practice of oncology,* ed. DeVita, V.T., Jr., Hellman, S., and Rosenberg, S.A. 3rd ed. J.B. Lippincott, Philadelphia. pp. 1612–1670.

Rubin, E., and Farber, J.L., eds. 1990. *Essential pathology.* J.B. Lippincott, Philadelphia.

Brain Tumors

Black, P.M. 1991. Brain tumors (two parts). *N. Engl. J. Med.* 324:1471–1476 and 1555–1564.

Levin, V.A., Sheline, G.E., and Gutin, P.H. 1989. Neoplasms of the central nervous system. In *Cancer: principles and practice of oncology,* ed. DeVita, V.T., Jr., Hellman, S., and Rosenberg, S.A. 3rd ed. J.B. Lippincott, Philadelphia. pp. 1557–1611.

Malkin, D., Li, F.P., Strong, L.C., Fraumeni, J.F., Jr., Nelson, C.E., Kim, D.H., Kassel, J., Gryka, M.A., Bischoff, F.Z., Tainsky, M.A., and Friend, S.H. 1990. Germ line *p53* mutations in a familial syndrome of breast cancer, sarcomas, and other neoplasms. *Science* 250:1233–1238.

Nigro, J.M., Baker, S.J., Preisinger, A.C., Jessup, J.M., Hostetter, R., Cleary, K., Bigner, S.H., Davidson, N., Baylin, S., Devilee, P., Glover, T., Collins, F.S., Weston, A., Modali, R., Harris, C.C., and Vogelstein, B. 1989. Mutations in the *p53* gene occur in diverse human tumour types. *Nature* 342:705–708.

Neuroblastoma

Look, A.T., Hayes, F.A., Shuster, J.J., Douglass, E.C., Castleberry, R.P., Bowman, L.C., Smith, E.I., and Brodeur, G.M. 1991. Clinical relevance of tumor cell ploidy and N-*myc* gene amplification in childhood neuroblastoma: a pediatric oncology group study. *J. Clin. Oncol.* 9:581–591.

Retinoblastoma

Friend, S.H., Bernards, R., Rogelj, S., Weinberg, R.A., Rapaport, J.M., Albert, D.M., and Dryja, T.P. 1986. A human DNA segment with properties of the gene that predisposes to retinoblastoma and osteosarcoma. *Nature* 323:643–646.

Fung, Y.-K., Murphree, A.L., T'Ang, A., Qian, J., Hinrichs, S.H., and Benedict, W.F. 1987. Structural evidence for the authenticity of the retinoblastoma gene. *Science* 236:1657–1661.

Knudson, A.G., Jr. 1971. Mutation and cancer: statistical study of retinoblastoma. *Proc. Natl. Acad. Sci. USA* 68:820–823.

Lee, W.-H., Bookstein, R., Hong, F., Young, L.-J., Shew, J.-Y., and Lee, E.Y.-H.P. 1987. Human retinoblastoma susceptibility gene: cloning, identification, and sequence. *Science* 235:1394–1399.

Wilms' Tumor

Matsunaga, E. 1981. Genetics of Wilms' tumor. *Hum. Genet.* 57:231–246.

National Cancer Institute. 1989. *Research report: adult kidney cancer and Wilms' tumor.* National Institutes of Health, Bethesda.

National Wilms' Tumor Study Committee. 1991. Wilms' tumor: status report, 1990. *J. Clin. Oncol.* 9:877–887.

Rose, E.A., Glaser, T., Jones, C., Smith, C.L., Lewis, W.H., Call, K.M., Minden, M., Champagne, E., Bonetta, L., Yeger, H., and Housman, D.E. 1990. Complete physical map of the WAGR region of 11p13 localizes a candidate Wilms' tumor gene. *Cell* 60:495–508.

Bone Tumors

Friend, S.H., Horowitz, J.M., Gerber, M.R., Wang, X.-F., Bogenmann, E., Li, F.P., and Weinberg, R.A. 1987. Deletions of a DNA sequence in retinoblastomas and mesenchymal tumors: organization of the sequence and its encoded protein. *Proc. Natl. Acad. Sci. USA* 84:9059–9063.

Malkin, D., Li, F.P., Strong, L.C., Fraumeni, J.F., Jr., Nelson, C.E., Kim, D.H., Kassel, J., Gryka, M.A., Bischoff, F.Z., Tainsky, M.A., and Friend, S.H. 1990. Germ line *p53* mutations in a familial syndrome of breast cancer, sarcomas, and other neoplasms. *Science* 250:1233–1238.

Mulligan, L.M., Matlashewski, G.J., Scrable, H.J., and Cavenee, W.K. 1990. Mechanisms of *p53* loss in human sarcomas. *Proc. Natl. Acad. Sci. USA* 87:5863–5867.

Soft Tissue Sarcomas

Malkin, D., Li, F.P., Strong, L.C., Fraumeni, J.F., Jr., Nelson, C.E., Kim, D.H., Kassel, J., Gryka, M.A., Bischoff, F.Z., Tainsky, M.A., and Friend, S.H. 1990. Germ line *p53* mutations in a familial syndrome of breast cancer, sarcomas, and other neoplasms. *Science* 250:1233–1238.

Mulligan, L.M., Matlashewski, G.J., Scrable, H.J., and Cavenee, W.K. 1990. Mechanisms of *p53* loss in human sarcomas. *Proc. Natl. Acad. Sci. USA* 87:5863–5867.

National Cancer Institute. 1989. *Research report: soft tissue sarcomas in adults and children.* National Institutes of Health, Bethesda.

Chapter 17
Common Solid Tumors of Adults

THE MAJORITY (ABOUT 90%) OF ADULT CANCERS are carcinomas arising from the epithelial cells that cover the surface of the body and line the internal organs. Most of the remainder are leukemias and lymphomas, which constitute approximately 8% of all adult cancers and were discussed in chapter 15. Sarcomas of bone and soft tissues are very rare in adults, altogether accounting for less than 1% of adult cancers. This chapter discusses the common solid tumors of adults in order of their incidence in the United States population (Table 17.1).

LUNG CANCER

Risk Factors

Cancer of the lung (Fig. 17.1) is responsible for approximately 15% of cancer cases, and 28% of cancer deaths, in the United States. As discussed in chapters 3 and 12, 80–90% of lung cancers are caused by cigarette smoking and could therefore be prevented by avoidance of this single carcinogenic agent. Additional risk factors for lung cancer include exposure to excess radon levels in the home and to certain industrial carcinogens, such as asbestos. The effect of these carcinogens combines with that of smoking, imparting an extremely high lung cancer risk to smokers exposed to these added carcinogenic insults. As discussed in chapter 5, lung cancer risk may also be affected by inherited differences in sensitivity to carcinogens.

Types, Oncogenes, and Tumor Suppressor Genes

Lung cancers are classified as small cell and non-small cell carcinomas. The most common non-small cell types are adenocarcinomas, squamous cell carcinomas, and large cell carcinomas. All types of lung cancer are associated with loss or inactivation of *p53* and a tumor suppressor gene on chromosome 3. Small cell carcinomas are also associated with loss or inactivation of the *RB* tumor suppressor gene and amplification of members of the *myc* oncogene family (c-*myc*, L-*myc*, and N-*myc*). *Ras*K oncogenes are frequently involved in adenocarcinomas.

Diagnosis and Treatment

Lung cancer is difficult to detect in early stages of the disease process. Symptoms, which include persistent cough, chest pain, and sputum streaked with blood, do not usually appear until the cancer has reached an advanced stage at which metastasis has already occurred. Treatment is usually surgery, combined with radiation and chemotherapy. However, lung cancers are not very responsive, and the five-year survival rate is only about 13%.

Table 17.1 Adult Solid Tumors

Type of Cancer	Cases per Year (United States)		Deaths per Year (United States)	
Lung	157,000	(15%)	142,000	(28%)
Colon/rectum	155,000	(15%)	61,000	(12%)
Breast	151,000	(14%)	44,000	(9%)
Prostate	106,000	(10%)	30,000	(6%)
Urinary				
Bladder	49,000	(5%)	10,000	(2%)
Kidney	24,000	(2%)	10,000	(2%)
Uterus				
Cervix	13,500	(1%)	6,000	(1%)
Endometrium	33,000	(3%)	4,000	(1%)
Oral cavity	31,000	(3%)	8,000	(2%)
Pancreas	28,000	(3%)	25,000	(5%)
Skin	28,000	(3%)	9,000	(2%)
Stomach	23,000	(2%)	14,000	(3%)
Ovary	21,000	(2%)	12,000	(2%)
Brain	16,000	(1.5%)	11,000	(2%)
Liver	15,000	(1.5%)	12,000	(2%)
Larynx	12,000	(1%)	4,000	(1%)
Thyroid	12,000	(1%)	1,000	(0.2%)
Esophagus	11,000	(1%)	10,000	(2%)
Testis	6,000	(0.6%)	400	(0.1%)
	892,000	(85%)	413,000	(81%)

Percentages refer to total United States cancer incidence and mortality. Nonmelanoma skin cancers and cervical carcinomas *in situ* are not included in incidence figures. The remaining cancers, not included in this table, are the leukemias and lymphomas (8% of incidence, 9% of mortality), sarcomas (1% of incidence and mortality), and relatively rare carcinomas of other and unspecified sites.

COLON AND RECTUM CANCER

Risk Factors

In combination, colon and rectum cancers (Fig. 17.2) account for approximately 15% of United States cancer incidence and 12% of mortality. Cancers of these sites are adenocarcinomas. Colon carcinoma is about twice as common as rectal carcinoma. Rare forms of colon cancer are inherited—for example, familial adenomatous polyposis, as discussed in chapter 5. In addition, the risk of developing colon and rectum cancers is about twice as high for individuals with an

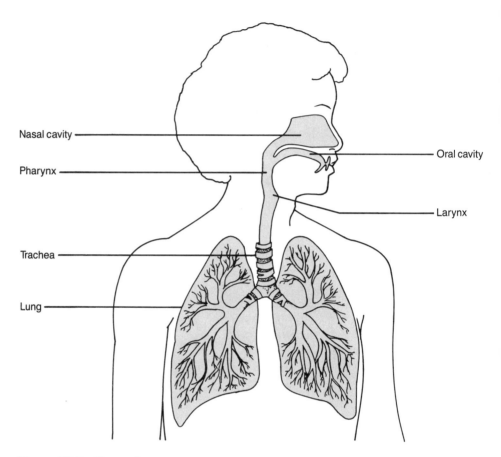

Figure 17.1 The respiratory system.

immediate family member who has had the disease as it is for the general population. It has been estimated that increased susceptibility to colon cancer is inherited by 10–20% of the population and may contribute to a substantial fraction of cases. Inflammatory bowel disease (e.g., ulcerative colitis) is also associated with a high risk of developing colon cancer. An individual who has had one polyp or carcinoma is also at increased risk of developing a second. Dietary factors are thought to be important determinants of the risk of colon and rectum cancers. As discussed in chapters 3 and 12, an increased incidence of these cancers appears to be associated with diets that are high in fat and low in fiber or other components of fruits and vegetables. The clearest risk factor is a high-fat diet (greater than 40% of total calories), which appears to result in about a twofold increase in colon cancer incidence.

Oncogenes and Tumor Suppressor Genes

Colon and rectum carcinomas are the best characterized cancers with respect to the roles of oncogenes and tumor suppressor genes in tumor development. The *ras*K oncogene and the *MCC* tumor suppressor gene appear to be involved in

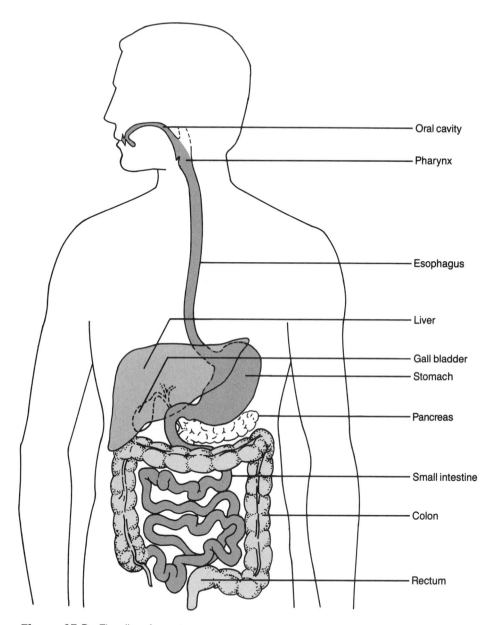

Figure 17.2   *The digestive system.*

early stages of the disease process, contributing to the development of pre-malignant adenomas (polyps). Progression to malignant carcinomas then involves loss or inactivation of the *p53* and *DCC* tumor suppressor genes.

Early Detection

In part because of the gradual progression of colon and rectum carcinomas, early detection is a feasible approach to reducing mortality from these cancers. As discussed in chapter 13, early stages of colon and rectum tumors can be detected, albeit with varying degrees of sensitivity and reliability, by digital rectal examination, sigmoidoscopy, and fecal occult blood testing. The use of all three of these screening tests is recommended by the American Cancer Society, particularly after age 50.

Treatment

The benefits of early detection are apparent in the treatment of colon and rectum cancers, which primarily relies on surgical removal of the tumor, sometimes combined with radiation and chemotherapy. Overall survival is about 50%, but this is largely dependent upon the extent of tumor progression at the time of diagnosis. Premalignant polyps are readily cured by removal during endoscopy. Cure rates for early-stage localized colon and rectum carcinomas are also high—close to 90 and 80%, respectively. After the cancers have spread regionally (to lymph nodes and adjacent organs), however, survival rates drop to about 50%, and survival rates for patients with distant metastases are less than 10%. Postsurgical chemotherapy with fluorouracil plus levamisole has recently been found to reduce mortality by about 30% for patients with colon cancer that has spread to regional lymph nodes; this is now offered as standard treatment to these patients (see chapter 14).

BREAST CANCER

Breast cancer accounts for approximately 14% of cancer incidence, and 9% of cancer mortality, in the United States. It is the most common cancer among women, and it is estimated that nearly one in every ten women will develop breast cancer at some point in life. Breast cancer also occurs in men, but is more than 100-fold less frequent than in women.

Risk Factors

Some rare forms of breast cancer are directly inherited, as discussed in chapter 5. In addition, the overall risk of developing breast cancer is increased two- or threefold for women whose mothers or sisters have had the disease, presumably reflecting inherited disease susceptibility (see chapter 5). Further, women who have had one breast cancer are at higher than average risk of developing a second. Other risk factors for breast cancer relate principally to the effects of hormones on the breast tissue. Breast cancer risk is increased about threefold for women who have never had children or who had their first child after age 35. In addition, early menarche (before age 12) or late menopause (after age 55) are associated with up to twofold increases in breast cancer risk. Oral contraceptive use is not associated with a significant overall increase in breast cancer incidence, although use of birth control pills for several years prior to first pregnancy may result in a modest increase in breast cancer risk. Long-term postmenopausal estrogen replacement therapy may also be associated with a modest increase (approximately 30%) in breast cancer risk, but this has not been definitively established. In addition, obesity may increase the risk of breast cancer by up to 50%.

Types of Breast Cancer

There are several histological types of breast carcinomas. The majority (nearly 90%) arise in the ducts (Fig. 17.3) and are called **ductal carcinomas.** They are further distinguished according to cell types. The most common is called invasive ductal, NOS (for "not otherwise specified"). Other types of ductal carcinomas are medullary, tubular, and mucinous. About 5% of breast cancers arise in

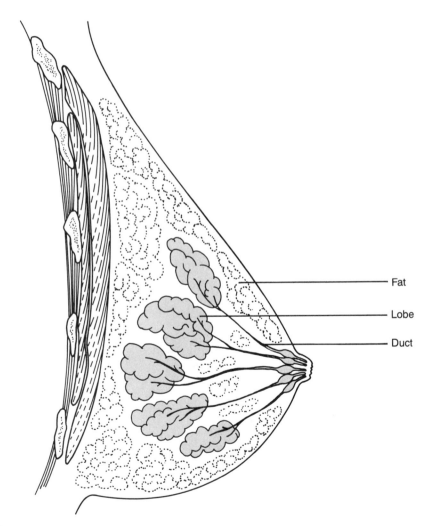

Fat

Lobe

Duct

Figure 17.3 Cross-section of a breast.

the lobules and are called **lobular carcinomas.** The remainder are classified as Paget's disease, which involves the nipple, and inflammatory carcinomas, which are associated with apparent inflammation of the breast.

Oncogenes and Tumor Suppressor Genes

Breast cancers frequently involve loss or inactivation of the *RB* and *p53* tumor suppressor genes; loss of other putative tumor suppressor genes on chromosomes 1, 3, 11, and 18; and amplification of the c-*myc* and *erb*B-2 oncogenes. Amplification of *erb*B-2, in particular, is more frequent in advanced-stage tumors and may be correlated with a less favorable prognosis (see chapter 10).

Early Detection

Early detection is of major importance in reducing breast cancer mortality, and screening for breast cancer by breast self-examination, examination by a physician, and mammography is recommended. It is estimated that annual screening by mammography for women over age 40 reduces breast cancer mortality by about 30%.

Treatment

Breast cancers are treated by surgery and radiation, sometimes combined with chemotherapy and hormone therapy. The benefits of early detection are apparent in terms of survival rates. The cure rate for carcinoma *in situ* is virtually 100%, and for localized invasive cancer it is 90%. The survival rate drops to about 70% for disease that has spread to regional lymph nodes, however, and to less than 20% once metastasis to distant body sites has occurred.

PROSTATE CANCER

Cancer of the prostate (Fig. 17.4) accounts for approximately 10% of all cancer cases and about 6% of total cancer mortality. It is the most common cancer in men, and approximately 1 out of every 11 men will develop this disease.

Risk Factors and the *RB* Tumor Suppressor Gene

The risk of prostate cancer (adenocarcinoma) increases with age, and over 80% of prostate cancers occur in men past age 65. Possible causes and risk factors for the disease have not been established, although development of these cancers may involve stimulation of prostate cell proliferation by testosterone. Prostate carcinomas frequently involve loss or inactivation of the *RB* tumor suppressor gene.

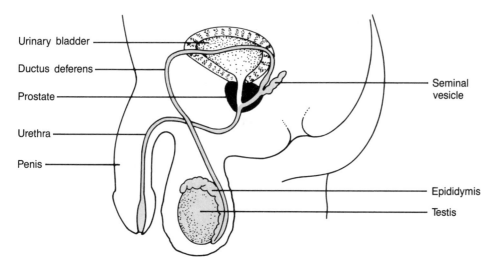

Urinary bladder

Ductus deferens

Prostate

Urethra

Penis

Seminal
vesicle

Epididymis

Testis

Figure 17.4   *The male reproductive tract.*

Early Detection

Early detection plays a major role in reducing mortality from prostate cancer. The most effective screen for prostate cancer is a rectal exam, which should be performed annually for men over the age of 40. Ultrasonography and blood tests for prostate-specific antigen (see chapter 13) are being evaluated as additional early detection methods. Symptoms, which may not occur in early stages of the disease, include problems in urination and persistent pain in the lower back, hips, or pelvis.

Treatment

The overall survival rate for prostate cancer is about 70%. More than half of prostate cancers are diagnosed while the disease is still localized, and the cure rate for these patients is over 80%. Except for very small tumors, surgical treatment usually involves removal of the entire prostate gland and surrounding tissue. Impotence, resulting from nerve damage, used to be a nearly universal side effect of this procedure, but improved surgical techniques now result in the recovery of potency in 50–80% of patients. An alternative for localized prostate cancer is radiation therapy, which usually preserves potency. More advanced stages of prostate cancer are treated by hormone therapy. As discussed in chapter 14, the growth of prostate cancer cells is dependent on testosterone, and several kinds of hormone therapy may be employed to block hormonal stimulation of the cancer cells. These treatments include removal of the testes (orchiectomy) and administration of estrogen, analogs of gonadotropin-releasing hormone, or antiandrogens. Hormone therapy is not curative but usually slows disease progression, relieves symptoms, and prolongs the lives of most patients.

URINARY CANCERS

Urinary cancers account for a total of approximately 7% of cancer incidence and 4% of mortality. About two-thirds of these cancers are bladder carcinomas, with most of the remainder being cancers of the kidney (Fig. 17.5).

Bladder Cancer: Risk Factors and Tumor Suppressor Genes

Almost all bladder cancers originate from the transitional epithelium of the bladder; they are therefore called **transitional cell carcinomas.** Bladder cancer occurs primarily at age 60 and above and is nearly three times more frequent in men than in women. Smoking results in about a twofold increase in bladder cancer risk. Occupational exposure to dyes, rubber, and leather are additional risk factors. In contrast to early concerns, artificial sweeteners or coffee drinking do not appear to be associated with increased bladder cancer incidence. Loss or inactivation of the *RB* and *p53* tumor suppressor genes occurs frequently in bladder carcinomas.

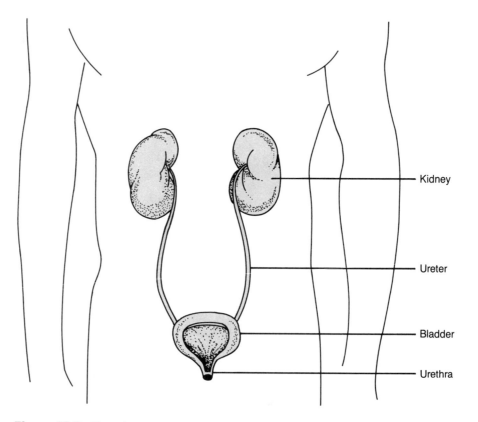

Kidney

Ureter

Bladder

Urethra

Figure 17.5 The urinary system.

Diagnosis and Treatment of Bladder Cancer

The primary symptom of bladder cancer is blood in the urine. Diagnosis is usually by examination of the bladder by **cystoscopy.** The primary treatment is surgery, either alone or in combination with radiation or chemotherapy. The cure rate is nearly 90% for patients with localized disease. For patients with regional disease, survival is about 40%, and it is about 10% for patients with distant metastases. It is noteworthy that favorable responses in over 50% of patients with advanced disease have recently been obtained by combination chemotherapy with the drugs methotrexate, vinblastine, doxorubicin, and cisplatin. The efficacy of this regimen, which may represent a significant treatment advance, is undergoing further evaluation.

Kidney Cancer

Kidney cancer, like bladder cancer, is more frequent among men than women, and smoking is associated with about a twofold increase in risk. Obesity is also a risk factor for women, possibly associated with excess estrogen production. **Renal cell carcinoma,** the most common form of kidney cancer, is associated with loss of a putative tumor suppressor gene on chromosome 3.

Early detection of kidney cancer is difficult, since early stages of the disease produce no clear signs or symptoms. The disease may be suggested by a variety of symptoms, including blood in the urine; pain in the side, abdomen, or back; weight loss; and weakness. Diagnosis is made by a variety of imaging techniques. The overall survival rate is about 40%, with surgery being the primary treatment. Prognosis is good for disease that remains localized to the kidney, but poor for patients with disseminated disease.

UTERINE CANCERS

Cancers of the uterus (Fig. 17.6) account for approximately 4% of total cancer incidence and 2% of mortality. The cancers of this organ include two distinct types, carcinomas of the cervix and carcinomas of the endometrium.

Papillomaviruses and Cervical Cancer Risk Factors

Cervical carcinoma (usually squamous cell carcinoma) is responsible for about 30% of uterine cancer cases and about 60% of deaths. As discussed in chapter 4, most cases of cervical carcinoma are associated with human papillomaviruses, which are sexually transmitted. The major risk factors for cervical carcinoma are therefore associated with sexual practices, such as intercourse with multiple partners. Cigarette smoking also appears to be associated with an increased disease incidence.

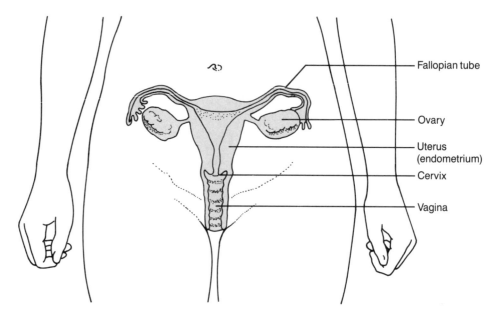

Figure 17.6 The female reproductive tract.

Early Detection of Cervical Cancer

Early detection by the Pap smear has been extremely effective against cervical carcinoma. As discussed in chapter 13, very early stages of disease can be reliably detected by this method. At present, about 50,000 cases of cervical carcinoma *in situ* are detected annually by the Pap test. These cases are not included in the incidence and mortality figures given above, which are based on 13,500 cases of cervical carcinoma per year that remain undetected until more advanced stages. Thus, screening by the Pap smear currently detects over 75% of cervical carcinomas before they become invasive, and the use of regular screening would be expected to prevent development of most of the remaining invasive cases of this disease.

Treatment of Cervical Cancer

Cervical cancer is treated by surgery or radiation. The cure rate for carcinoma *in situ* is virtually 100% following minor procedures, and localized carcinoma remains curable in nearly 90% of cases. Survival rates for disease that has spread regionally, however, drop to about 50%, and rates fall to below 20% for metastatic cancers. Therefore, the key to effective reduction in mortality from this disease remains early detection.

Endometrial Cancer and Estrogens

Endometrial cancer (usually adenocarcinoma) accounts for about 70% of uterine cancers, excluding cervical carcinomas *in situ*. The risk factors for endometrial cancer involve excess stimulation of endometrial cell proliferation by estrogen in the absence of progesterone. Obesity increases endometrial cancer risk several-fold, most likely as a result of estrogen production by fat cells. Other risk factors are failure to ovulate and a history of infertility, which may also be associated with an estrogen imbalance. In addition, the risk of endometrial cancer is increased by high-dose postmenopausal estrogen replacement therapy, particularly in the absence of progesterone to counteract the stimulatory effect of estrogen on endometrial cell proliferation. Currently available birth control pills, however, contain progesterone as well as estrogen, and they appear to result in a reduced risk of endometrial cancer.

Diagnosis and Treatment of Endometrial Cancer

Unfortunately, the Pap smear is only partially effective in diagnosis of endometrial cancer, and this disease is usually not detected until symptoms are evident. The disease is most common in women between the ages of 55 and 70, and the American Cancer Society recommends that women at increased risk (e.g., with a history of infertility or obesity) have an endometrial biopsy at menopause. Patients receiving estrogen replacement therapy should have endometrial biopsies repeated periodically. The most common symptom of the disease after menopause is vaginal bleeding.

Diagnosis of endometrial cancer is by biopsy or dilatation and curettage (D and C). Surgery and radiotherapy are effective in treatment of localized disease, and survival rates for these patients are over 90%. Regional spread of disease outside of the uterus is still treatable by radiation, with survival rates in excess of 65%. Advanced metastatic endometrial cancer is generally treated by administration of progesterone, which results in prolonged survival, although not cures, in about one-third of patients. The overall survival rate for endometrial cancer is about 85%.

ORAL CANCERS

Risk Factors

Oral cancers, including cancers of the lip, tongue, mouth, and pharynx (Figs. 17.1 and 17.2), account for approximately 3% of cancer incidence and 1.6% of mortality. Cancers at these sites are usually squamous cell carcinomas. The major risk factors for oral cancers are tobacco use and excessive consumption of alcoholic beverages, particularly in combination. As a consequence of such carcinogen exposure, patients with one such cancer are at high risk of developing a second, independent tumor. Excessive exposure to sunlight is also an important risk factor for lip cancer.

The genetic alterations involved in oral cancers have not been well characterized. However, given the relationship of these tumors to tobacco carcinogens, they might be expected to involve some of the same types of carcinogen-related events found in lung cancers, such as activation of *ras* oncogenes and inactivation of the *p53* tumor suppressor gene.

Diagnosis and Treatment

Early detection is important in the prognosis of oral cancers; it is best accomplished by regular examination of the mouth and throat by dentists and physicians. Leukoplakia, which may be observed during such an examination, is a white patch (*leuko* = white and *plakia* = patch) on a mucous membrane of the mouth. Sometimes, but not always, leukoplakia represents either a preneoplastic cell proliferation or a more advanced stage of carcinoma development. Following biopsy, approximately 10% of patients with leukoplakia are diagnosed with preneoplastic dysplasias or carcinoma *in situ*, and approximately 5% with invasive carcinoma.

Treatment of oral cancers is primarily by surgery or radiotherapy. The overall survival rate for these cancers is about 50%, depending on the site and stage of disease. The survival rate for lip cancer, for example, is about 90%, while that for cancer of the pharynx is approximately 30%.

PANCREATIC CANCER

Cancer of the pancreas (adenocarcinoma) accounts for approximately 3% of cancer incidence and 5% of mortality. Most cases occur after age 65. Cigarette smoking, which increases pancreatic cancer incidence about twofold, is the only known risk factor. Nearly 90% of these cancers involve *ras*K oncogenes.

There is no known early detection method for pancreatic cancer, and the disease produces no symptoms until it reaches advanced stages. Treatment is ineffective, and five-year survival rates are only 3%.

MELANOMA AND OTHER SKIN CANCERS

Melanoma: Risk Factors and Oncogenes

Melanoma is a carcinoma arising from pigment-producing cells in the skin. It accounts for approximately 2.6% of cancer incidence and 1.2% of mortality. The incidence of melanoma throughout the world has been steadily increasing for the last 40 years.

The major risk factor for melanoma is ultraviolet radiation from sunlight. It is about ten times more frequent among whites than blacks, presumably because the greater pigmentation of black skin protects against radiation. In rare cases, familial susceptibility to melanoma may be inherited—for example in individuals with dysplastic nevus syndrome or xeroderma pigmentosum (see chapter 5). Melanomas frequently involve *ras*N oncogenes, which may be mutated directly by the action of ultraviolet light (see chapter 10).

Detection and Treatment of Melanoma

Early detection is critical to the outcome of melanoma and is best accomplished by self-examination of the skin. Melanomas may develop within a mole or as a new mole-like growth. They are characterized by increasing size and changes in color (Fig. 17.7). The American Cancer Society emphasizes four warning signs of melanoma: (1) assymetry, meaning the shape of one half of a mole is different from the other; (2) border irregularities, such as uneven, ragged, or notched edges; (3) different colors within a mole; and (4) a mole greater than 6 mm (about ¼ inch) in diameter.

Treatment of melanoma is primarily surgical removal, possibly including excision of regional lymph nodes. Survival rates are about 90% for localized melanoma, but metastatic disease is not responsive to therapy. Since melanomas can metastasize quickly, early detection is the major determinant of prognosis.

Nonmelanoma Skin Cancers

The nonmelanoma skin cancers, basal and squamous cell carcinomas, are extremely common but seldom lethal. For this reason, these cancers are generally not included in calculations of cancer incidence, including those in this chapter. The nonmelanoma skin cancers, like melanoma, are caused by solar ultraviolet radiation. In contrast to melanoma, however, basal and squamous cell carcinomas metastasize very slowly. They are consequently readily cured by surgery or radiation. There are estimated to be more than 600,000 cases of nonmelanoma skin cancers diagnosed each year, but they result in only about 2,500 deaths—a cure rate of approximately 99.5%.

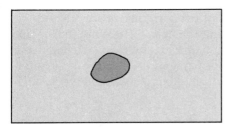

Normal mole

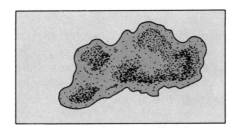

Melanoma

Figure 17.7 Comparison of a normal mole and a melanoma. Melanomas are characterized by increasing size and irregularities in shape, border, and coloration.

STOMACH CANCER

Stomach cancer (adenocarcinoma) accounts for approximately 2% of United States cancer incidence and 2.7% of mortality, although it is much more common in other parts of the world. For example, stomach cancer in Japan is about five times more common than in the United States. It is also noteworthy that, as discussed in chapter 1, stomach cancer in the United States has declined more than fivefold since 1930.

Risk Factors

Excessive consumption of cured, smoked, and pickled foods, which contain large amounts of salt, nitrates, and nitrites, is thought to be a major risk factor for stomach cancer. As discussed in chapters 3 and 12, nitrates and nitrites are converted to potent carcinogens, nitrosamines. Vitamin C blocks the formation of nitrosamines and may be responsible for the protective effect of fresh fruits and vegetables against stomach cancer. The decline in stomach cancer is thought to be the result of the use of refrigeration for food preservation, correlated with a decreased use of cured and smoked foods as well as increased consumption of fresh fruits and vegetables.

Diagnosis and Treatment

Early detection of stomach cancer is difficult, since symptoms frequently do not develop until the disease is relatively advanced. Symptoms include indigestion, abdominal discomfort, loss of appetite, and weight loss. Diagnosis involves X-rays and **gastroscopy.** Surgery is the primary treatment, which may be combined with radiation or chemotherapy. Only surgery is potentially curative of localized disease, however, and overall survival rates are only 15%.

OVARIAN CANCER

Types of Ovarian Tumors

Cancer of the ovary (see Fig. 17.6) accounts for approximately 2% of cancer incidence and 2.4% of mortality. About 90% of ovarian cancers are adenocarcinomas arising from the epithelial cells covering the ovary. There are four major types of these tumors: serous, mucinous, endometrioid, and clear cell. Relatively infrequent ovarian tumors arise from germ cells (**dysgerminomas, yolk sac carcinomas, teratomas,** and **choriocarcinomas**) or from other ovarian cell types, such as granulosa, theca, and Sertoli-Leydig cells.

Risk Factors and Oncogenes

The risk of ovarian cancer appears to be affected primarily by reproductive history, indicating that hormonal factors are important in cancer development.

Women who have not had children have about a twofold increased risk of developing ovarian cancer, and the risk decreases further for women who have had several pregnancies. Birth control pills (which, like pregnancy, prevent ovulation) also appear to decrease ovarian cancer risk. In addition, women who have breast or endometrial cancers are about twice as likely to develop ovarian cancer. Ovarian cancers frequently involve inactivation of the *p53* tumor suppressor gene. Amplification of the *erb*B-2 oncogene is also a frequent event, which appears to correlate with more aggressive tumor growth. The *gip* oncogene is often involved in the relatively rare ovarian tumors arising from granulosa and theca cells, which are hormone-producing cells that surround the egg in ovarian follicles.

Diagnosis and Treatment

Ovarian cancer is usually asymptomatic until relatively advanced stages, and it is not detected by the Pap test. Early detection is best accomplished through periodic pelvic examinations, but the majority of ovarian carcinomas have already reached an advanced stage by the time of diagnosis. Ultrasonography and tests for an ovarian tumor marker (**CA-125**) are being evaluated as possible screening methods. Common symptoms are abdominal swelling and bloating. Treatment includes surgery, radiation, and chemotherapy. The survival rate for localized disease is over 80%, but it drops to 45% for cancers that have spread beyond the ovary and to less than 20% once metastasis to distant sites has occurred. Overall, the five-year survival rate for ovarian cancer is 38%.

BRAIN TUMORS

Types of Adult Brain Tumors and Oncogenes

Brain tumors account for approximately 1.5% of adult cancer incidence and 2.2% of mortality. As discussed for childhood brain tumors in chapter 16, causes and risk factors are unknown. There are many different types of brain cancers, the most prevalent in adults being **astrocytomas, ependymomas,** and **meningiomas.** Astrocytomas and ependymomas are malignant, whereas meningiomas are benign. The astrocytomas, which include **glioblastomas,** account for about 50% of adult brain tumors. These tumors frequently involve loss or inactivation of the *p53* tumor suppressor gene. The *gsp* oncogene is frequently involved in pituitary tumors, benign neoplasms which constitute about 5% of brain tumors.

Diagnosis and Treatment

Headaches are the most common symptom of adult brain tumors, which are then diagnosed by a variety of imaging techniques, such as computed tomography (CT scan) and magnetic resonance imaging (MRI), as well as electroencephalography (EEG). The standard treatment is surgery, often combined with radiation and chemotherapy. The overall survival rate is under 30%, but this varies considerably according to tumor type and location. Even benign brain tumors, such as meningiomas, are not always curable, since their location may preclude complete surgical removal.

LIVER CANCER

Risk Factors, Hepatitis B Virus, and *p53*

Cancers of the liver and gall bladder (see Fig. 17.2) constitute about 1.4% of United States cancer incidence and 2.3% of mortality. The primary cause of the most common type of liver cancer, hepatocellular carcinoma, is infection with hepatitis B virus, and the disease is much more common in other parts of the world where infection with this virus is prevalent (see chapter 4). Aflatoxin is also a potent liver carcinogen, which is produced by a fungus that frequently contaminates poorly stored peanuts and other foods. The level of aflatoxin allowed in peanut butter is carefully controlled in the United States, but aflatoxin contamination of foodstuffs may be associated with high rates of hepatocellular carcinoma in other countries. Cirrhosis, which can be caused by excessive alcohol consumption, is also associated with an increased risk of liver cancer. Inactivation or loss of the *p53* tumor suppressor gene, possibly resulting from aflatoxin-induced mutations (see chapter 10), is a frequent event in hepatocellular carcinomas.

Diagnosis and Treatment

Symptoms of liver cancer are pain, weight loss, loss of appetite, and fatigue. The disease may be treated by surgery, radiation, and chemotherapy. Prognosis is poor, however, and the five-year survival rate is only 4%.

LARYNGEAL CANCER

Risk Factors

Cancer of the larynx (see Fig. 17.1) accounts for 1.2% of cancer incidence and 0.7% of mortality. Laryngeal cancers are squamous cell carcinomas which, like lung cancer, are primarily caused by cigarette smoking. As in the case of oral cancers, excessive alcohol consumption, particularly in combination with smoking, may also increase laryngeal cancer risk.

Diagnosis and Treatment

Hoarseness is an early symptom of laryngeal cancer, and persistent hoarseness should be evaluated by a physician. Additional symptoms include pain or sore throat. Diagnosis is by examination with a laryngoscope. The primary treatment for laryngeal cancer is surgery or radiotherapy. Radiation is the preferred treatment for localized laryngeal cancers, since loss of speech is prevented. The overall survival rate for patients with laryngeal cancer is 68%.

THYROID CANCER

Types, Risk Factors, and Oncogenes

Cancers of the thyroid gland account for about 1% of cancer incidence and 0.2% of mortality. There are four major histological types: papillary, follicular, medullary, and anaplastic. Papillary carcinomas are the most common, accounting for over 50% of all thyroid cancers. Some medullary carcinomas, which arise from cells secreting the hormone calcitonin, are inherited as part of multiple endocrine neoplasia types 2A and 2B (see chapter 5). The major environmental risk factor for thyroid cancer is radiation. In particular, childhood exposure to radiation of the head and neck as therapy for such conditions as enlarged thymus or tonsils is associated with a subsequent increased incidence of thyroid carcinomas. Radiation was a common treatment for these conditions prior to the 1950s, although it is no longer used. Several oncogenes are regularly involved in papillary thyroid carcinomas, including all three members of the *ras* family (*ras*H, *ras*K, and *ras*N), *ret*, and *trk*.

Diagnosis and Treatment

Thyroid carcinomas are usually detected as a lump during physical examination. The primary treatment is surgery, frequently in combination with administration of radioactive iodine, a source of radiation that localizes in the thyroid gland and may serve to eliminate residual cancer cells. The prognosis for most patients with thyroid cancer is excellent, and survival rates are greater than 90%. However, this is not the case for anaplastic carcinoma; five-year survival rates for patients with this disease (which represents approximately 20% of all thyroid carcinomas) are less than 10%.

ESOPHAGEAL CANCER

Risk Factors and *p53*

Cancer of the esophagus (see Fig. 17.2) accounts for about 1% of cancer incidence and nearly 2% of mortality. Cancers of this site are usually squamous cell carcinomas. The major risk factors for esophageal cancer are the same as for oral cancers: tobacco and excessive alcohol consumption. The *p53* tumor suppressor gene is frequently lost or inactivated in esophageal carcinomas.

Diagnosis and Treatment

The most common symptom of esophageal cancer is difficulty in swallowing, but the disease has frequently progressed to an advanced state before symptoms are evident. Surgery and radiotherapy are the principal treatments, but the prognosis for patients with esophageal cancer is poor, corresponding to a five-year survival rate of 8%.

TESTICULAR CANCER

Cancer of the testes (see Fig. 17.4) accounts for about 0.6% of total cancer incidence and less than 0.1% of mortality. The causes of testicular cancer are unknown, but men between the ages of 20 and 35 are at the greatest risk. The disease is usually detected as a lump on the testis, either during self-examination or by a physician.

Types of Testicular Cancer

There are several histologic types of testicular cancers, nearly all of which arise from germ cells within the testis. **Seminomas,** which account for approximately 40% of all testicular cancers, are composed of morphologically undifferentiated cells. The nonseminomatous testicular cancers, which include **choriocarcinomas, embryonal carcinomas, teratomas,** and **yolk sac carcinomas,** contain more specialized cell types.

Diagnosis and Treatment

Seminomas are usually detected at an early stage, when they are still localized. In addition, these tumors are particularly sensitive to radiation. Consequently, treatment of seminomas with a combination of surgery plus radiotherapy is curative for more than 90% of patients.

The nonseminomatous testicular cancers, on the other hand, are usually not detected until a more advanced stage of disease, and in most cases these tumors have already metastasized by the time of diagnosis. Nonetheless, recent advances in chemotherapy have made this a highly treatable form of cancer. In particular, these tumors are unusually sensitive to the drug cisplatin. Therefore, treatment of nonseminomatous testicular cancers usually consists of surgery plus combination chemotherapy. A common drug combination is cisplatin, bleomycin, and vinblastine. Survival rates for nonseminomatous testicular carcinomas are in excess of 75%, and overall survival rates for all testicular carcinomas are in excess of 90%.

SUMMARY

The solid tumors reviewed above account for about 85% of United States cancer incidence and 81% of mortality. Together with the leukemias and lymphomas (8% of incidence and 9% of mortality), these diseases account for over 90% of the cancer burden in this country, the remainder being sarcomas (approximately 1% of both incidence and mortality) and comparatively infrequent carcinomas of other sites. The major risk factors for these cancers include tobacco, alcohol, diet, radiation, estrogen imbalances, and viruses. Alterations in a variety of oncogenes and tumor suppressor genes are involved in these tumors, often with multiple genes contributing in a cumulative fashion to the development of malignancy.

In most cases, the survival rates for adult patients with solid tumors are determined by the stage at which cancer is diagnosed. In general, localized cancers can be successfully treated, but therapy is much less effective once the cancer has spread from its site of origin. Hence, early diagnosis is often the major determinant of disease outcome.

KEY TERMS

small cell carcinoma
squamous cell carcinoma
adenocarcinoma
adenoma
polyp
ductal carcinoma
lobular carcinoma
transitional cell carcinoma
renal cell carcinoma
leukoplakia
dysgerminoma
yolk sac carcinoma
teratoma
choriocarcinoma
CA-125
astrocytoma
ependymoma
meningioma

glioblastoma
papillary carcinoma
follicular carcinoma
medullary carcinoma
anaplastic carcinoma
seminoma
embryonal carcinoma

REFERENCES AND FURTHER READING

General References

American Cancer Society. 1990. *Cancer facts and figures—1990.* American Cancer Society, Atlanta.

Bos, J.L. 1989. *ras* oncogenes in human cancer: a review. *Cancer Res.* 49:4682–4689.

Cooper, G.M. 1990. *Oncogenes.* Jones and Bartlett, Boston.

Crowley, L.V. 1988. *Introduction to human disease.* 2nd ed. Jones and Bartlett, Boston.

DeVita, V.T., Jr., Hellman, S., and Rosenberg, S.A., eds. 1989. *Cancer: principles and practice of oncology.* 3rd ed. J.B. Lippincott, Philadelphia.

National Cancer Institute. 1991. *Physician data query.* BRS Colleague, New York. (*Note:* This is a computer database that contains frequently updated information on the diagnosis, staging, and treatment of each different type of cancer. It is available at most medical libraries and through several software vendors in addition to BRS Colleague.)

Page, H.S., and Asire, A.J. 1985. *Cancer rates and risks.* 3rd ed. National Institutes of Health, Bethesda.

Rubin, E., and Farber, J.L., eds. 1990. *Essential pathology.* J.B. Lippincott, Philadelphia.

Lung Cancer

Dillman, R.O., Seagren, S.L., Propert, K.J., Guerra, J., Eaton, W.L., Perry, M.C., Carey, R.W., Frei, E.F. III., and Green, M.R. 1990. A randomized trial of induction chemotherapy plus high-dose radiation versus radiation alone in stage III non-small-cell lung cancer. *N. Engl. J. Med.* 323:940–945.

Harbour, J.W., Lai, S.-L., Whang-Peng, J., Gazdar, A.F., Minna, J.D., and Kaye, F.J. 1988. Abnormalities in structure and expression of the human retinoblastoma gene in SCLC. *Science* 241:353–357.

Johnson, B.E., Ihde, D.C., Makuch, R.W., Gazdar, A.F., Carney, D.N., Oie, H., Russell, E., Nau, M.M., and Minna, J.D. 1987. *myc* family oncogene amplification in tumor cell lines established from small cell lung cancer patients and its relationship to clinical status and course. *J. Clin. Invest.* 79:1629–1634.

National Cancer Institute. 1989. *Research report: cancer of the lung.* National Institutes of Health, Bethesda.

Nigro, J.M., Baker, S.J., Preisinger, A.C., Jessup, J.M., Hostetter, R., Cleary, K., Bigner, S.H., Davidson, N., Baylin, S., Devilee, P., Glover, T., Collins, F.S., Weston, A.,

Modali, R., Harris, C.C., and Vogelstein, B. 1989. Mutations in the *p53* gene occur in diverse human tumour types. *Nature* 342:705–708.

Slebos, R.J.C., Kibbelaar, R.E., Dalesio, O., Kooistra, A., Stam, J., Meijer, C.J.L.M., Wagenaar, S.S., Vanderschueren, R.G.J.R.A., van Zandwijk, N., Mooi, W.J., Bos, J.L., and Rodenhuis, S. 1990. K-*ras* oncogene activation as a prognostic marker in adenocarcinoma of the lung. *N. Engl. J. Med.* 323:561–565.

Takahashi, T., Nau, M.M., Chiba, I., Birrer, M.J., Rosenberg, R.K., Vinocour, M., Levitt, M., Pass, H., Gazdar, A.F., and Minna, J.D. 1989. *p53*: a frequent target for genetic abnormalities in lung cancer. *Science* 246:491–494.

Colon and Rectum Cancer

Fearon, E.R., and Vogelstein, B. 1990. A genetic model for colorectal tumorigenesis. *Cell* 61:759–767.

Moertel, C.G., Fleming, T.R., MacDonald, J.S., Haller, D.G., Laurie, J.A., Goodman, P.J., Underleider, J.S., Emerson, W.A., Tormey, D.C., Glick, J.H., Veeder, M.H., and Mailliard, J.A. 1990. Levamisole and fluorouracil for adjuvant therapy of resected colon carcinoma. *N. Engl. J. Med.* 322:352–358.

NIH Consensus Conference. 1990. Adjuvant therapy for patients with colon and rectal cancer. *J. Amer. Med. Assoc.* 264:1444–1450.

Willett, W.C., Stampfer, M.J., Colditz, G.A., Rosner, B.A., and Speizer, F.E. 1990. Relation of meat, fat, and fiber intake to the risk of colon cancer in a prospective study among women. *N. Engl. J. Med.* 323:1664–1672.

Breast Cancer

Case, C., ed. 1984. *The breast cancer digest.* 2nd ed. National Institutes of Health, Bethesda.

Lee, E.Y.-H.P., To, H., Shew, J.-Y., Bookstein, R., Scully, P., and Lee, W.-H. 1988. Inactivation of the retinoblastoma susceptibility gene in human breast cancers. *Science* 241:218–221.

Nigro, J.M., Baker, S.J., Preisinger, A.C., Jessup, J.M., Hostetter, R., Cleary, K., Bigner, S.H., Davidson, N., Baylin, S., Devilee, P., Glover, T., Collins, F.S., Weston, A., Modali, R., Harris, C.C., and Vogelstein, B. 1989. Mutations in the *p53* gene occur in diverse human tumour types. *Nature* 342:705–708.

Slamon, D.J., Godolphin, W., Jones, L.A., Holt, J.A., Wong, S.G., Keith, D.E., Levin, W.J., Stuart, S.G., Udove, J., Ullrich, A., and Press, M.F. 1989. Studies of the *HER*-2/*neu* proto-oncogene in human breast and ovarian cancer. *Science* 244:707–712.

Willett, W.C. 1989. The search for the causes of breast and colon cancer. *Nature* 338:389–394.

Prostate Cancer

Bookstein, R., Rio, P., Madreperla, S.A., Hong, F., Allred, C., Grizzle, W.E., and Lee, W.H. 1990. Promoter deletion and loss of retinoblastoma gene expression in human prostate carcinoma. *Proc. Natl. Acad. Sci. USA* 87:7762–7766.

Gittes, R.F. 1991. Carcinoma of the prostate. *N. Engl. J. Med.* 324:236–245.

Urinary Cancers

National Cancer Institute. 1989. *Research report: adult kidney cancer and Wilms' tumor.* National Institutes of Health, Bethesda.

National Cancer Institute. 1987. *Research report: cancer of the bladder.* National Institutes of Health, Bethesda.

Raghavan, D., Shipley, W.U., Garnick, M.B., Russell, P.J., and Richie, J.P. 1990. Biology and management of bladder cancer. *N. Engl. J. Med.* 322:1129–1138.

Sidransky, D., Von Eschenbach, A., Tsai, Y.C., Jones, P., Summerhayes, I., Marshall, F., Paul, M., Green, P., Hamilton, S.R., Frost, P., and Vogelstein, B. 1991. Identification of *p53* gene mutations in bladder cancers and urine samples. *Science* 252:706–709.

Uterine Cancers

Henderson, B.E., Ross, R., and Bernstein, L. 1988. Estrogens as a cause of human cancer: The Richard and Hinda Rosenthal Foundation Award Lecture. *Cancer Res.* 48:246–253.

National Cancer Institute. 1987. *Research report: cancer of the uterus.* National Institutes of Health, Bethesda.

Zur Hausen, H. 1989. Papillomaviruses in anogenital cancer as a model to understand the role of viruses in human cancers. *Cancer Res.* 49:4677–4681.

Oral Cancers

National Cancer Institute. 1988. *Research report: oral cancer.* National Institutes of Health, Bethesda.

Pancreatic Cancer

Almoguera, C., Shibata, D., Forrester, K., Martin, J., Arnheim, N., and Perucho, M. 1988. Most human carcinomas of the exocrine pancreas contain mutant c-Ki-*ras* genes. *Cell* 53:549–554.

National Cancer Institute. 1987. *Research report: cancer of the pancreas.* National Institutes of Health, Bethesda.

Melanoma and Other Skin Cancers

National Cancer Institute. 1988. *Research report: melanoma.* National Institutes of Health, Bethesda.

National Cancer Institute. 1988. *Research report: nonmelanoma skin cancers.* National Institutes of Health, Bethesda.

Van't Veer, L.J., Burgering, B.M.T., Versteeg, R., Boot, A.J.M., Ruiter, D.J., Osanto, S., Schrier, P.I., and Bos, J.L. 1989. N-*ras* mutations in human cutaneous melanoma from sun-exposed body sites. *Mol. Cell. Biol.* 9:3114–3116.

Stomach Cancer

National Cancer Institute. 1988. *Research report: cancer of the stomach.* National Institutes of Health, Bethesda.

Ovarian Cancer

Lyons, J., Landis, C.A., Harsh, G., Vallar, L., Grunewald, K., Feichtinger, H., Duh, Q.-Y., Clark, O.H., Kawasaki, E., Bourne, H.R., and McCormick, F. 1990. Two G protein oncogenes in human endocrine tumors. *Science* 249:655–659.

Marks, J.R., Davidoff, A.M., Kerns, B.J., Humphrey, P.A., Pence, J.C., Dodge, R.K., Clarke-Pearson, D.L., Iglehart, J.D., Bast, R.C., Jr., and Berchuck, A. 1991. Overexpression and mutation of *p53* in epithelial ovarian cancer. *Cancer Res.* 51:2979–2984.

National Cancer Institute. 1989. *Research report: cancer of the ovary.* National Institutes of Health, Bethesda.

Slamon, D.J., Godolphin, W., Jones, L.A., Holt, J.A., Wong, S.G., Keith, D.E., Levin, W.J., Stuart, S.G., Udove, J., Ullrich, A., and Press, M.F. 1989. Studies of the *HER-2/neu* proto-oncogene in human breast and ovarian cancer. *Science* 244:707–712.

Brain Tumors

Black, P.M. 1991. Brain tumors (two parts). *N. Engl. J. Med.* 324:1471–1476 and 1555–1564.

Lyons, J., Landis, C.A., Harsh, G., Vallar, L., Grunewald, K., Feichtinger, H., Duh, Q.-Y., Clark, O.H., Kawasaki, E., Bourne, H.R., and McCormick, F. 1990. Two G protein oncogenes in human endocrine tumors. *Science* 249:655–659.

Nigro, J.M., Baker, S.J., Preisinger, A.C., Jessup, J.M., Hostetter, R., Cleary, K., Bigner, S.H., Davidson, N., Baylin, S., Devilee, P., Glover, T., Collins, F.S., Weston, A., Modali, R., Harris, C.C., and Vogelstein, B. 1989. Mutations in the *p53* gene occur in diverse human tumour types. *Nature* 342:705–708.

Liver Cancer

Bressac, B., Kew, M., Wands, J., and Ozturk, M. 1991. Selective G to T mutations of *p53* gene in hepatocellular carcinoma from southern Africa. *Nature* 350:429–431.

Harris, C.C. 1990. Hepatocellular carcinogenesis: recent advances and speculation. *Cancer Cells* 2:146–148.

Hsu, I.C., Metcalf, R.A., Sun, T., Welsh, J.A., Wang, N.J., and Harris, C.C. 1991. Mutational hotspot in the *p53* gene in human hepatocellular carcinomas. *Nature* 350:427–428.

Thyroid Cancer

Bongarzone, I., Pierotti, M.A., Monzine, N., Mondellini, P., Manenti, G., Donghi, R., Pilotti, S., Grieco, M., Santoro, M., Fusco, A., Vecchio, G., and Della Porta, G. 1989. High frequency of activation of tyrosine kinase oncogenes in human papillary thyroid carcinoma. *Oncogene* 4:1457–1462.

Esophageal Cancer

Hollstein, M.C., Metcalf, R.A., Welsh, J.A., Montesano, R., and Harris, C.C. 1990. Frequent mutation of the *p53* gene in human esophageal cancer. *Proc. Natl. Acad. Sci. USA* 87:9958–9961.

Testicular Cancer

National Cancer Institute. 1987. *Research report: testicular cancer.* National Institutes of Health, Bethesda.

PART V

PROSPECTS FOR
THE FUTURE

THE PRECEDING PARTS OF THIS BOOK have described our understanding of the development of cancer and present methods of
cancer prevention and treatment. In this, the concluding section,
our progress against cancer is summarized, and possible future
developments, based on current directions in cancer research, are
discussed.

Chapter 18

The War on Cancer— Progress and Promises

CANCER HAS BEEN WITH US throughout the history of mankind and has long been a focus of medical practice. Indeed, the term "carcinoma" was coined by Hippocrates in the fourth century B.C., and efforts to deal with the cancer problem have been ongoing since that time. Some of the notable events in the battle against cancer during the last two centuries have included the first association between the use of tobacco (snuff) and cancer in 1761, identification of the first occupational carcinogen (chimney soot) in 1775, the first surgical cure of an abdominal tumor in 1809 (the patient survived 30 years after removal of a 22-pound ovarian carcinoma), the discovery of tumor viruses in 1908, identification of the first chemical carcinogen (coal tar) in 1915, the development of the Pap test in 1928, the first successful chemotherapy of childhood leukemia in 1947, and the report of the Surgeon General on Smoking and Health in 1964.

It was against this historical background that President Richard M. Nixon signed the National Cancer Act in 1971, declaring the start of our current War on Cancer. The prevention and treatment of cancer were obviously already areas of intense research, but the notion of a War on Cancer focused public attention on the cancer problem as a national priority. Twenty years later, it seems reasonable to ask how we are doing. Put simply, does the War on Cancer more closely resemble our efforts in Vietnam or in the Persian Gulf?

The War on Cancer has been fought on two fronts, which to date have remained largely distinct efforts. Practical efforts have focused on cancer prevention and treatment, while major basic research efforts have been directed at understanding cancer at the cellular and molecular levels. It is all too clear that cancer has not been conquered, but significant advances in the prevention and treatment of cancer have been made. Moreover, progress in understanding the fundamental mechanisms responsible for the development of cancer has been dramatic. Major improvements in our ability to deal with cancer may eventually result from applying our increasing understanding of cancer's molecular basis to the development of new strategies for cancer prevention and treatment.

PROGRESS IN CANCER PREVENTION AND TREATMENT

Some critics of the War on Cancer have asserted that there has been little progress in dealing with cancer, since there has been no significant decrease in overall cancer deaths. Although the likelihood of dying from cancer is the ultimate bottom line, this seems to be too dismissive a position, which ignores the progress that has been made. Although far from representing the successful conquest of cancer, such progress is significant and needs to be understood in the context of the overall cancer problem.

Avoidance of Known Risk Factors Would Substantially Reduce Current Cancer Mortality

First, it is important to emphasize that we already have knowledge at hand that could prevent a significant fraction of cancer deaths, but these preventive measures are not being effectively applied. Significant progress has been made both in the identification of the causes of certain cancers and in the detection of some cancers at early, more treatable, stages of the disease process. The translation of these findings into a significant reduction in cancer mortality, however, involves changes in lifestyle that are difficult to implement. In addition, it is important to realize that, because of the characteristic long lag time between carcinogen exposure and tumor development, the effects of preventive measures on overall cancer mortality will often not be evident for several decades after their implementation.

Most notably, about one-third of cancer deaths in the United States could be eliminated by avoidance of tobacco. This realization has led to a reduction in the prevalence of smoking, but hardly to its elimination. About 40% of American adults smoked in 1965, and about 30% still smoke today, including many teenagers. Even 20 years after its identification as a major cause of cancer, the use of tobacco remains a problem. Thus, although the elimination of tobacco has the potential of making a major impact against cancer, this has not been realized. Apparently, the prospect of cancer prevention is not a strong enough motivation for many members of our society to avoid tobacco use.

Given the limited impact of identifying tobacco as a major carcinogen, it is difficult to be optimistic about the possibility of cancer prevention based on voluntary changes in lifestyle. Thus, although alcohol, obesity, high-fat diets, and sexually transmitted viruses are clearly risk factors for some cancers, it seems unlikely that cancer prevention is a sufficient motivation for many people to alter their behavior so as to reduce cancer risk accordingly. Recent studies of dietary factors, for example, indicate that colon cancer incidence might be halved by reducing dietary fat intake from 40% to 30% of total calories. Such a change in eating habits throughout the United States would be expected to reduce total cancer mortality by about 5%, but the widespread adoption of a major dietary modification to achieve this relatively modest reduction in cancer risk must be considered unlikely. Similar motivational problems appear to apply to the widespread acceptance of early screening tests, which could reduce overall cancer mortality by about 10%.

It appears that the major exposures to carcinogens are determined by personal lifestyle choices, rather than by factors such as industrial pollution. Consequently, voluntary modifications of behavior will be needed to achieve any significant reduction in cancer mortality based on preventive measures. Nonetheless, this should not obscure the fact that significant progress has been made in identifying the causes of some cancers and in detecting some cancers at early, readily treatable stages. If current recommendations for cancer prevention and early detection were put into general practice, they would result in about a twofold reduction in total cancer mortality. This would clearly represent substantial progress, suggesting that the most effective step that can currently be taken against cancer is to increase public awareness and motivation in order to take advantage of what we already know in the area of cancer prevention.

Vaccination Against Hepatitis B Virus

The discovery of viruses as the causes of some cancers affords an alternative approach to prevention—namely, the development of antiviral vaccines. Indeed, a safe and apparently effective vaccine against hepatitis B virus is already in use and can be reasonably expected to have a significant impact on the incidence of liver cancer. Although rare in the United States, hepatocellular carcinoma is extremely common in parts of Asia and Africa, and probably accounts for about 10% of total worldwide cancer incidence. Its potential prevention by a vaccination program thus represents a major step in the international cancer effort.

Some Cancers Can Be Successfully Treated by Chemotherapy

Progress has also been made in the treatment of some cancers, although not for those diseases that are most common in our society. For the majority of cancers, the success of treatment is primarily determined by early diagnosis. Localized cancers can frequently be cured by surgery and radiation, but chemotherapy of metastatic disease usually fails. However, drug combinations have now been developed that are capable of curing most patients suffering from acute lymphocytic leukemia, Hodgkin's disease, some non-Hodgkin's lymphomas, and testicular cancer. Together, these diseases account for less than 5% of the total cancer burden, so their successful treatment has not made a significant impact on overall cancer statistics. On the other hand, these successes are real, and they suggest that similar progress against other cancers is within the realm of possibility. Indeed, the activity of fluorouracil plus levamisole against colon carcinoma (see chapter 14) represents a recent example of at least modest progress in the treatment of one of the most common cancers of adults.

The Development of Immunotherapy

Significant progress in some types of immunotherapy has also been made, although this has not yet been translated into general practice. Specific populations of anti-tumor lymphocytes (tumor-infiltrating lymphocytes) have been identified, and they have been shown to have significant therapeutic effects in early-stage clinical trials against melanomas and renal cell carcinomas. Moreover, experiments are currently underway to use genetic engineering to make these lymphocytes even more potent killers of cancer cells. Success along these avenues could lead to significant future advances in cancer treatments.

In summary, real progress against cancer has been made, but practical advances have been slow, and efforts to date have not led to any significant decrease in cancer deaths. Public education, to implement currently understood preventive measures effectively (particularly avoidance of tobacco), would have the greatest immediate impact on cancer mortality. Advances in chemotherapy have led to successes against a few malignancies, but not against the majority of common cancers.

PROGRESS IN UNDERSTANDING CANCER

The Discovery of Oncogenes and Tumor Suppressor Genes

In contrast to the relatively slow progress in cancer prevention and treatment, the last 20 years have seen dramatic advances in our understanding of cancer at the cellular and molecular levels. The discoveries of oncogenes and tumor suppressor genes have provided a conceptual framework for understanding the mechanisms that control normal cell growth and differentiation and the ways in which breakdowns of these normal cellular controls lead to the development of cancer. Some of the major historical milestones in this part of the War on Cancer included the discoveries of viral oncogenes in 1970, cellular proto-oncogenes in 1976, human tumor oncogenes in 1981, and the first human tumor suppressor gene in 1987.

A large number of oncogenes and tumor suppressor genes have now been identified, and it is clear that alterations in these genes are fundamental to the development of human cancers. Many of these genes have been characterized with respect to their roles in regulating normal cell growth, and we have begun to understand how their malfunction can lead to the uncontrolled proliferation of cancer cells. Moreover, pictures of the ways in which cumulative damage to multiple oncogenes and tumor suppressor genes contributes to the development of different types of human cancers are beginning to emerge.

The Molecular and Cellular Basis of Cancer

Our understanding of the molecular and cellular basis of cancer has thus advanced tremendously, and progress in this area continues at a rapid pace. In addition to illuminating the mechanisms involved in abnormal neoplastic cell growth, studies at this level have also resulted in substantial gains in our understanding of the basic mechanisms that regulate normal cell growth and differentiation. The current challenge is to apply these newfound insights to the practical arenas of cancer prevention and treatment.

THE FUTURE'S PROMISE: WILL OUR UNDERSTANDING OF CANCER YIELD PRACTICAL BENEFITS?

Oncogenes and Tumor Suppressor Genes in Diagnosis

There is little question that our increasing understanding of cancer will have an impact on diagnosis and treatment, but it is uncertain how profound this impact will be. It is already evident that oncogenes and tumor suppressor genes will be useful markers for the diagnosis of some cancers, and that this will contribute to improvements in early detection and therapy. Several examples are readily apparent and are already being put into clinical practice. Analysis of tumor suppressor genes will allow identification of individuals at high risk for inherited cancers.

Detection of the *abl* oncogene translocation in chronic myelogenous leukemia provides a sensitive assay for leukemic cells, which is useful in monitoring the response of patients to therapy. Amplification of the N-*myc* oncogene in neuroblastomas, and of the *erb*B-2 oncogene in breast and ovarian carcinomas, is predictive of rapid disease progression; it may contribute to choices among treatment options for patients with these diseases. Ongoing research in this area will undoubtedly identify diagnostic roles for additional genes in other cancers. However, although this information will be useful and may well lead to improvements in diagnosis and subsequent treatment, it is not likely to have a major impact on the overall cancer problem. In the end, we will still be limited by the problems inherent in current treatment methods.

Oncogenes and Tumor Suppressor Genes as Targets for Chemotherapy

As discussed in chapter 14, the limitation of current chemotherapeutic drugs is that they are not specific for cancer cells. Instead, the present chemotherapeutic agents interfere nonspecifically with cell division and are therefore toxic to rapidly proliferating normal cells as well as to cancer cells. This toxicity to normal cells limits the effectiveness of chemotherapy—a problem that could be overcome if drugs specific for cancer cells could be found. Can the discoveries of oncogenes and tumor suppressor genes be exploited to provide such specific targets, against which a new generation of anticancer drugs might act? Is it possible, for example, to design drugs that specifically interfere with the function of oncogene proteins, or that augment the activity of tumor suppressor gene products?

Unfortunately (from the standpoint of cancer chemotherapy), oncogenes and tumor suppressor genes are important in normal cells as well as in cancer cells. Since the products of these genes are critical regulators of normal cell proliferation, they do not provide chemotherapeutic targets that are unique to cancer cells. Consequently, the possible exploitation of oncogenes and tumor suppressor genes in the treatment of cancer is not a straightforward proposition, but there are also reasons to hope that it will not ultimately be an impossible one.

The *ras* Oncogenes: An Example of Approaches and Problems

The *ras* oncogenes are a good example of the potentials and problems inherent in a therapeutic strategy targeted against an oncogene. In about 20% of human cancers, including many lung and colon carcinomas, point mutations generate *ras* oncogenes that drive abnormal cell proliferation. It is reasonable to anticipate that interference with *ras* oncogenes would effectively inhibit the growth of these cancers, and this has in fact been demonstrated in experimental systems. However, normal *ras* genes (proto-oncogenes) are expressed in apparently all normal cells and appear to be required for normal cell growth, as well as for some pathways of cell differentiation. Consequently, a drug that abolished all *ras* expression or function might be expected to have severe toxic effects, much like our present chemotherapeutic agents. Nonetheless, while such putative inhibitors would not be entirely specific for cancer cells, they might still provide useful alternatives to current chemotherapy protocols. It has been proposed, for example, that drugs that block the function of *ras* oncogenes by inhibiting the addition of lipid to *ras* proteins (a modification required for the association of *ras* proteins with the plasma membrane) might be useful chemotherapeutic agents. Clearly, such drugs would inhibit both *ras* oncogenes in cancer cells and *ras* proto-oncogenes in normal cells. However, it is possible that anticancer activity would outweigh toxicity, and the development of this class of drugs is an area of active research.

An alternative possibility is the design of drugs that would specifically inhibit oncogene, but not proto-oncogene, function. For some oncogenes (for example, the *myc* genes), this is not a possibility at all, since the oncogene proteins are the same as those encoded by the proto-oncogenes—their activity as oncogenes results simply from abnormal gene expression. However, other oncogene proteins, such as *ras*, do differ structurally from those encoded by the corresponding proto-oncogenes. In these cases, the development of oncogene-specific drugs is a theoretical possibility, which would achieve real specificity against cancer cells. However, the structural differences between these oncogenes and proto-oncogenes are subtle—for example, the alteration of only a single amino acid in *ras* oncogene proteins. The design of drugs that would be specific for *ras* oncogene proteins, while allowing the closely related proto-oncogene proteins to function normally, is therefore an unquestionably difficult and challenging problem. Ultimately, it may not be an impossible one, but a great deal more research is needed before such a challenge can be met.

The Possibility of Gene Therapy

A distinct possibility is the use of gene therapy to eliminate or block oncogene expression or, more likely, to restore functional tumor suppressor genes to cancer cells in which they have been lost or inactivated. Both blocking oncogene expression and restoring tumor suppressor gene function have been shown in experimental culture systems to inhibit cancer cell growth effectively. The problem of applying these methods to the treatment of clinical cancer, however, is that the desired gene would have to be introduced into virtually all of the cancer cells. At present, there is no technically feasible method for even approaching this objective, although it is obviously not possible to predict what technological advances may be seen in years to come.

The Potential of Manipulating Signal Transduction Pathways

Yet another possibility for exploiting our understanding of oncogenes and tumor suppressor genes is the potential of using drugs that stimulate or inhibit specific intracellular signaling pathways that regulate cell growth. In fact, such drugs are already being used in both cancer prevention and therapy, and this kind of approach would seem to have increasing potential as we learn more about the opposing actions of oncogenes and tumor suppressor genes in regulating cell growth and differentiation.

One example of such an oncogene-directed therapy, discussed in chapters 11 and 14, is the treatment of acute promyelocytic leukemia with retinoic acid. In this disease, the retinoic acid receptor is disrupted by a chromosomal translocation. The precise effects of this disruption are not yet known, but it appears that the translocation yields an altered oncogene protein that most likely interferes with the ability of the normal retinoic acid receptor to induce cell differentiation. Strikingly, the disease can be treated by administration of retinoic acid, which appears to compensate for the aberrant receptor and induces differentiation of the leukemic cells. Although the mechanisms involved remain to be fully elucidated, this seems to be a clear case of effective cancer treatment directed against a specific oncogene.

Other examples are provided by the activity of retinoic acid in chemoprevention and by the use of tamoxifen in both prevention and treatment of breast cancer. In their target cells, both of these steroid hormones induce expression of the growth inhibitory factor TGFβ. Although not all of the actions of TGFβ have been delineated, at least one pathway by which TGFβ inhibits cell proliferation involves activation of the *RB* tumor suppressor gene product, which in turn inhibits expression of the c-*myc* proto-oncogene. Our understanding of such relationships between hormones, extracellular growth factors, tumor suppressor genes, and oncogenes is growing rapidly, and further research is likely to suggest specific manipulations of hormones and growth factors for the treatment of some tumors. For example, might administration of retinoic acid or tamoxifen be particularly useful for breast cancers in which c-*myc* oncogenes are amplified or overexpressed? Is it possible that interference with the activity of specific growth factor receptors might inhibit the proliferation of certain cancers, such as breast and ovarian carcinomas in which the *erb*B-2 oncogene (a growth factor receptor) is amplified?

The development of drugs that affect specific intracellular signaling pathways, and their rational use based on the oncogenes and tumor suppressor genes involved in particular types of cancer, seems to represent a plausible area of future developments in cancer therapy. A great deal of further work will clearly be required, but this approach appears at least to offer the promise of developing drugs that are specifically targeted towards cancer cells. If such drugs were nontoxic, their use in chemoprevention could also be considered. Indeed, as discussed in chapter 12, chemoprevention trials of retinoic acid derivatives and tamoxifen are already in progress.

SUMMARY

Although we are far from the ultimate conquest of cancer, major advances have been made. The greatest immediate impact on cancer mortality could be attained by public acceptance of the lifestyle changes required to eliminate major carcinogens, most notably tobacco, and to maximize the benefits of early detection.

Chemotherapy has proven successful against a few relatively rare cancers, but not so far against the more common malignancies. On the other hand, striking progress has been made in understanding cancer at the molecular and cellular levels. The challenge of the future, which may well determine the ultimate success of the War on Cancer, is to apply this understanding to the practical matters of cancer prevention and treatment.

REFERENCES AND FURTHER READING

Baker, S.J., Markowitz, S., Fearon, E.R., Willson, J.K.V., and Vogelstein, B. 1990. Suppression of human colorectal carcinoma cell growth by wild-type *p53*. *Science* 249:912–915.

Biondi, A., Rambaldi, A., Alcalay, M., Pandolfi, P.P., Lo Coco, F., Diverio, D., Rossi, V., Mencarelli, A., Longo, L., Zangrilli, D., Masera, G., Barbui, T., Mandelli, F., Grignani, F., and Pelicci, P.G. 1991. *RAR*-α gene rearrangements as a genetic marker for diagnosis and monitoring in acute promyelocytic leukemia. *Blood* 77:1418–1422.

Bishop, J.M. 1991. Molecular themes in oncogenesis. *Cell* 64:235–248.

Davis, D.L., Hoel, D., Fox, J., and Lopez, A. 1990. International trends in cancer mortality in France, West Germany, Italy, Japan, England and Wales, and the USA. *Lancet* 336:474–481.

Doll, R. 1989. Progress against cancer: are we winning the war? *Acta Oncol.* 28:611–621.

Goldstein, J.L., and Brown, M.S. 1990. Regulation of the mevalonate pathway. *Nature* 343:425–430.

Loeb, L.A. 1989. Endogenous carcinogenesis: molecular oncology into the twenty-first century—Presidential Address. *Cancer Res.* 49:5489–5496.

Look, A.T., Hayes, F.A., Shuster, J.J., Douglass, E.C., Castleberry, R.P., Bowman, L.C., Smith, E.I., and Brodeur, G.M. 1991. Clinical relevance of tumor cell ploidy and N-*myc* gene amplification in childhood neuroblastoma: a pediatric oncology group study. *J. Clin. Oncol.* 9:581–591.

Marshall, E. 1990. Experts clash over cancer data. *Science* 250:900–902.

Marx, J. 1990. Oncogenes evoke new cancer therapies. *Science* 249:1376–1378.

Muir, C.S. 1990. Epidemiology, basic science, and the prevention of cancer: implications for the future. *Cancer Res.* 50:6441–6448.

Rosenberg, S.A., Aebersold, P., Cornetta, K., Kasid, A., Morgan, R.A., Moen, R., Karson, E.M., Lotze, M.T., Yang, J.C., Topalian, S.L., Merino, M.J., Culver, K., Miller, A.D., Blaese, R.M., and Anderson, W.F. 1990. Gene transfer into humans—immunotherapy of patients with advanced melanoma, using tumor-infiltrating lymphocytes modified by retroviral gene transduction. *N. Engl. J. Med.* 323:570–578.

Sawyers, C.L., Timson, L., Kawasaki, E.S., Clark, S.S., Witte, O.N., and Champlin, R. 1990. Molecular relapse in chronic myelogenous leukemia patients after bone marrow transplantation detected by polymerase chain reaction. *Proc. Natl. Acad. Sci. USA* 87:563–567.

Schafer, W.R., Trueblood, C.E., Yang, C.-C., Mayer, M.P., Rosenberg, S., Poulter, C.D., Kim, S.-H., and Rine, J. 1990. Enzymatic coupling of cholesterol intermediates to a mating pheromone precursor and to the *ras* protein. *Science* 249:1133–1139.

Appendix: Information and Resources for Cancer Patients

American Cancer Society, Inc.
1599 Clifton Road, N.E.
Atlanta, Georgia 30329
800-ACS-2345

The American Cancer Society is a national voluntary organization that supports cancer research and offers a variety of services for cancer patients, including information, counseling, and rehabilitation programs. Educational programs include distribution of a number of pamphlets and booklets on cancer. Local American Cancer Society units provide counseling, home care items, and transportation for cancer patients. Additional national rehabilitation and education programs are:

CanSurmount: A short-term visitor program for cancer patients and their families. One-on-one visits offer support by an individual who has experienced the same type of cancer as the patient.

I Can Cope: A series of lectures and group discussions covering the concerns of cancer patients and their families.

Laryngectomy Rehabilitation: The International Association of Laryngectomees is a group of more than 250 clubs that provide support for laryngectomy patients.

Ostomy Rehabilitation: A support program, in cooperation with the United Ostomy Association, to help patients with urinary or intestinal cancers cope with ostomies. One-on-one counseling is provided by volunteers who have undergone the same type of surgery.

Look Good—Feel Better: A program designed to help patients develop cosmetic skills to improve their appearance and deal with side effects of cancer treatment.

Reach to Recovery: A support program for women who have had mastectomies.

Chartered Divisions of the American Cancer Society

Alabama Division, Inc.
504 Brookwood Boulevard
Homewood, Alabama 35209
205–879–2242

Alaska Division, Inc.
406 West Fireweed Lane
Suite 204
Anchorage, Alaska 99503
907–277–8696

Arizona Division, Inc.
2929 East Thomas Road
Phoenix, Arizona 85016
602–224–0524

Arkansas Division, Inc.
901 North University
Little Rock, Arkansas 72207
501–664–3480

California Division, Inc.
1710 Webster Street
P.O. Box 2061
Oakland, California 94612
415–893–7900

Colorado Division, Inc.
2255 South Oneida
P.O. Box 24669
Denver, Colorado 80224
303–758–2030

Connecticut Division, Inc.
Barnes Park South
14 Village Lane
Wallingford, Connecticut 06492
203–265–7161

Delaware Division, Inc.
92 Read's Way
New Castle, Delaware 19720
302–324–4227

District of Columbia Division, Inc.
1825 Connecticut Avenue, N.W.
Suite 315
Washington, D.C. 20009
202–483–2600

Florida Division, Inc.
1001 South MacDill Avenue
Tampa, Florida 33629
813–253–0541

Georgia Division, Inc.
46 Fifth Street, N.E.
Atlanta, Georgia 30308
404–892–0026

Hawaii/Pacific Division, Inc.
Community Services Center Building
200 North Vineyard Boulevard
Honolulu, Hawaii 96817
808–531–1662

Idaho Division, Inc.
2676 Vista Avenue
P.O. Box 5386
Boise, Idaho 83705
208–343–4609

Illinois Division, Inc.
77 East Monroe
Chicago, Illinois 60603
312–641–6150

Indiana Division, Inc.
8730 Commerce Park Place
Indianapolis, Indiana 46268
317–872–4432

Iowa Division, Inc.
8364 Hickman Road, Suite D
Des Moines, Iowa 50325
515–253–0147

Kansas Division, Inc.
1315 SW Arrowhead Road
Topeka, Kansas 66604
913–273–4114

Kentucky Division, Inc.
701 West Muhammad Ali Blvd.
P.O. Box 1807
Louisville, Kentucky 40201–1807
502–584–6782

Louisiana Division, Inc.
Fidelity Homestead Bldg.
837 Gravier Street
Suite 700
New Orleans, Louisiana 70112–1509
504–523–4188

Maine Division, Inc.
52 Federal Street
Brunswick, Maine 04011
207–729–3339

Maryland Division, Inc.
8219 Town Center Drive
White Marsh, Maryland 21162–0082
301–931–6868

Massachusetts Division, Inc.
247 Commonwealth Avenue
Boston, Massachusetts 02116
617–267–2650

Michigan Division, Inc.
1205 East Saginaw Street
Lansing, Michigan 48906
517–371–2920

Minnesota Division, Inc.
3316 West 66th Street
Minneapolis, Minnesota 55435
612–925–2772

Mississippi Division, Inc.
1380 Livingston Lane
Lakeover Office Park
Jackson, Mississippi 39213
601–362–8874

Missouri Division, Inc.
3322 American Avenue
Jefferson City, Missouri 65102
314–893–4800

Montana Division, Inc.
313 N. 32nd Street
Suite #1
Billings, Montana 59101
406–252–7111

Nebraska Division, Inc.
8502 West Center Road
Omaha, Nebraska 68124–5255
402–393–5800

Nevada Division, Inc.
1325 East Harmon
Las Vegas, Nevada 89119
702–798–6857

New Hampshire Division, Inc.
360 Route 101, Unit 501
Bedford, New Hampshire 03102–6800
603–472–8899

New Jersey Division, Inc.
2600 Route 1, CN 2201
North Brunswick, New Jersey 08902
201–297–8000

New Mexico Division, Inc.
5800 Lomas Boulevard, NE
Albuquerque, New Mexico 87110
505–260–2105

New York Division, Inc.
6725 Lyons Street
P.O. Box 7
East Syracuse, New York 13057
315–437–7025

• Long Island Division, Inc.
  145 Pidgeon Hill Road
  Huntington Station, New York 11746
  516–385–9100

• New York City Division, Inc.
  19 West 56th Street
  New York, New York 10019
  212–586–8700

• Queens Division, Inc.
  112–25 Queens Boulevard
  Forest Hills, New York 11375
  718–263–2224

• Westchester Division, Inc.
  30 Glenn St.
  White Plains, New York 10603
  914–949–4800

North Carolina Division, Inc.
11 South Boylan Avenue
Suite 221
Raleigh, North Carolina 27603
919–834–8463

North Dakota Division, Inc.
123 Roberts Street
P.O. Box 426
Fargo, North Dakota 58107
701–232–1385

Ohio Division, Inc.
5555 Frantz Road
Dublin, Ohio 43017
614–889–9565

Oklahoma Division, Inc.
3000 United Founders Boulevard
Suite 136
Oklahoma City, Oklahoma 73112
405–843–9888

Oregon Division, Inc.
0330 SW Curry
Portland, Oregon 97201
503–295–6422

Pennsylvania Division, Inc.
P.O. Box 897
Route 422 & Sipe Avenue
Hershey, Pennsylvania 17033–0897
717–533–6144

• Philadelphia Division, Inc.
  1422 Chestnut Street
  Philadelphia, Pennsylvania 19102
  215–665–2900

Puerto Rico Division, Inc.
Calle Alverio #577
Esquina Sargento Medina
Hato Rey, Puerto Rico 00918
809–764–2295

Rhode Island Division, Inc.
400 Main Street
Pawtucket, Rhode Island 02860
401–722–8480

South Carolina Division, Inc.
128 Stonemark Lane
Columbia, South Carolina 29210
803–750–1693

South Dakota Division, Inc.
4101 Carnegie Place
Sioux Falls, South Dakota 57106–2322
605–361–8277

Tennessee Division, Inc.
1315 Eighth Avenue, South
Nashville, Tennessee 37203
615–255–1227

Cancer Care, Inc.
1180 Avenue of the Americas
New York, New York 10036
212–221–3300

Texas Division, Inc.
2433 Ridgepoint Drive
Austin, Texas 78754
512–928–2262

Utah Division, Inc.
610 East South Temple
Salt Lake City, Utah 84102
801–322–0431

Vermont Division, Inc.
13 Loomis Street, Drawer C
P.O. Box 1452
Montpelier, Vermont 05601–1452
802–223–2348

Virginia Division, Inc.
4240 Park Place Court
Glen Allen, Virginia 23060
804–270–0142

Washington Division, Inc.
2120 First Avenue North
Seattle, Washington 98109–1140
206–283–1152

West Virginia Division, Inc.
2428 Kanawha Boulevard East
Charleston, West Virginia 25311
304–344–3611

Wisconsin Division, Inc.
615 North Sherman Avenue
Madison, Wisconsin 53704
608–249–0487

Wyoming Division, Inc.
2222 House Avenue
Cheyenne, Wyoming 82001
307–638–3331

The service arm of the National Cancer Foundation, which provides information, counseling, and support for patients and family members.

Candlelighters Childhood Cancer Foundation, Inc.
1901 Pennsylvania Avenue, N.W.
Washington, D.C. 20006
202–659–5136

An international organization of parents whose children have or have had cancer. The society provides guidance and support services.

The Concern for Dying
250 West 57th Street
New York, New York 10107
212–246–6962

An educational organization that distributes information on "the living will," a document that records patient wishes on treatment, death, and dying.

Corporate Angel Network
Westchester County Airport
Hangar F
White Plains, New York 10604
914–328–1313

An organization to alleviate the costs of cancer patients undergoing therapy in National Cancer Institute-approved centers by arranging free transportation on corporate aircraft.

Leukemia Society of America, Inc.
733 Third Avenue
New York, New York 10017
212–573–8484

A national organization that supports research on leukemia and provides consultation and financial assistance for patients with leukemia and related diseases.

Make-a-Wish Foundation of America
Suite 936
2600 North Central Avenue
Phoenix, Arizona 85004
602–240–6600

An organization to cover expenses and arrange granting a "special wish" for terminally ill children.

Make Today Count
10 ½ South Union Street
Alexandria, Virginia 22314
703–548–9674

An association of cancer patients and other people with life-threatening diseases, family members, nurses, physicians, and community members. The organization provides emotional self-help.

National Cancer Institute
National Institutes of Health
Bethesda, Maryland 20892

The National Cancer Institute is the division of the National Institutes of Health that supports programs of cancer research and treatment, as well as providing a variety of information services to patients and physicians.

Cancer Information Service
800–4-CANCER

A toll-free number by which patients or family members are connected to staff members trained to answer a variety of questions about cancer.

Office of Cancer Communications
National Cancer Institute
Building 31, Room 10A24
Bethesda, Maryland 20892

This office provides a wide variety of written materials on cancer for both patients and physicians.

PDQ (Physician Data Query) Service

A computerized database providing current treatment information for most types of cancer, descriptions of ongoing clinical trials, and names of organizations and physicians involved in cancer treatment. The Cancer Information Service (1-800-4-CANCER) provides free PDQ searches.

Comprehensive Cancer Centers

Comprehensive cancer centers are National Cancer Institute-recognized centers for cancer research and treatment.

Alabama
University of Alabama
Comprehensive Cancer Center
205–934–5077

California
Kenneth Norris, Jr. Comprehensive
   Cancer Center
University of Southern California
213–224–6600

Jonsson Comprehensive Cancer Center
University of California at Los Angeles
213–825–3181

Connecticut
Yale Comprehensive Cancer Center
203–785–4095

Florida
Sylvester Comprehensive Cancer Center
University of Miami
305–548–4800

Illinois
Illinois Cancer Council
312–346–9813

Maryland
Johns Hopkins Oncology Center
301–955–8822

Massachusetts
Dana-Farber Cancer Institute
617–732–3000

Michigan
Wayne State University
Comprehensive Cancer Center
313–745–8870

Minnesota
Mayo Comprehensive Cancer Center
507–284–4718

New York
Columbia University Cancer Center
212–305–6921

Memorial Sloan-Kettering Cancer Center
212–639–6561

Roswell Park Memorial Institute
716–845–5770

North Carolina
Duke Comprehensive Cancer Center
919–684–3377

University of North Carolina
Lineberger Cancer Research Center
919–966–3036

Cancer Center
Wake Forest University
Bowman Gray School of Medicine
919–748–4464

Ohio
Ohio State University
Comprehensive Cancer Center
614–293–3302

Pennsylvania
Fox Chase Cancer Center
215–728–2781

University of Pittsburgh
Pittsburgh Cancer Institute
412–647–2072

Texas
University of Texas
M.D. Anderson Cancer Center
713–792–6000

Washington
Fred Hutchinson Cancer Research Center
206–467–4302

Wisconsin
University of Wisconsin
Clinical Cancer Center
608–263–8610

National Hospice Organization
Suite 901
1901 North Fort Myer Drive
Arlington, Virginia 22209
703–243–5900

An organization of groups and institutions dealing with care for the terminally ill and their families. The organization provides information and referrals.

United Ostomy Association, Inc.
36 Executive Park
Suite 120
Irvine, California 92714
714–660–8624

The association provides counseling and support for patients who have had colostomy, ileostomy, or urostomy surgery.

Glossary

abl   An oncogene that encodes a protein-tyrosine kinase and is translocated in chronic myelogenous and acute lymphocytic leukemias.

actinomycin D   A chemotherapeutic drug that binds to DNA and inhibits RNA synthesis.

acute lymphocytic leukemia (ALL)   A rapidly progressing leukemia of immature lymphocytes.

acute monocytic leukemia   A rapidly progressing leukemia of immature monocytes.

acute myelocytic leukemia (AML)   A rapidly progressing leukemia of myeloblastic cells (immature granulocytes).

acute myelomonocytic leukemia   A rapidly progressing leukemia of precursors to both monocytes and granulocytes.

acute promyelocytic leukemia   A rapidly progressing leukemia of promyelocytes (granulocyte precursors).

acutely transforming virus   A retrovirus that rapidly induces tumors in infected animals and transforms cells in culture. The genomes of acutely transforming viruses contain one or more oncogenes.

adenocarcinoma   A carcinoma arising from glandular epithelium.

adenoma   A benign tumor arising from glandular epithelium.

aflatoxin   A potent liver carcinogen found in contaminated food supplies.

alkylating agents   A group of chemotherapeutic drugs that react with DNA.

anchorage dependence   The requirement of fibroblasts and epithelial cells for attachment to a surface in order to proliferate.

androgen   A steroid hormone that stimulates male sex characteristics.

aneuploidy   An abnormal number of chromosomes.

angiogenesis   The formation of new blood vessels.

angiography   X-ray examination of the blood vessels.

antimetabolite   A chemotherapeutic drug that interferes with one or more steps in DNA synthesis.

AP-1   A transcription factor composed of the *fos* and *jun* oncogene products.

apoptosis   *See* programmed cell death.

asparaginase   An enzyme, used in chemotherapy, that degrades the amino acid asparagine.

astrocytoma   A type of malignant brain tumor.

ataxia telangiectasia An inherited disease associated with genetic instability, immunodeficiency, and an increased susceptibility to leukemias and lymphomas.

ATP (adenosine triphosphate) A high-energy phosphate compound used by cells in energy-transfer reactions.

autocrine growth stimulation Production of a growth factor by a responsive cell, resulting in continual stimulation of cell proliferation.

basal cell carcinoma A common type of skin cancer.

basement membrane A layer of extracellular material underlying epithelia.

B cell *See* B lymphocyte.

bcl-2 An oncogene translocated in follicular B-cell lymphomas which acts to inhibit programmed cell death.

bcr Breakpoint cluster region, a gene on chromosome 22 that is fused to *abl* as a result of the Philadelphia translocation in chronic myelogenous and acute lymphocytic leukemias.

benign tumor A tumor that remains confined to its original location and does not invade adjacent tissue or metastasize to distant body sites.

bischloroethylnitrosourea (BCNU) An alkylating agent used in chemotherapy.

bleomycin A chemotherapeutic drug that reacts with DNA.

blocking agent A chemopreventive drug that interferes with the action of carcinogens.

Bloom's syndrome An inherited disease associated with genetic instability and an increased incidence of leukemias and lymphomas.

B lymphocyte An antibody-producing lymphocyte.

BNLF-1 An oncogene of Epstein-Barr virus.

bone marrow transplantation A procedure in which patients receive high doses of chemotherapeutic drugs or radiation, followed by a transplant of new bone marrow to bypass toxicity to the blood-forming cells.

Burkitt's lymphoma A non-Hodgkin's B-cell lymphoma associated with Epstein-Barr virus infection and translocation of the c-*myc* oncogene.

CA-125 A tumor marker for ovarian carcinomas.

cancer A malignant tumor.

carcinoembryonic antigen (CEA) A tumor marker frequently secreted by gastrointestinal carcinomas.

carcinogen A cancer-inducing agent.

carcinogenesis Development of cancer.

carcinoma A malignant tumor of epithelial cells.

carcinoma *in situ* A small carcinoma that has not yet invaded surrounding normal tissue.

carotenoids A group of compounds, including β-carotene, which are dietary sources of vitamin A and may display chemopreventive activity.

cell cycle The sequence of events that occur during the process of cell reproduction.

cellular oncogene An oncogene formed by mutation or rearrangement of a proto-oncogene in a tumor.

chemoprevention The administration of drugs to reduce cancer risk.

chemotherapy The treatment of cancer with drugs.

chlorambucil An alkylating agent used in chemotherapy.

choriocarcinoma A malignancy of trophoblastic tissue arising from male or female germ cells.

chromosome A DNA molecule bound to proteins.

chromosome translocation Exchange of segments between nonhomologous chromosomes.

chronic lymphocytic leukemia (CLL) A slowly progressing leukemia of immature lymphocytes.

chronic myelogenous leukemia (CML) A slowly progressing leukemia originating in the pluripotent stem cell of the bone marrow.

cirrhosis A disease characterized by liver cell degeneration.

cisplatin A platinum compound used in chemotherapy.

clinical staging *See* tumor staging.

clinical trial Testing of new drugs in cancer patients.

clonality Origin of a tumor from a single progenitor cell.

clonal selection The process of tumor progression in which new clones of tumor cells with increased proliferative capacity, invasiveness, or metastatic potential become dominant in the tumor cell population.

collagen The major extracellular component of connective tissue.

colonoscopy Examination of the colon with a flexible lighted tube.

colony-stimulating factor-1 (CSF-1) A growth factor that stimulates proliferation and differentiation of macrophage progenitors.

combination chemotherapy The use of drug combinations for cancer treatment.

computed tomography (CT) scan A scanning X-ray technique that employs computer analysis to generate cross-sectional images of the body.

contact inhibition The cessation of movement of normal fibroblasts that results from cell contact.

cruciferous vegetables A group of vegetables (including broccoli, Brussels sprouts, and cauliflower) containing several compounds that inhibit cancer development.

cyclohexylchloroethylnitrosourea (CCNU) An alkylating agent used in chemotherapy.

cyclophosphamide An alkylating agent used in chemotherapy.

cystoscopy Examination of the bladder with a thin lighted tube.

cytokine A factor produced by lymphocytes or other blood cells during the immune response.

cytosine arabinoside An antimetabolite that is incorporated into DNA in place of cytosine and inhibits further DNA replication.

cytoskeleton Structural framework of the cell.

daunomycin A chemotherapeutic drug that binds to DNA and leads to strand breakage by inhibiting topoisomerase II.

DCC A tumor suppressor gene that is frequently deleted in colon and rectum carcinomas and encodes a protein related to cell adhesion molecules.

deletion Loss of genetic material.

density-dependent inhibition The characteristic cessation of normal cell proliferation at a finite cell density in culture.

deoxyribonucleic acid (DNA) The genetic material.

diacylglycerol A compound that activates protein kinase C.

diethylstilbestrol (DES) A synthetic estrogen.

dihydrofolate reductase An enzyme that catalyzes the formation of tetrahydrofolate, which is required for DNA synthesis. The target enzyme for methotrexate.

DNA *See* deoxyribonucleic acid.

doxorubicin   A chemotherapeutic drug that binds to DNA and leads to strand breakage by inhibiting topoisomerase II.

drug resistance   Failure of cancer cells to respond to a chemotherapeutic drug.

ductal carcinoma   A type of breast cancer that arises in the ducts of the mammary gland.

dysgerminoma   An ovarian tumor of undifferentiated germ cells.

dysplasia   An early preneoplastic stage of the development of carcinomas, characterized by loss of cellular regularity.

E2A   An oncogene translocated in acute lymphocytic leukemias.

E6 and *E7*   Oncogenes of human papillomaviruses.

ectoderm   Germ layer giving rise to tissues that include the skin and nervous system.

embryonal carcinoma   A malignancy of embryonic stem cells.

endoderm   Germ layer giving rise to tissues including the epithelial lining of the gastrointestinal and respiratory tracts, as well as the epithelial portions of the liver and other internal organs.

endometrium   The lining of the uterus.

endoplasmic reticulum   An array of membranes within the cell.

endoscopy   The use of thin lighted tubes for examination of internal body cavities.

endothelial cell   A cell lining the interior of blood vessels.

ependymoma   A type of malignant brain tumor.

epidemiology   Study of the incidence of disease in different population groups.

epithelial cells   Cells that form continuous sheets covering the surface of the body and lining the internal organs.

Epstein-Barr virus   A virus associated with Burkitt's lymphoma and nasopharyngeal carcinoma.

*erb*B-2   An oncogene frequently amplified in mammary and ovarian carcinomas, which encodes a receptor protein-tyrosine kinase related to the EGF receptor.

erythrocyte   Red blood cell.

erythroleukemia   A leukemia of immature red blood cells.

erythropoietin   A growth factor that promotes proliferation and differentiation of red blood cell precursors.

estrogen   A steroid hormone that stimulates female sex characteristics.

estrogen replacement therapy   Administration of estrogen to relieve symptoms of menopause.

etoposide   A chemotherapeutic drug that causes DNA breakage by inhibiting topoisomerase II.

Ewing's sarcoma   A childhood bone tumor.

extracellular domain   The part of a receptor molecule that is exposed to the external environment on the outside of the cell surface.

extracellular matrix   Insoluble meshwork of secreted material that fills the space between cells in tissues.

familial adenomatous polyposis   A rare inherited form of colon cancer in which affected individuals develop multiple colon adenomas (polyps).

Fanconi's anemia   A hereditary disease characterized by genetic instability and a high incidence of leukemias and lymphomas.

fecal occult blood test   A colon cancer screening test in which the presence of small amounts of blood in the stool is detected.

fibroblast A common type of connective tissue cell involved in wound healing.

fibroblast growth factors A family of growth factors that stimulate proliferation of a variety of cell types.

fluorouracil An antimetabolite that inhibits DNA replication by blocking the synthesis of thymidine monophosphate.

fos An oncogene that encodes a component of AP-1 transcription factor.

G0 A quiescent state in which cells are not proliferating.

G1 The stage of the cell cycle between the end of mitosis and the beginning of DNA synthesis.

G2 The stage of the cell cycle between the end of DNA synthesis and the beginning of mitosis.

gastroscopy Examination of the stomach with a thin lighted tube.

gene A region of DNA that encodes a single protein.

gene amplification Increased number of gene copies in a cell, resulting from repeated DNA replication.

gene transfer Transfer of genetic information, usually in the form of DNA, from one cell to another; also called transfection.

genome The complete genetic information of a species.

gip An oncogene involved in adrenal cortical and ovarian tumors.

glioblastoma A highly malignant type of astrocytoma.

glucocorticoid A steroid hormone produced by the adrenal cortex.

Golgi complex Subcellular organelle that functions in secretion of material out of the cell.

gonadotropin A hormone produced by the pituitary that controls the function of the gonads.

gonadotropin-releasing hormone (GnRH) A hormone produced by the hypothalamus that acts on the pituitary to signal gonadotropin release.

graft versus host disease The principal complication of bone marrow transplantation, in which lymphocytes from the donor marrow react against cells of the recipient.

granulocyte A type of white blood cell that functions in inflammation.

granulocyte colony-stimulating factor (G-CSF) A growth factor that stimulates the proliferation and differentiation of granulocyte precursors.

granulocyte-macrophage colony-stimulating factor (GM-CSF) A growth factor that stimulates the proliferation and differentiation of precursors to both granulocytes and macrophages.

growth factor A secreted protein that stimulates cell proliferation or differentiation.

growth factor receptor A cell surface protein to which a growth factor binds.

gsp An oncogene involved in some pituitary tumors.

GTPase activating protein (GAP) A protein that stimulates hydrolysis of bound GTP by *ras* proto-oncogene proteins.

hematopoietic cells Cells giving rise to mature blood cells.

hepatitis B viruses A family of viruses that can cause hepatocellular carcinoma.

hepatocellular carcinoma The most common type of liver cancer.

Hodgkin's disease A type of lymphoma.

hormone A chemical produced by one type of cell that affects the activity of a second type of cell.

human immunodeficiency virus (HIV) A retrovirus that causes AIDS.

human T-cell lymphotropic virus (HTLV) A retrovirus that causes adult T-cell leukemia.

hydroxyurea A chemotherapeutic drug that inhibits DNA replication by blocking the synthesis of deoxyribonucleotides.

imaging The use of noninvasive methods, such as X-rays, to view the inside of the body.

immortality Indefinite lifespan of some cells in culture.

immunoglobulin An antibody.

immunosuppressive Suppressing the activity of the immune system.

immunotherapy Stimulating the immune system to eliminate cancer cells.

initiating agent A carcinogen that acts to initiate development of a tumor by inducing mutations.

initiation *See* tumor initiation.

inositol phospholipids A group of lipids involved in intracellular signal transduction.

inositol triphosphate (IP₃) A compound that causes the concentration of intracellular calcium to increase.

integrins A family of cell surface receptors that mediate cell-cell and cell-matrix interactions.

interferon A cytokine that inhibits virus infection and displays antitumor activity.

interleukin-2 (IL-2) A growth factor that stimulates proliferation of T lymphocytes.

interleukin-3 (IL-3) A growth factor that stimulates proliferation of hematopoietic stem cells.

intracellular domain The part of a cell surface receptor that is inside of the cell.

intracellular signal transduction The process by which a signal initiated by growth factor binding to a cell surface receptor is conveyed to the nucleus.

ionizing radiation High-energy forms of radiation such as X-rays and radiation produced by the decay of radioactive particles.

jun An oncogene that encodes a component of AP-1 transcription factor.

Kaposi's sarcoma A vascular neoplasm that is common in AIDS patients.

keratinocyte An epithelial cell of the skin.

kinase An enzyme that transfers phosphate groups, usually from ATP, to another molecule.

latent period Time between exposure to a carcinogen and development of a tumor.

leukemia A cancer arising from the blood-forming cells in bone marrow.

levamisole A nonspecific immune stimulant used in tumor therapy.

Li-Fraumeni cancer family syndrome A rare hereditary cancer susceptibility, leading to the development of multiple kinds of tumors, which results from inherited mutations of the *p53* tumor suppressor gene.

lipoma A benign tumor of fat cells.

liposarcoma A malignant tumor of fat cells.

lobular carcinoma A type of breast cancer that arises in the lobules of the mammary gland.

luprolide An analog of gonadotropin-releasing hormone used in tumor therapy.

lymphatic system   The structural system involved in the immune response, including lymphatic vessels, lymph nodes, and the spleen, tonsils, and thymus.

lymph node   An aggregate of lymphatic tissue.

lymphocyte   A white blood cell that functions in the immune response.

lymphoid   The blood cell lineage giving rise to lymphocytes.

lymphoma   A cancer arising from lymphocytes or macrophages in lymphatic tissue.

Lynch cancer family syndrome   A rare inherited cancer susceptibility leading to the development of breast and ovarian carcinomas.

lysosome   A subcellular organelle containing digestive enzymes.

M phase   The stage of the cell cycle during which cell division takes place.

macrophage   A white blood cell that functions in inflammatory reactions and the immune response.

magnetic resonance imaging (MRI)   An imaging technique that employs computer analysis of the movement of molecules in a strong magnetic field.

malignant tumor   A tumor that is capable of invading surrounding normal tissue and metastasizing to distant body sites.

mammography   A low-dose X-ray of the breast.

MCC   A tumor suppressor gene frequently inactivated in colon and rectum carcinomas.

mechlorethamine   An alkylating agent, nitrogen mustard.

medulloblastoma   A type of malignant brain tumor, most common in children.

melanoma   A cancer of the pigment-producing cells of the skin.

melphalan   An alkylating agent used in chemotherapy.

meningioma   A type of benign brain tumor.

mercaptopurine   An antimetabolite that inhibits DNA replication by blocking purine synthesis.

mesoderm   Germ layer giving rise to connective tissues, muscle, bone, and the cells of the blood and immune system.

metastasis   The spread of cancer cells to distant body sites.

methotrexate   An antimetabolite that interferes with DNA replication by inhibiting dihydrofolate reductase.

mitochondria   Subcellular organelles that generate energy within the cell.

mitomycin C   A chemotherapeutic drug that reacts with DNA.

mitosis   The type of cell division in which daughter cells receive chromosomes identical to those of the parental cell.

monoclonal antibody   An antibody produced by a single clone of B lymphocytes.

monocyte   A type of white blood cell involved in inflammatory reactions.

mRNA   Messenger RNA (ribonucleic acid). An RNA molecule copied from DNA that serves as a template for protein synthesis.

multidrug resistance   Simultaneous resistance of a tumor to multiple chemotherapeutic drugs.

multiple myeloma   Cancer of mature antibody-producing B lymphocytes (plasma cells).

mutagen   An agent that induces mutations.

mutation   An alteration in the base sequence of DNA.

myc A family of oncogenes (c-*myc*, L-*myc*, and N-*myc*) that are activated by chromosome translocation in Burkitt's and other B-cell lymphomas and by amplification in a variety of tumors, including neuroblastomas and small cell lung carcinomas.

myeloid The blood cell lineage giving rise to platelets, erythrocytes, granulocytes, monocytes, and macrophages.

neoplasm An abnormal growth of cells.

neuroblastoma A childhood cancer of embryonic neural cells.

neurofibroma A benign tumor arising from connective tissue of a nerve.

NF1 A tumor suppressor gene responsible for inheritance of type 1 (von Recklinghausen) neurofibromatosis.

nitrogen mustard An alkylating agent used in chemotherapy.

nitrosamines A class of carcinogens that can be formed from nitrates and nitrites in the digestive tract.

N-**nitroso compounds** *See* nitrosamines.

non-Hodgkin's lymphoma Any type of lymphoma other than Hodgkin's disease.

nuclear magnetic resonance (NMR) *See* magnetic resonance imaging.

nucleoli The sites of ribosome synthesis in the nucleus.

occupational carcinogen A carcinogen to which workers are exposed.

oncogene A gene capable of inducing one or more characteristics of cancer cells.

oncogenic virus *See* tumor virus.

osteosarcoma A type of bone tumor.

oxygen free radical An oxygen molecule with a reactive unshared electron.

p53 A tumor suppressor gene that is lost or inactivated in a variety of neoplasms, including breast, colon, and lung carcinomas, sarcomas, and leukemias.

Pap smear (Pap test) A cytological screening test for early detection of cervical carcinoma.

papilloma A benign tumor projecting from an epithelial surface.

papillomaviruses A family of tumor viruses that induce papillomas and carcinomas in a variety of species, including anogenital carcinomas in humans.

P-glycoprotein A membrane protein that acts to "pump" foreign compounds, including some chemotherapeutic drugs, out of cells.

Philadelphia chromosome An abnormal human chromosome 22, formed by translocation of the *abl* oncogene from chromosome 9 in chronic myelogenous leukemias.

phosphatidylinositol kinase An enzyme that phosphorylates phosphatidylinositol.

phospholipase C An enzyme that catalyzes the formation of diacylglycerol and inositol triphosphate.

plasma membrane A bilayer of lipid molecules, with proteins inserted into it, that surrounds the cell.

platelet A blood cell that functions in coagulation.

platelet-derived growth factor (PDGF) A growth factor that stimulates proliferation of connective tissue cells.

point mutation Alteration of a single nucleotide of DNA.

polyp A benign tumor projecting from an epithelial surface.

prednisone A glucocorticoid used in treatment of leukemias and lymphomas.

premalignant   A neoplastic cell that has not yet acquired the ability to invade surrounding normal tissue.

preneoplastic   A cell that displays increased proliferative potential and is capable of progressing to the full neoplastic phenotype as a result of further alterations.

procarbazine   An alkylating agent used in chemotherapy.

progesterone   A steroid hormone secreted by the corpus luteum that functions to prepare the uterine endometrium for implantation of a developing embryo.

programmed cell death   An active process in which cell death occurs as part of the normal developmental program of a cell lineage.

promoting agent   A compound that leads to tumor development by stimulating cell proliferation.

promotion   *See* tumor promotion.

prostate-specific antigen (PSA)   An enzyme, secreted by prostate cells, that is a marker for prostate carcinoma.

protease   An enzyme that degrades proteins.

protein   A polymer of amino acids that is encoded by a gene.

protein kinase   An enzyme that phosphorylates a protein.

protein kinase C   A family of protein-serine/threonine kinases that are activated by diacylglycerol and function in intracellular signal transduction.

protein-serine/threonine kinase   A protein kinase that phosphorylates serine or threonine residues on its substrate proteins.

protein-tyrosine kinase   A protein kinase that phosphorylates tyrosine residues on its substrate proteins.

proto-oncogene   A normal cell gene that can be converted to an oncogene.

quiescent cell   A cell that is not proliferating.

rad   Radiation absorbed dose. The amount of radiation absorbed by tissue.

radioisotope scanning   An imaging method in which radioactive isotopes are administered and then detected in tissues of the patient.

radon   A natural source of radiation formed as a decay product of uranium.

raf   A protein-serine/threonine kinase oncogene involved in signal transduction from receptor protein-tyrosine kinases.

RAR   The retinoic acid receptor, activated as an oncogene by chromosome translocation in acute promyelocytic leukemia.

ras   A family of oncogenes (*ras*H, *ras*K, and *ras*N) involved in a variety of human tumors, including colon, lung, pancreatic, and thyroid carcinomas, leukemias, and lymphomas.

RB   A tumor suppressor gene identified by genetic studies of retinoblastoma, and also frequently inactivated in osteosarcomas, rhabdomyosarcomas, and bladder, breast, lung, and prostate carcinomas.

Reed-Sternberg cell   The characteristic cell of Hodgkin's disease.

rem   Radiation equivalent man. Amount of radiation absorbed by tissue, corrected for the biological effectiveness of the type of radiation being considered.

renal cell carcinoma   The most common form of kidney cancer.

ret   A protein-tyrosine kinase oncogene frequently activated by DNA rearrangement in thyroid carcinomas.

retinoblastoma   A childhood eye tumor.

retinoic acid (vitamin A)   A steroid-related hormone that induces differentiation, and inhibits proliferation, of a variety of epithelial cells.

retinoids   Compounds related to retinoic acid (vitamin A).

retroviruses   A family of viruses that includes HIV and HTLV.

rhabdomyosarcoma   A cancer of skeletal muscle cells.

ribosome   Subcellular organelle that carries out protein synthesis.

risk factor   A factor affecting the likelihood that an individual will develop cancer.

RNA   Ribonucleic acid. *See* mRNA.

Rous sarcoma virus   An acutely transforming retrovirus, in which the first oncogene was identified.

S   The stage of the cell cycle during which DNA replication occurs.

sarcoma   A cancer of connective tissue.

screening   Testing to detect early stages of cancer development in healthy asymptomatic individuals.

secondary prevention   Screening to detect early stages of cancer development that are readily treatable.

seminoma   A cancer of undifferentiated germ cells.

senescence   Limited proliferative capacity of normal cells.

sigmoidoscopy   Endoscopic examination of the rectum and lower part of the colon.

signal transduction   *See* intracellular signal transduction.

sis   A retroviral oncogene that encodes platelet-derived growth factor.

small cell carcinoma   A type of lung cancer.

squamous cell carcinoma   A cancer of flat epithelial cells.

src   The oncogene of Rous sarcoma virus.

staging   *See* tumor staging.

stem cell   A cell that divides to form one new stem cell and another cell that differentiates.

steroid hormones   A group of hormones, including estrogens, androgens, progesterone, glucocorticoids, thyroid hormone, and retinoic acid, consisting of small lipid-soluble molecules that pass directly through the plasma membrane.

suppressing agent   A chemopreventive agent that acts to inhibit cell proliferation.

tamoxifen   An estrogen antagonist.

tax   The oncogene of HTLV.

T cell   *See* T lymphocyte.

teniposide   A chemotherapeutic drug that causes DNA breakage by inhibiting topoisomerase II.

teratoma   A tumor of embryonic stem cells.

testosterone   A steroid hormone produced by the testes that stimulates male sex characteristics.

TGFβ   A growth factor that stimulates differentiation of a variety of epithelial cells.

thioguanine   An antimetabolite that inhibits DNA replication by blocking purine synthesis.

thiotepa   An alkylating agent used in chemotherapy.

thymidylate synthetase   An enzyme that catalyzes the formation of thymidine monophosphate. The target enzyme for fluorouracil.

thyroid hormone   A steroid-related hormone produced by the thyroid gland.

T lymphocyte   A lymphocyte that functions in cell-mediated immune responses.

TNM system   A system of tumor staging based on tumor size, invasion of surrounding tissue, lymph node involvement, and metastasis.

topoisomerase II An enzyme that catalyzes the breakage and rejoining of DNA strands.

transcription Synthesis of RNA from a DNA template.

transcription factor A protein that regulates transcription.

transformation Conversion of a normal cell in culture to a cell displaying one or more of the characteristics of a tumor cell.

transitional cell carcinoma A carcinoma arising from the transitional epithelium of the bladder.

translation Protein synthesis directed by mRNA.

translocation *See* chromosome translocation.

transmembrane domain The part of a protein that passes through the plasma membrane.

trk A protein-tyrosine kinase oncogene frequently activated by DNA rearrangement in thyroid carcinomas.

tumor An abnormal growth of cells.

tumor grading Histologic examination of tumor cell morphology and rate of cell division.

tumor-infiltrating lymphocyte A lymphocyte with antitumor activity.

tumor initiation The first step in development of a tumor.

tumor marker A substance, produced by cancer cells, that can be used diagnostically as a test for tumor growth.

tumor necrosis factor A cytokine with antitumor activity.

tumor progression The continuing development of increasing malignancy during growth of a tumor.

tumor promoter A compound that leads to tumor development by stimulating cell proliferation.

tumor promotion The second stage in development of a tumor, during which cell proliferation leads to formation of a tumor cell population.

tumor staging Assessment of the extent of tumor growth, invasion, and metastasis.

tumor suppressor gene A gene that inhibits tumor development.

tumor virus A virus that induces cancer.

ultrasound An imaging method in which the echoes of high-frequency sound waves are used to reveal tissue masses.

vinblastine A chemotherapeutic drug that blocks cell division.

vincristine A chemotherapeutic drug that blocks cell division.

viral oncogene An oncogene present in a tumor virus.

virus An infectious particle that reproduces inside cells.

VP-16 *See* etoposide.

Warthin cancer family syndrome A rare inherited cancer susceptibility leading to the development of colon and endometrial carcinomas.

Wilms' tumor A childhood kidney cancer.

WT1 A tumor suppressor gene that is inactivated in Wilms' tumor.

xeroderma pigmentosum An inherited skin disease, associated with a high incidence of skin cancer, in which patients are unable to repair DNA damage resulting from ultraviolet light.

X gene A candidate oncogene of hepatitis B virus.

yolk sac carcinoma A type of embryonic tumor arising from male or female germ cells.

Index